T0155663

NEUROMETHODS

For further volumes:
http://www.springer.com/series/7657

Advanced Patch-Clamp Analysis for Neuroscientists

Edited by

Alon Korngreen

The Leslie and Susan Gonda Multidisciplinary Brain Research Center
The Mina and Everard Goodman Faculty of Life Sciences
Bar-Ilan University, Ramat-Gan, Israel

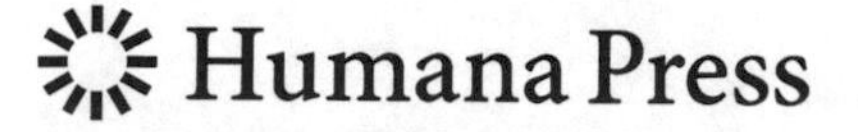

Editor
Alon Korngreen
The Leslie and Susan Gonda
Multidisciplinary Brain Research Center
The Mina and Everard Goodman
Faculty of Life Sciences
Bar-Ilan University
Ramat-Gan, Israel

ISSN 0893-2336 ISSN 1940-6045 (electronic)
Neuromethods
ISBN 978-1-4939-8043-7 ISBN 978-1-4939-3411-9 (eBook)
DOI 10.1007/978-1-4939-3411-9

Springer New York Heidelberg Dordrecht London

Softcover reprint of the hardcover 1st edition 2016

Printed on acid-free paper

Humana Press is a brand of Springer
Springer Science+Business Media LLC New York is part of Springer Science+Business Media (www.springer.com)

Series Preface

Experimental life sciences have two basic foundations: concepts and tools. The *Neuromethods* series focuses on the tools and techniques unique to the investigation of the nervous system and excitable cells. It will not, however, shortchange the concept side of things as care has been taken to integrate these tools within the context of the concepts and questions under investigation. In this way, the series is unique in that it not only collects protocols but also includes theoretical background information and critiques which led to the methods and their development. Thus it gives the reader a better understanding of the origin of the techniques and their potential future development. The *Neuromethods* publishing program strikes a balance between recent and exciting developments like those concerning new animal models of disease, imaging, in vivo methods, and more established techniques, including, for example, immunocytochemistry and electrophysiological technologies. New trainees in neurosciences still need a sound footing in these older methods in order to apply a critical approach to their results.

Under the guidance of its founders, Alan Boulton and Glen Baker, the *Neuromethods* series has been a success since its first volume published through Humana Press in 1985. The series continues to flourish through many changes over the years. It is now published under the umbrella of Springer Protocols. While methods involving brain research have changed a lot since the series started, the publishing environment and technology have changed even more radically. Neuromethods has the distinct layout and style of the Springer Protocols program, designed specifically for readability and ease of reference in a laboratory setting.

The careful application of methods is potentially the most important step in the process of scientific inquiry. In the past, new methodologies led the way in developing new disciplines in the biological and medical sciences. For example, Physiology emerged out of Anatomy in the nineteenth century by harnessing new methods based on the newly discovered phenomenon of electricity. Nowadays, the relationships between disciplines and methods are more complex. Methods are now widely shared between disciplines and research areas. New developments in electronic publishing make it possible for scientists that encounter new methods to quickly find sources of information electronically. The design of individual volumes and chapters in this series takes this new access technology into account. Springer Protocols makes it possible to download single protocols separately. In addition, Springer makes its print-on-demand technology available globally. A print copy can therefore be acquired quickly and for a competitive price anywhere in the world.

Saskatoon, Canada ***Wolfgang Walz***

Preface

What is patch-clamp for you? For me, patch-clamp is my primary research instrument. I was introduced to patch clamping over 20 years ago as a graduate student at Ben-Gurion University of the Negev. Bert Sakmann and Erwin Neher received the Nobel prize a few years earlier and the second edition of their book *Single Channel Recording* was about to come out. Using the new Axopatch-200B I recorded my first ionic currents from lung ciliary epithelium. I remember my excitement when I got my first gigaseal, holding my breath as the oscilloscope trace slowly moved and suddenly jumped to zero. Even today it is fun to watch the membrane suddenly seal to the glass pipette. With some surprise I realize that I am still using the Axopatch-200B; it is a great amplifier. You can argue that using the same type of amplifier for more than 20 years means that I am old fashioned or that the patch-clamp technique has been frozen for two decades. I refuse to address the first possibility on the grounds that it may incriminate me. As for the patch-clamp technique, it is not frozen and it is adapting.

Scientific techniques evolve and mature. At the beginning there is excitement; many rush to use the new tool. It becomes the primary research tool for many laboratories. Slowly, the tide ebbs. Commercial companies start producing parts and turnkey systems. It is no longer required to handcraft components; they are ready for purchase. In parallel, the limitations of the technique are discovered and the technique's shine is tarnished. It becomes just another tool in the scientist's arsenal. Once the low hanging fruit is consumed, the herd moves to greener pastures. Some stay, modifying the technique combining it with other techniques to go where no one has gone before.

Patch-clamp is a mature, even middle aged, technique. The 1980s were the years of the single-channel recording gold rush. During the 1990s it has been discovered that the whole-cell configuration of the technique can be extremely useful for recording from single neurons. Using this configuration allowed the first peek at unitary synaptic transmission and dendritic integration in the mammalian central nervous system. During the first decade of the twenty-first century, many studies combined the patch-clamp technique with other techniques especially fluorescence microscopy and two-photon imaging. A novel application has been the use of patch electrodes for various experiments in vivo. The new, literally shiny, kid on the block, optogenetics, has been immediately combined with patch-clamp. Here I gathered representatives of current adaptations of the patch-clamp technique to neuroscience. Thus, chapters are devoted to in vivo recordings, voltage-gated channel recording and analysis, dendritic and axonal recordings, synaptic current recording and analysis, advanced fluorescent techniques, optogenetics and voltage-sensitive dye imaging, and finally channel and neuronal modeling. The chapters also display the highly varying points of view of the patch-clamp technique. Some authors wrote a completely technical chapter; others view patch-clamp as a standard tool and detailed how it was used in a specific scientific settings. Naturally, the chapters dealing with analysis and theory only addressed the products of the technique. The chapters thus provide a snapshot of patch-clamp application and analysis. They also provide a window at the developmental stage of a mature scientific tool that is, and will continue to be, a major tool in neuroscience.

Ramat-Gan, Israel *Alon Korngreen*

Contents

Contributors

COREY D. ACKER • *R.D. Berlin Center for Cell Analysis and Modeling, University of Connecticut Health Center, Farmington, CT, USA*

MARA ALMOG • *The Mina and Everard Goodman, Faculty of Life Sciences, Bar-Ilan University, Ramat-Gan, Israel; Leslie and Susan Gonda Multidisciplinary Brain Research Center, Bar-Ilan University, Ramat-Gan, Israel*

SRDJAN D. ANTIC • *Department of Neuroscience, Institute for Systems Genomics, University of Connecticut Health Center, Farmington, CT, USA*

DIRK FELDMEYER • *Institute of Neuroscience and Medicine, INM-2, Research Centre Jülich, Jülich, Germany; Department of Psychiatry, Psychotherapy and Psychosomatics, Medical School, RWTH Aachen University, Aachen, Germany; Jülich-Aachen Research Alliance-Brain, Translational Brain Medicine, Aachen, Germany*

SONIA GASPARINI • *Neuroscience Center of Excellence, LSU Health Sciences Center, New Orleans, LA, USA*

XIAOLONG JIANG • *Department of Pharmacology, University of Virginia School of Medicine, Charlottesville, VA, USA*

GEORGE KASTELLAKIS • *Institute of Molecular Biology and Biotechnology (IMBB), Foundation for Research and Technology, Hellas (FORTH), Heraklion, Crete, Greece; Department of Biology, University of Crete, Heraklion, Crete, Greece*

GERGELY KATONA • *Institute of Experimental Medicine of the Hungarian Academy of Sciences, Budapest, Hungary; Pázmány Péter University, Budapest, Hungary*

CHRISTIAAN P.J. DE KOCK • *Department of Integrative Neurophysiology, Center for Neurogenomics and Cognitive Research, VU University Amsterdam, Amsterdam, The Netherlands*

BOJANA KOKINOVIC • *Molecular Biology Department, Umeå University, Umeå, Sweden; Integrative Medical Biology Department, Umeå University, Umeå, Sweden; Neuroscience and Brain Technologies Department, Italian Institute of Technology, Genova, Italy*

MAARTEN H. P. KOLE • *Department of Axonal Signaling, Royal Netherlands Academy of Arts and Sciences, Netherlands Institute for Neuroscience, Amsterdam, The Netherlands; Department of Cell Biology, Faculty of Science, Utrecht University, Utrecht, The Netherlands*

ALON KORNGREEN • *The Mina and Everard Goodman, Faculty of Life Sciences, Bar-Ilan University, Ramat-Gan, Israel; Leslie and Susan Gonda Multidisciplinary Brain Research Center, Bar-Ilan University, Ramat-Gan, Israel*

JANA KUSCH • *Institut für Physiologie II, Universitätsklinikum Jena, Jena, Germany*

ANGELIKA LAMPERT • *Institute of Physiology, University Hospital RWTH Aachen, Aachen, Germany*

FREDERIC LANORE • *Department of Neuroscience, Physiology and Pharmacology, University College London, London, UK*

TREVOR C. LARRY • *Department of Pharmacology, University of Virginia School of Medicine, Charlottesville, VA, USA; Department of Neuroscience Undergraduate Program, University of Virginia School of Medicine, Charlottesville, VA, USA*

PAOLO MEDINI • *Molecular Biology Department, Umeå University, Umeå, Sweden; Integrative Medical Biology Department, Umeå University, Umeå, Sweden*

JANNIS E. MEENTS • *Institute of Physiology, University Hospital RWTH Aachen, Aachen, Germany*
LORIN S. MILESCU • *Division of Biological Sciences, University of Missouri, Columbia, MO, USA*
MARCO A. NAVARRO • *Division of Biological Sciences, University of Missouri, Columbia, MO, USA*
LUCY M. PALMER • *Florey Institute of Neuroscience and Mental Health, University of Melbourne, Melbourne, VIC, Australia*
STYLIANOS PAPAIOANNOU • *Molecular Biology Department, Umeå University, Umeå, Sweden; Integrative Medical Biology Department, Umeå University, Umeå, Sweden*
HENRIKE PLANERT • *Institute of Neurophysiology, Charité Universitätsmedizin Berlin, Berlin, Germany*
PANAYIOTA POIRAZI • *Department of Biology, University of Crete, Heraklion, Crete, Greece; Computational Biology Lab, IMBB-FORTH, Heraklion, Crete, Greece*
MARKO POPOVIC • *Department of Axonal Signaling, Royal Netherlands Academy of Arts and Sciences, Netherlands Institute for Neuroscience, Amsterdam, The Netherlands*
GABRIELE RADNIKOW • *Institute of Neuroscience and Medicine, INM-2, Research Centre Jülich, Jülich, Germany*
BALÁZS RÓZSA • *Institute of Experimental Medicine of the Hungarian Academy of Sciences, Budapest, Hungary; Pázmány Péter University, Budapest, Hungary*
AUTOOSA SALARI • *Division of Biological Sciences, University of Missouri, Columbia, MO, USA*
GILAD SILBERBERG • *Department of Neuroscience, Karolinska Institute, Stockholm, Sweden*
R. ANGUS SILVER • *Department of Neuroscience, Physiology and Pharmacology, University College London, London, UK*
MANDAKINI B. SINGH • *Department of Neuroscience, Institute for Systems Genomics, University of Connecticut Health Center, Farmington, CT, USA*
STEFANOS S. STEFANOU • *Institute of Molecular Biology and Biotechnology (IMBB), Foundation for Research and Technology, Hellas (FORTH), Heraklion, Crete, Greece; Department of Biology, University of Crete, Heraklion, Crete, Greece*
GERGELY SZALAY • *Institute of Experimental Medicine of the Hungarian Academy of Sciences, Budapest, Hungary*
GUANGFU WANG • *Department of Pharmacology, University of Virginia School of Medicine, Charlottesville, VA, USA*
DANIEL R. WYSKIEL • *Department of Pharmacology, University of Virginia School of Medicine, Charlottesville, VA, USA; Department of Neuroscience Graduate Program, University of Virginia School of Medicine, Charlottesville, VA, USA*
J. JULIUS ZHU • *Department of Pharmacology, University of Virginia School of Medicine, Charlottesville, VA, USA*
GIOVANNI ZIFARELLI • *Department of Physiology, Anatomy and Genetics, University of Oxford, Oxford, United Kingdom*

Chapter 1

In Vivo Whole-Cell Recordings

Bojana Kokinovic, Stylianos Papaioannou, and Paolo Medini

Abstract

The introduction of whole-cell, patch clamp recordings in vivo has allowed measuring the synaptic (excitatory and inhibitory) inputs and the spike output from molecularly or anatomically identified neurons. Combining this technique with two-photon microscopy also allows to optically target such recordings to the different subtypes of inhibitory cells (e.g., soma- and dendrite targeting), as well as to measure dendritic integration of synaptic inputs in vivo. Here we summarize the potentialities and describe the critical steps to successfully apply such an informative technique to the study of the physiology and plasticity of brain microcircuits in the living, intact brain.

Key words In vivo whole cell, Patch clamp, Brain microcircuits, Two-photon microscopy, Synaptic physiology

1 Introduction

Patch clamp pipettes have initially been used to measure the tiny currents that flow through single channels in voltage clamp [1]. Thereafter, patch clamp recordings have been used to record intracellular currents and voltages with the patch pipette in physical continuity with the cell cytoplasm (whole-cell configuration) [2]. Whole-cell recording remains the technique of choice to study synaptic connectivity and dynamics among visually identified cells in slices, by causing the occurrence of a spike in the presynaptic neurons while recording the postsynaptic response (paired whole-cell recordings) [3]. The use of patch electrodes to perform in vivo recordings has been carefully described by a seminal work done at Bert Sakmann's Lab [4]. Compared to sharp microelectrode recordings, whole-cell recordings have mainly the advantage that the access resistance is significantly lower, mostly due to the larger diameter of the pipette tip and the significantly shorter taper: this in turn allows smaller temporal distortions of fast electrical events, smaller voltage drops during current injections, and easier dialysis of the cytoplasm. The latter is a characteristic that turned

Alon Korngreen (ed.), *Advanced Patch-Clamp Analysis for Neuroscientists*, Neuromethods, vol. 113,
DOI 10.1007/978-1-4939-3411-9_1, © Springer Science+Business Media New York 2016

out to be useful for proper filling of cells with dyes (with the aim of neuroanatomical reconstructions) or with functional indicators (allowing for example dendritic calcium imaging in vivo—[5]). Another advantage in comparison with the sharp intracellular electrodes is that whole-cell electrodes do not rupture the neuronal membrane and hence cause less initial leakage of ions through it.

Here we will review the main potentials and the current limitations of the in vivo whole-cell recording techniques. We will stress the advantages of combining electrophysiological recordings with two-photon microscopy and optogenetics in the exploration of the physiology of cortical microcircuits in the living brain, with a concluding note on possible future developments of this technique. Finally, we present two experimental protocols related to the current use of this technique in our Lab: one to perform whole-cell recordings followed by neuronal anatomy in a functionally defined cortical area and one to perform two-photon targeted loose-patch recordings.

2 Potentialities

2.1 Simultaneous Measurements of Synaptic Inputs and Spike Outputs

In vivo whole-cell recordings allow simultaneous measurements of the synaptic inputs received by a neuron and of its spike output, during both spontaneous and sensory-driven activity. Indeed, the functional response properties of neurons are often different at sub- and suprathreshold level (Fig. 1). In general, response selectivity for stimulus features are sharper at action potential level: for example, in the primary visual cortex this is the case for orientation selectivity [6], binocularity [7], and segregation of ON and OFF subfields [8]. Also, comparison of synaptic and spike visual responses indicated that the reliability of the sensory responses is generally higher at the level of synaptic inputs [9, 10].

Blind in vivo whole-cell recordings are usually somatic [4]. However, recent work [11] demonstrated that it is possible to measure synaptic inputs with a patch pipette directly at the level of dendrites in vivo: importantly such work indicated that local, dendritic spikes contribute to the subthreshold orientation selectivity of V1 neurons.

2.2 Anatomical Reconstructions

By filling cells with biocytin it is possible to reveal the morphology and hence the anatomical identity of recorded cells by means of standard histochemical, peroxidase-based methods—e.g., [12]—see also Fig. 1. Slices can also be counterstained with cytochrome oxidase or with Nissl or myelin staining to reveal layering or other cytoarchitectonic features respectively. Both the dendrites and the axonal arbors of biocytin-filled cells can be reconstructed in 3D using a computer-assisted tracing system under 40× or

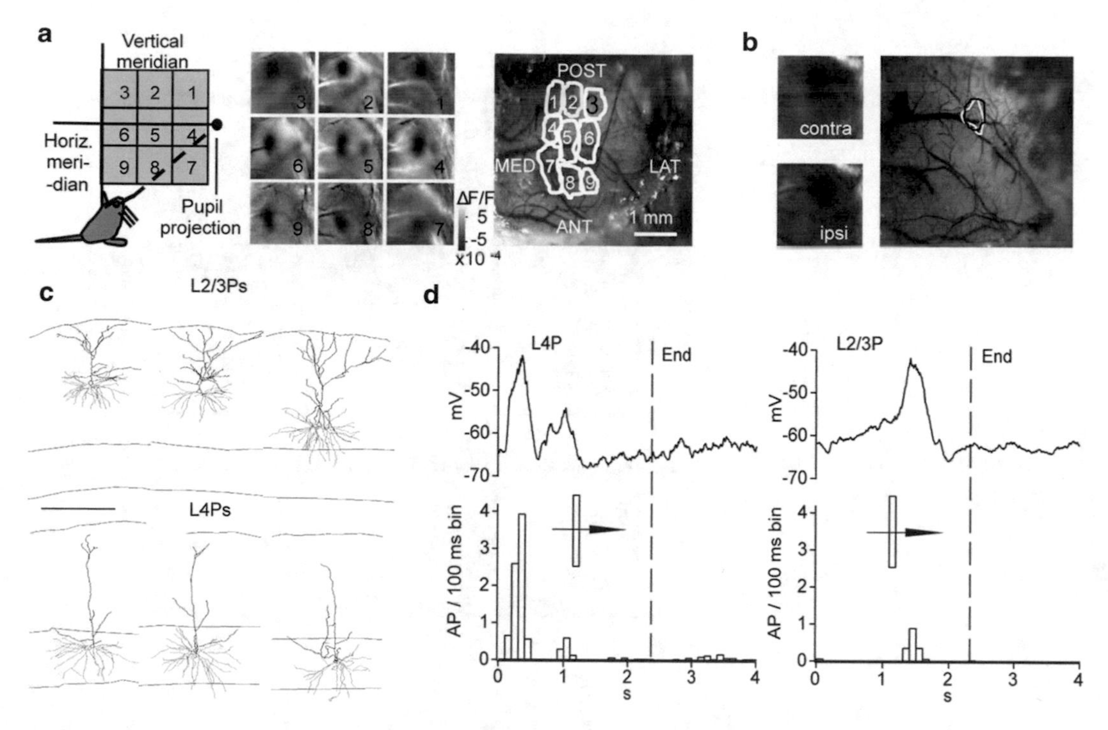

Fig. 1 Intrinsic optical imaging (IOI)-targeted in vivo whole-cell recordings of anatomically reconstructed layer 4 and layer 2/3 pyramids (L4Ps and L2/3Ps, respectively) in rat binocular V1. (**a**) Twenty-degree square spots displaying drifting gratings were randomly presented in different visual field positions (*left*). The corresponding IOI spots, representing the visually evoked focal decrease of light reflectance, were averaged over 40 stimulus presentations (*middle*), and overlaid on the vasculature image acquired before imaging with a green filter (*right*). All craniotomies were made over the IOI spot centered at 20° of elevation and neighboring the vertical meridian in the contralateral visual field (position 3, *red*), which corresponds to the center of the binocular region in the rat. (**b**) The two eyes were independently stimulated to confirm that the region of interest was located within the binocular portion of V1. Note the correspondence of the spots obtained through contralateral and ipsilateral eye stimulations in position 3. (**c**) Examples of coronal projections of the 3D reconstructions of basal (*red*) and apical (*black*) dendritic arbors of recorded L2/3Ps and L4Ps. Pial and layer 4 borders are outlined. Bar: 300 μm. (**d**) *L2/3Ps had similar PSP but smaller AP responses.* Examples of averaged PSP (*top* traces) and AP (*bottom* histograms) visual responses of a L4P (*left*) and of a L2/3P (*right*) to an optimally oriented moving light bar presented to the dominant eye. The bar starts sweeping the screen at time 0. *Dashed lines* indicate the beginning of the interstimulus period. From [10]

100× objectives. Such reconstructions can include also dendritic spines and axonal boutons quantification.

2.3 Estimate of Inhibitory and Excitatory Conductances

It is possible to estimate the excitatory and inhibitory components of synaptic responses by using in vivo whole-cell recordings [13–15]. This cannot be easily done by recording synaptic currents at the calculated reversal potentials for excitatory and inhibitory currents because of the space clamp problems—that prevent a homogeneous voltage clamping of all neuronal branches, and because depolarizing the cell at the reversal potential for excitatory currents can lead to uncontrolled activation of voltage-gated

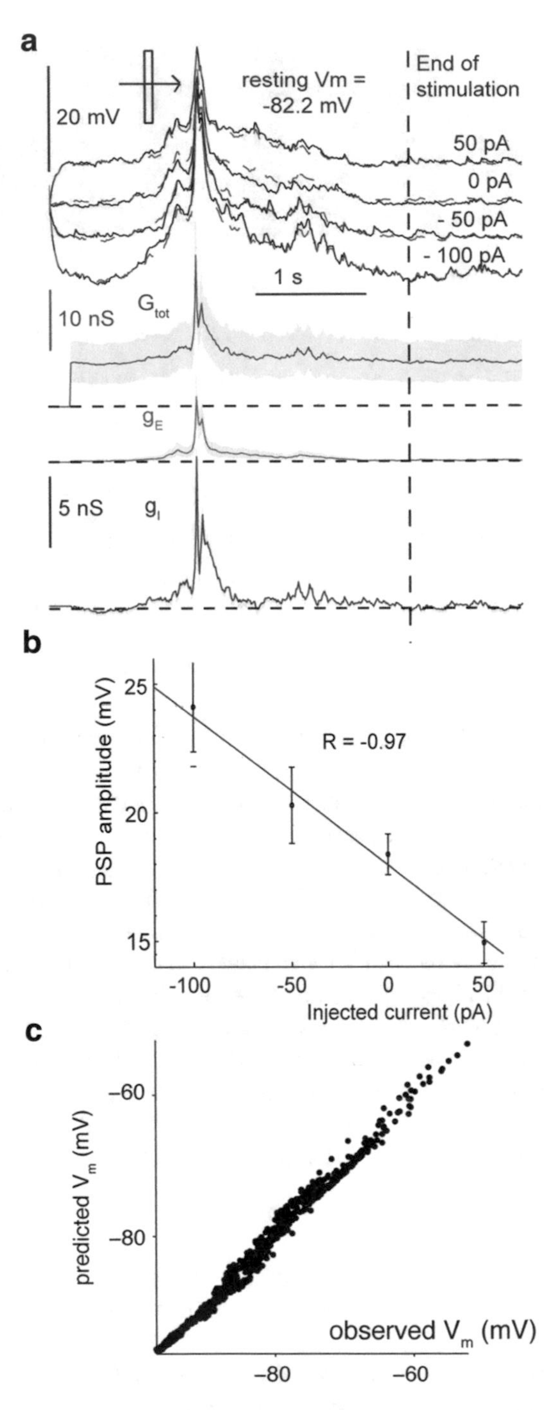

Fig. 2 Estimate of visually driven excitatory and inhibitory synaptic conductances. (**a**) Visual responses to an optimally oriented moving light bar of a layer 4 regular spiking neuron (4RSN) recorded under 1 mM QX314 while injecting different steady DC currents. The *black*, *continuous lines* are the recorded V_m values, whereas the *blue*, *dashed trace* shows the reconstructed V_m values obtained by inserting back the estimated g_E and g_I values into the fundamental membrane equation. The instantaneous total synaptic conductance

channels. For these reasons, excitatory and inhibitory conductances are usually estimated using a linear model of the membrane that assumes that the synaptic excitatory and inhibitory, as well as the nonsynaptic resting conductances, are largely voltage independent. This assumption must be carefully controlled, as there are clearly extrasynaptic, voltage-dependent conductances (e.g., I_h currents though HCN channels) as well as voltage-dependent synaptic conductances (e.g., NMDA channels). In the case of current clamp, the voltage response to sensory stimulation is recorded upon intracellular injection of different, mostly hyperpolarizing, current levels, a strategy to minimize the contributions of NMDA-receptor currents [16] as well as the activation of voltage-dependent channels (Fig. 2). A crucial point is controlling for the linearity of the current to voltage relation at the various injection steps, a prerequisite to apply this method. Preliminary to this calculation, the distortion of the voltage reading due to the series resistance must be corrected offline (Series resistance compensation) [17, 18]. The fundamental membrane equation is solved to extract the excitatory and inhibitory synaptic conductances as described in [19]. A further control for the linearity assumption is to reinsert back the values of excitatory and inhibitory conductances obtained in every point in time to see whether this predicts the actual membrane voltage values (membrane potential reconstruction). There are no significant differences in doing conductance estimates in current and voltage clamp, both in vitro and in vivo [16]. Also, blocking sodium and potassium voltage-active channels with intracellular QX-314 and Cs^+ helps in rendering neurons electrotonically more compact; however works in current clamp do not show significant differences

Fig. 2 (continued) is calculated based on the instantaneous slope of the current–voltage relation (G_{tot}, *blue*). The time-dependent excitatory (g_E, *green*) and inhibitory (g_I, *red*) conductances are plotted below. Gray traces represent the 95 % confidence intervals obtained by bootstrapping of the data. Conductance measurements began after the response to the injected current was at steady state (after 200 ms). Resting conductances were calculated in absence of visual stimulation (*dashed line*: stimulus end). (**b**) Visually driven PSPs vary linearly with the injected current. Plot showing the linearity of the relationship between the amplitude of the visually driven PSP response and the value of the injected current ($r = -0.97$) for a 4RSN (this plot refers to the example shown in Fig. 2 of the Main Text). Means ± standard errors are shown. The median of the correlation coefficients for all the recorded neurons was −0.94 (25th–75th percentiles: −0.88 to −0.99). (**c**) Plot of the recorded vs. reconstructed V_m values obtained by inserting back the estimated g_E and g_I into the membrane equation. The linearity of the cell and the accuracy of the V_m reconstruction are shown by the fact that data points align along the line of steepness 1 and intercept 0 in the plot. From [17]

between the conductance estimates when these blockers are added or omitted [17, 19].

When accurately controlled, the estimates of the excitatory and inhibitory conductances allow exploring the *temporal dynamics* of the excitatory and inhibitory synaptic inputs, and to put this in relation to the neuronal response. For example, work in the auditory cortex showed that after tone presentation excitation rises first and inhibition rises later, and that neuronal spiking occurs between the excitatory and inhibitory peaks, suggesting that the later inhibitory peak temporally confines the spiking, reducing its jitter [20]. Second, such methodology also allows the evaluation of the changes of the *relative strength* of the excitatory and inhibitory conductances, occurring for example, when a neuron is presented with the preferred or non-preferred orientation [13, 15], or after a change in the sensory environment, like monocular deprivation in area V1 [17, 21]. Indeed, the absolute conductance values obtained for inhibition and excitation can be affected by the different spatial distribution of the two types of synapses. For example, dendritic excitatory inputs are electrotonically more distant compared to perisomatic inhibitory contacts, a fact that is relevant because in vivo whole-cell recordings are mostly obtained from the soma.

2.4 Two-Photon Targeted Patch

Blind in vivo whole-cell patch recordings have a lower yield of inhibitory, non-pyramidal cells as compared to blind juxtasomal recordings and also in relation to the actual proportion of non-pyramidal, inhibitory cells (about 20 % of cortical neurons) (expected yield) [22, 23]. Two-photon targeted patch clamp has been introduced also for this purpose. The initial work indeed targeted GFP-labeled, parvalbumin-positive fast spiking cells in the cortex of transgenic mice using a patch pipette filled with a red fluorescent indicator [24]. This technique has been extensively used by the group of Carl Petersen to study the role of specific subpopulations of cortical interneurons in sensory processing in the primary somatosensory cortex—e.g., [25]. For example, such an approach showed that somatostatin-positive, dendritic targeting cells respond to whisker deflection in the barrel cortex with hyperpolarizations, a phenomenon that is not observed in any other cell type [26]. Several other Labs have used two-photon-targeted patching to perform loose-patch (juxtasomal) recordings from specific types of inhibitory neurons in both primary and association cortices. For example, this technique confirmed that the majority of parvalbumin-positive cells have scarcer orientation tuning in V1 compared to all remaining cortical neurons [27–29]. Also, in the multisensory (visuo-tactile) association cortex RL (Fig. 3), this technique showed a scarce multisensory integration in parvalbumin-positive cells, as compared to neighboring excitatory pyramidal neurons [30].

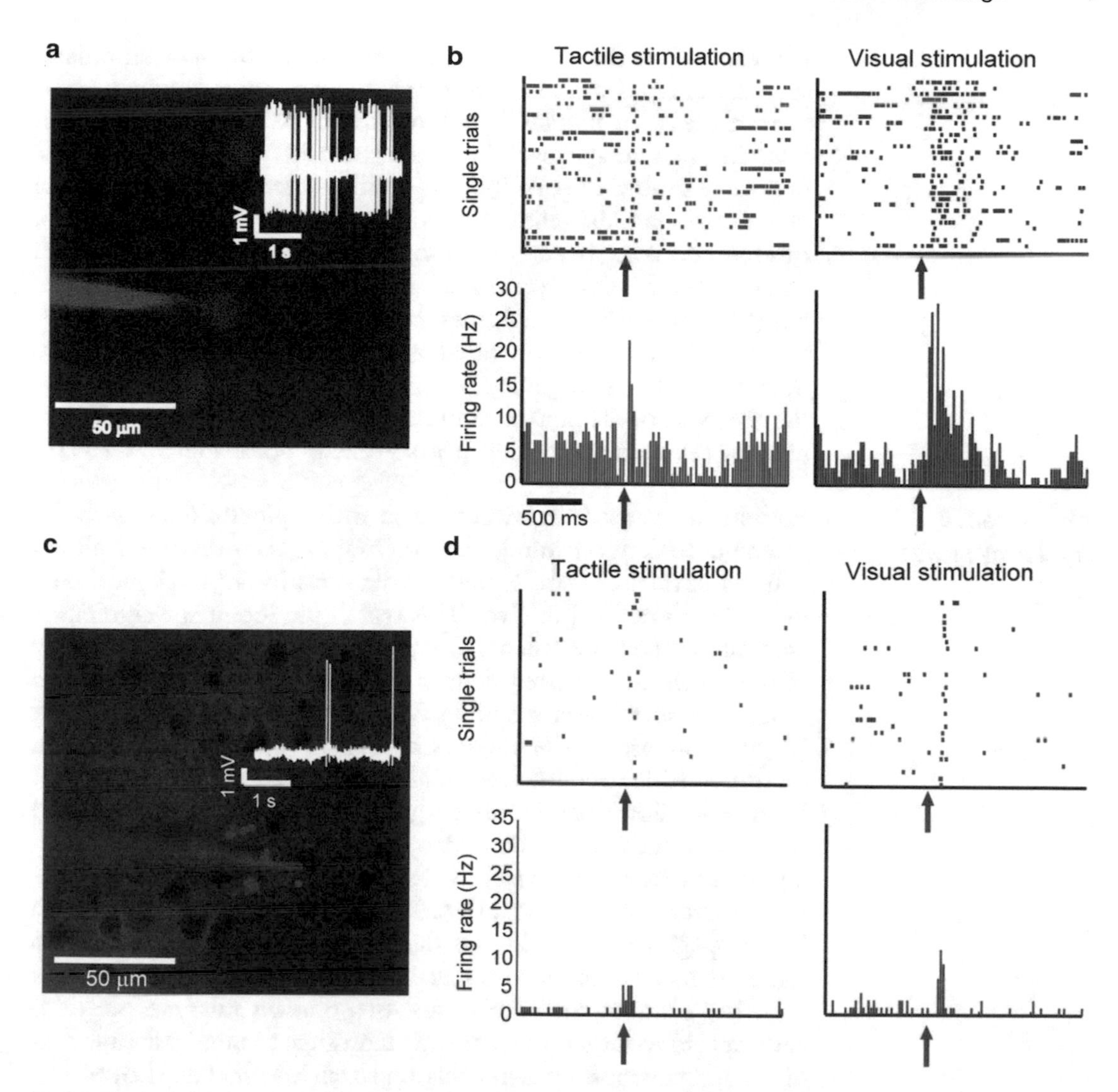

Fig. 3 Two-photon-targeted loose-patch recordings from inhibitory and excitatory neurons in the multisensory, visuo-tactile, mouse area RL. (**a**, **b**) Example raster plots (*top*) and peristimulus time histograms (*bottom*) for a two-photon-targeted juxtasomal recording of a bimodal Pv-IN upon tactile and visual stimulation (*blue* and *red*, respectively) from a mouse expressing the red protein tdTomato selectively in PV-INs (the pipette is filled with the green Na^+-salt dye Alexa488). Note the high frequency bursts of APs with pronounced after hyperpolarizations typical of PV-INs (*white*) *Arrows* are stimulus onsets. (**c**, **d**) Same as in (**a**, **b**) but for a regular-spiking pyramid targeted under the two-photon with the "shadow patching" technique (Alexa 488 being gently ejected in the extracellular matrix to visualize pyramidal cell bodies as *dark* structures). From [30]

2.5 Combining In Vivo Whole-Cell Recordings with Optogenetics: The Optopatcher

A second, interesting approach to record from specific cell types is the use of the so-called "optogenetic tags" realized with a conditional expression technique. Channelrhodopsin can be expressed in specific cell types by crossing mice that express the Cre recombinase in the aforementioned cell types with mice that bear a floxed-channelrhodopsin construct [31] (the latter construct can be also transduced into neurons by means of viral particles). In this case, illumination of the cortical surface with an optic fiber stimulates specific cell types and this is used as a guide to evaluate whether the neuron facing the pipette tip belongs or not to the group of cells of interest. Such an approach has been recently used to compare the frequency tuning of parvalbumin-positive interneurons and of the excitatory pyramidal cells in the primary auditory cortex [32]. Also, this methodology is very promising to record from identified cell types in deep cortical layers or in deep subcortical structures that are not (yet) accessible to multiphoton microscopy. A recent development in this direction is provided by a modified version of the pipette holder that allows to guide the light coming from an optic fiber to the glass wall of the pipette until the very tip so to selectively illuminate the cell under recording (the so-called "optopatcher"—[33]).

2.6 Dendritic Physiology In Vivo

Patching a layer 2/3 cortical neuron with a pipette filled with the green fluorescent calcium indicator Oregon green BAPTA-1 allows to monitor calcium transients along the dendritic arbor [5] and even in single spines [34] in vivo. This technique requires a significant amount of time (more than 30 min) to allow proper dendritic filling of the previously patched neuron with the dye and benefits from using high-speed laser scanning system such as a resonant scanner [5], but can also be done with standard galvanometric mirrors on single dendritic stretches (Fig. 4). Patching the neuron is important because it allows the monitoring of the subthreshold (synaptic) response of the neuron (that are not captured by the calcium imaging) on one side, and also allows hyperpolarizing the cell to prevent the occurrence of somatic spikes. This is in turn important because a backpropagating spike in the dendrites can cause local calcium signals that can contaminate sensory-driven calcium transients.

Also, shadow patching under two-photon microscope guidance can aid to target patch clamp recordings to dendritic branches of cortical neurons: by using this approach Smith et al. showed the existence of sensory-driven dendritic spikes (independently of the occurrence of backpropagating action potentials) in vivo, and that such events influence the subthreshold tuning of visual cortical neurons as assessed by somatic recording [11].

2.7 Single-Cell, Intracellular Pharmacological and Genetic Manipulation

The relatively fast and efficient cytoplasmic dialysis of recorded neurons occurring during whole-cell recordings greatly facilitates intracellular perfusion with specific pharmacological agents or even with DNA plasmids for molecular manipulations. One example of the first case (pharmacological modulation) is given by experiments where intracellular NMDA blockers have been used to block dendritic spikes and to prove the role of these dendritic events in setting the angular tuning of layer 4 neurons in the barrel cortex [35], or by the intracellular perfusion of neurons with QX314 and Cs^+ to do conductance measurements—e.g., [19]. Another interesting application of genetic transduction of neurons with DNA plasmids

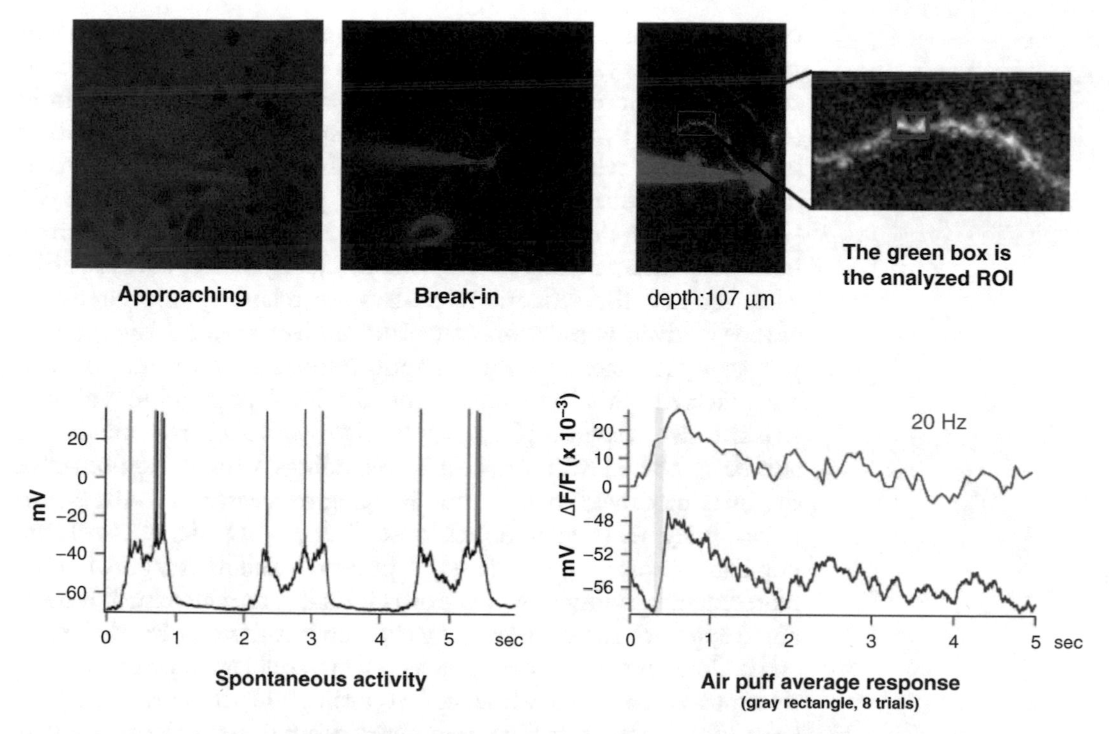

Fig. 4 Patching in vivo followed by dendritic calcium imaging in mouse barrel–somatosensory cortex to identify dendritic "hot spots." Top images: neurons are approached by using the shadow patching technique with the red Alexa 594 dye. The patch pipette is also filled with the green calcium indicator OGB-1, K+ salt, and the calcium imaging is done on the region of interest (ROI) on dendrites. *Bottom*: the *black* trace on the *left* represents the somatic voltage as recorded in current clamp. On the *right* plot the dendritic hot spot fluorescence response from the ROI (*green*) and the somatic voltage response (*red*) to whisker pad stimulation (Iurilli G and Medini P, unpublished data)

to study their microcircuits was given by a work of Rancz et al. [36]. In that work the authors used a set of plasmids in the recording pipette that restricted the retrograde transsynaptic labeling of afferent neurons only to the neurons that were monosynaptically connected with the recorded, transfected cell. Yet another application of single-cell labeling by means of plasmid transfection via the intracellular solution of the recording pipette is the ability to make time-lapsed in vivo whole-cell recordings from the same-tagged neuron. This technique brings the advantages of chronic recordings (follow the temporal evolution of response properties) into the armament of the intracellular electrophysiologist [37].

3 Current Limitations and Future Developments

The main current limitation of in vivo whole-cell recordings is represented by the limited sampling capability of the technique as compared to extracellular recordings or to two-photon population calcium imaging. However, two considerations are relevant in this regard: (1) the variability of the functional response properties at subthreshold level is significantly lower than at suprathreshold (spike) level [9, 10]. So, synaptic responses are usually rather homogeneous within a given neuronal population (e.g., sound-driven hyperpolarization in layer 2/3 pyramidal neurons in V1 [18]; or tactile-driven hyperpolarizations in somatostatin-positive inhibitory interneurons in the somatosensory barrel cortex [26]). The fact that the synaptic responses are relatively homogeneous among a given population of cells in a given cortical layer is also proven by the fact that the synaptic responses of a single neuron are reflected with high fidelity by the local population voltage-sensitive dye response [38]. (2) In vivo whole-cell recordings can also be combined with extracellular recordings, with voltage-sensitive dye imaging or with two-photon imaging to measure also the population integrated synaptic response (e.g., VSD signal) and the population spike output (e.g., 2-photon calcium imaging). One representative example of how powerful and informative such a combined approach can be is given by experiments done by Randy Bruno and colleagues where a combination of extracellular recordings in the thalamus and in vivo whole-cell recordings in the cortex allowed measuring for the first time the unitary postsynaptic efficacy of thalamocortical synapses [39] and the synaptic strength of thalamic innervation in the different cell types composing columnar, excitatory microcircuits [40].

A second limitation is that in vivo whole-cell recordings are rather time-consuming and have a yield of a few neurons per animal on average. Of relevance, an automated system to perform non-human-assisted in vivo whole-cell recordings has been recently described. Such a system could allow a single human operator to simultaneously perform different sets of experiments on different animals (and rigs) [41].

A third limitation is that in vivo whole-cell recordings are currently done in immobile, anesthetized or awake, head-fixed rodents, cats, and recently even primates [42]. Importantly, the feasibility of in vivo whole-cell recordings in freely moving rodents has been at least demonstrated [43]. Further evolution will be needed to record neuronal activity intracellularly and chronically in freely moving animals. A very interesting evolution with respect to this might be represented by gold mushroom shaped microelectrode array that can be internalized—at least partially—by neurons to obtain intracellular-like recordings in vitro [44]. Developing proper technical approaches to implement a similar approach in vivo might allow important advances in this direction in the future.

4 Protocol for Intrinsic Signal Imaging-Targeted In Vivo Whole-Cell Recordings Followed by Anatomical Neuronal Identification

4.1 Animal and Surgical Preparation

Anesthetize the mouse with 20 % urethane (0.8–1 g/kg) i.p. (intraperitoneal). It is a long-lasting anesthetic that provides stable anesthesia conditions allowing keeping animals in spontaneous breathing. However, since it is carcinogenic, it is only allowed to be used for terminal procedures and must be handled accordingly.

Important: provide additional supplements of 10 % of the initial dose until the anesthesia depth foreseen in your Ethical Permit is reached. Anesthesia level should be repeatedly checked at the beginning of the experiment, and then regularly throughout the experiment, by observing the appearance of automatic movements (whisking, chewing) and by checking pinch and corneal reflexes. Absolutely avoid alarming signs that the level of anesthesia is insufficient such as piloerection and extensive salivation. In the recording setup, heart and breathing rate together with O_2 saturation should be carefully checked.

1. Place the mouse onto a heating plate (37 °C). It has to be kept in mind that anesthesia also impairs thermoregulation so it is crucial to carefully monitor and keep the body temperature (37 °C) of the animal during the whole procedure.
2. Inject dexamethasone intramuscularly (0.01 mg/kg). This prevents formation of brain and mucosal edemas.
3. Provide animal with a continuous supply of humidified oxygen through a nasal cannula.
4. Before surgery, infiltrate the skin with local lidocaine solution (1 %). With fine scissors, cut and remove the skin. With a fine spatula or a delicate bone scrapper also remove periosteum.
5. Place the recording chamber onto the exposed skull and fix it with the help of acrylic glue and dental cement. Pay attention to fix it on the skull firmly.
6. Gently thin the skull by using a drill until the blood vessels are clearly visible under the microscope. Stop drilling when you perceive that the bone has a "papyraceous" (i.e., paper-like) consistence and avoid drilling continuously in order to prevent friction-generated heat.

Intrinsic signal imaging can be done in case the in vivo whole-cell recordings need to be done in a precise position in the cortex. This procedure is more advisable than opening a large craniotomy and doing extracellular mapping because, for getting a good mechanical stability of the preparation, it is essential to open relatively small craniotomies (from 0.5 to 1 mm).

4.2 Intrinsic Signal Imaging Procedure and Craniotomy

1. Transfer the animal to the recording setup on a heating plate.
2. Acquire a vasculature ("green") image under 540 nm light. The diaphragms of the macroscope lenses should be kept closed in order to increase the depth of focus, minimize optical aberrations, and get the best possible image quality. Start imaging session by illuminating the cortex with monochromatic 630 nm light provided by stable DC current source. All images are acquired with a cooled CCD camera defocused ca 500–600 μm below the pial surface. Light intensity should be adjusted just below pixel saturation. To achieve this, completely open the bottom macroscope lens diaphragm.
3. Monitor the signal on line: for this purpose as well as for offline data analysis, divide the averaged images after stimulus presentation (allowing ca 1 s latency for the intrinsic signal to peak) by the average image before stimulus onset (Resting State Normalization). To analyze the spot extension, the ratio of the two images should be coded on a grayscale. In setting the clipping values, please remember that the intrinsic signal amplitude is usually about 1/1000 of the absolute reflected light. The region of interest ("spot area") can be reasonably taken as the image area where the stimulus-evoked decrease in reflectance is higher than 50 % of the peak decrease. This region has to be then overlaid with the vasculature "green" image by which location of craniotomy is determined.
4. The craniotomy should be cut with sharp blades (#11). In doing this there should be no bleeding and also no bone left. In general, stop any minimal bleeding coming from the nearby bone or dura mater by gently pushing some clotting cellulose sponge (e.g., Sugi®, Kettenbach, Germany) close to the bleeding point (but not *on* the damaged vessel).
5. Keep the craniotomy always moist by filling the recording chamber with warmed Ringer solution. In mice it is not necessary to remove the dura matter, whereas in rats it is: for this purpose, use a 33 gauge needle.

Advice: Pull your patch pipettes before opening the craniotomy (e.g., during the intrinsic imaging). Check the size and shape of the very tip of all your pipettes using a 100× air objective with a convenient working distance: the shape of the tip should be convex in order to maximize surface adhesion with cell membrane and the very final opening should be around 1 μm of diameter. Keep the pipettes protected from dust and use them no longer than the end of the day (cracks can appear on the pipette).

4.3 Patching In Vivo

1. Fill the patch pipettes with intracellular solution (in mM: 135 K gluconate, 10 HEPES, 10 Na phosphocreatine, 4 KCl, 4 ATP-Mg, 0.3 GTP, pH 7.2, 291 mOsm). In order to be

able to perform anatomical identification of the cell biocytin (3 mg/ml) can be added to the intracellular solution. The ideal resistance of the pipette for patching should be between 5 and 9 MΩ: this is the best compromise between having a reasonable probability to make a seal and having a reasonably low access resistance (<100 MΩ).

2. Apply positive pressure (300–400 mbar) using a 10-ml syringe in order to prevent occlusion of the pipette before touching the brain surface.
3. Dry the recording chamber from the Ringer solution and carefully approach the craniotomy surface until you touch the skull with the pipette tip.
4. Fill in the chamber again with Ringer solution.
5. Penetrate the dura—in case of mice—and the pia (only a temporary increase of the pipette resistance should be observed) and navigate through the cortex slowly while maintaining a high pressure level till the depth of interest is reached. Always keep the pipette resistance controlled while the amplifier is operating in voltage-clamp mode.
6. Once you have reached the depth of interest, decrease the pipette pressure to +30 mbar. In this way cells are not pushed away from the tip of the pipette.
7. Search for neurons in voltage clamp, monitor the current and pipette resistance, and while stepping the *Z* axis (step size = 2 μm), repeatedly null the pipette current to zero.
8. As the pipette gets closer to the cell, causing dimpling of its surface, the pipette resistance will increase (amplitude of the current pulse in voltage clamp decreases).

   Note that a slow increase in resistance probably does not indicate that the cells are being approached but rather means the tip is becoming clogged. When the resistance is approximately doubled and it is possible to see in the current trace heart-beat-associated pulsations (strokes) usually it is a sign that the pipette is leaning against a cell. Try to release the pressure while hyperpolarizing the pipette potential: in ideal conditions this leads to the spontaneous formation of a gigaseal. However, it is usually necessary to apply transient negative pressure (50–100 mbar) to promote gigaseal formation. Forming a gigaseal is usually a relatively fast process and if the seal is not formed within 1–2 min it is best to try with a new pipette. In case of failure slowly come out and try a new pipette.
9. After having established a seal compensate the fast capacitive artifacts. Avoid overcompensation that can lead to amplifier oscillation.

10. Membrane rupturing is done by a quick, negative pressure ramp: suddenly the slow exponential due to membrane capacitance charging will appear. At that point immediately release pressure. Switch to current clamp in "voltage follower" or "bridge mode."
11. It is highly advisable to also compensate the series resistant in order to minimize the voltage drop across the access resistance when current is injected (e.g., to estimate synaptic conductances). Series resistance compensation can also be done offline. Note: it is advisable, in case of pulsed sweep acquisition, to have a brief (e.g., 200 ms), slightly hyperpolarizing pulse (e.g., −100 pA) at the beginning or end of each sweep, first to allow for series resistance estimates as well as for monitoring it repeatedly during the experiment. *Tip:* Once a whole-cell configuration has been achieved, the duration of stable recordings can be increased by gently moving the pipette backward a few μm approx. every 30 min. In this way one can compensate for the initial displacement of the cell by the pipette during approach and gigaseal formation.

5 Troubleshooting

- If gigaohm seals cannot be obtained, there are a couple of probable causes that can be checked:

  First of all the health state of the animal: Are all vital signs within a physiological range (i.e., PCO2) (35 ± 3 mmHg) if the mouse is ventilated, heart rate (545 ± 78 beats/min) and body temperature (36.5–37.5 °C)? Next step is to check the shape and the size of the pipette tip. The resistance needs to be kept in the right range (5–9 MΩ). The pressure line to the pipette holder should be checked for leaks as well. Also osmolarity and pH of used intracellular solution should have proper value (see above).
- If gigaohm seals can be obtained but cannot be broken, or last for very short time, the pressure line to the pipette holder should be checked for leaks or obstructions. Moreover tuning the puller settings to improve pipette resistance and tip shape can be of great importance.
- Finding and eliminating sources of vibration is crucial to get gigaseals and for an acceptable duration of the recordings.

In general it is crucial to monitor animal health condition during experiment, all the vital signs need to be in physiological range, osmolarity of used solution must be appropriate, shape and size of used pipettes should be proper, and pressure line has to be without any leaks or obstruction.

5.1 Expected Results

If the animal is handled correctly and if the preparation was done in such a way to keep the vital signs under physiological condition, recordings can be obtained for more than 12 h. Moreover stable recordings from single neurons can be obtained for up to 3 h. Per day in general two to eight cells can be recorded.

5.2 Histological Procedure

1. Deeply anesthetize the mouse with urethane. Check if there is any responses to tail/toe pinches and also if there is any corneal reflex. Proceed only after the mouse is unresponsive and all of the above reflexes are absent.
2. Put the mouse inside a chemical fume hood in the supine position.
3. Perform transcardial perfusion with cold phosphate buffer saline (PBS 0.1 M, 20 ml) followed by aldehyde-based fixative (4 % paraformaldehyde in 0.1 M PBS, 60 ml).
4. Carefully extract the brain from the skull (first remove the dura mater).
5. Put the brain in aldehyde-based fixative overnight for postfixation.

5.3 Biocytin Staining Procedure

1. Cut 100–150 μm tick sections of brain using a vibratome.
2. Wash the slices three times in PBS, 10 min each.
3. Incubate the slices at 37 °C for 1 h in a solution of Cytochrome-c (300 μg/ml), Catalase (200 μg/ml), and Diaminobenzidine (0.5 mg/ml).
4. For stopping reaction wash the wells with bleach.
5. Wash the slices four times in PBS, 10 min each.
6. Quenching of endogenous peroxidase is done by incubating slices in a solution of 3 % hydrogen peroxide in PBS for 10–15 min.
7. Wash the slices five times in PBS, 10 min each.
8. Put the slices for one hour in a solution of Triton X-100 2 % in PBS for permeabilization of the slices.
9. Leave slices in solution of VECTASTAIN® ABC kit (sensitive avidin/biotin-based peroxidase system) (20 drops of A and 20 drops of B) and Triton X-100 1 % at 4 °C overnight, or for 2 h at room temperature.
10. Wash the slices five times in PBS, 10 min each.
11. For 15 min incubate slices in a solution of Diaminobenzidine (0.5 mg/ml) and 3 % hydrogen peroxide.
12. Wash the slices five times in PBS, 10 min each.

13. Mount the slices according to the order of sectioning on slides with mounting medium based on Mowiol-488, as it does not require dehydration.
14. Analyze the slices by using bright-field microscope. Screening for the slides containing the cell should be done with a 5× or 10× objective. Fine neuroanatomical analysis (and eventually neuronal reconstructions) should be done with a 40× or 100× objective. For the purpose of neuroanatomical reconstructions, it is highly advisable to patch only one or two (better one) neuron per craniotomy.

5.4 Two-Photon-Targeted Juxtasomal Recording

Two-photon targeted patching is a method which uses two-photon microscopy to optically target neurons that express fluorescent markers such as GFP variants or red fluorescent protein. Nevertheless, this method is restricted to mice expressing reporter proteins in identified cell types. However, it is possible to optically target patch recordings to genetically non-labeled neurons by using the shadow patching technique. The latter consists in ejection of a fluorescent dye in the extracellular space to visualize and approach the neuron cell bodies (with the possibility to distinguish pyramidal shaped and non-pyramidal shaped cells), because the dye is not taken up by the neuronal cells which appear like black (non-stained) objects on light (stained) background of the extracellular space, hence the name shadow patching.

With the same pipette that is used for delivering the fluorescent dye and guided by the two-photon microscope to individual shadowed cells, patch clamp can be performed. Since there is a constant flow of dye from the pipette, detecting when the pipette comes in contact with a neuron is possible due to a fluorescent cleft made in the cell membrane. If the suction is applied promptly the gigaseal or patch clamp in the cell membrane is formed. Here we will describe how we do two-photon targeted juxtasomal recordings in our lab [30].

5.5 Procedure

1. Perform the surgery of the animal as described. Be particularly careful not to cause any bleeding and not to produce too much heat during the drilling. Always administer systemic cortisone, provide oxygen, and carefully monitor the animal physiological parameters.
2. Fill the patch pipette with a solution of dye Alexa 594 (25 μm) in Ringer solution. The resistance of the pipette for shadow patching should be between 5 and 9 MΩ and should remain constant when lowering the pipette in the brain. Do not apply high pressure (300 mbar) on the pipette to avoid excessive spillover of the dye, particularly on the cortical surface. Use 20–40 mbar instead.

3. Tune the laser to a wavelength that is optimal for the dye used (e.g., 780–800 nm for Alexa 594).
4. Lower the pipette until layer 2/3 is reached (150–300 μm). Take into account the pipette angle in calculating the absolute depth. During the descent, move the pipette along its axis and move the objective focus to keep the pipette tip on the focal plane. To this purpose, it is useful to keep ejecting some dye in the surrounding tissue.
5. Apply small puffs of the dye (40–100 mbar). Provided that the pipette is not clogged, every pulse should label a volume of about 300 μm diameter. Cell somata should be instantaneously counterstained.
6. By moving the pipette in the $x - y$ plane with a micromanipulator slowly approach the cell of interest within 50 μm of pipette tip to avoid excessive tissue damage.
7. Approach the cell while monitoring seal resistance, the spike shape, and the microscope frame: juxtasomal configuration is reached when: (a) seal resistance is in the order of 30–100 MΩ; (b) the spike shape is positive and has a height of about 2 mV. Check if you can drive the cell to fire using a range of currents 1–5 nA [45]. During the experiment monitor the spike shape and the absolute value of voltage: spike broadening and a DC shift towards hyperpolarized values usually indicate cell damage. In that case one can observe labeling of the soma.

Two-photon microscopy can also be used to optically target non-labeled neurons by counterstaining the extracellular space with a fluorescent dye such as Alexa 594. In this way, neuronal cell bodies are visualized as black "holes" or shadows and the pipette can be advanced to the cell membrane so to patch the neuron ("shadow" patching—[46]). Significantly, three-dimensional reconstruction of the optical sections taken through a volume of tissue allows to target recordings to pyramidal shaped, excitatory cells or to non-pyramidal shaped, putative inhibitory neurons, thus allowing targeting different cell types even in the case of the unstained mammalian neocortex.

References

1. Neher E, Sakmann B (1976) Single-channel currents recorded from membrane of denervated frog muscle fibres. Nature 260(5554):799–802
2. Sakmann B (2006) Patch pipettes are more useful than initially thought: simultaneous pre- and postsynaptic recording from mammalian CNS synapses in vitro and in vivo. Pflugers Arch 453(3):249–259
3. Feldmeyer D, Sakmann B (2000) Synaptic efficacy and reliability of excitatory connections between the principal neurones of the input (layer 4) and output layer (layer 5) of the neocortex. J Physiol 525(Pt 1):31–39
4. Margrie TW, Brecht M, Sakmann B (2002) In vivo, low-resistance, whole-cell recordings from neurons in the anaesthetized and awake mammalian brain. Pflugers Arch 444(4):491–498. doi:10.1007/s00424-002-0831-z
5. Jia H, Rochefort NL, Chen X, Konnerth A (2010) Dendritic organization of sensory

input to cortical neurons in vivo. Nature 464 (7293):1307–1312
6. Carandini M, Ferster D (2000) Membrane potential and firing rate in cat primary visual cortex. J Neurosci 20(1):470–484
7. Medini P (2011) Layer- and cell-type-specific subthreshold and suprathreshold effects of long-term monocular deprivation in rat visual cortex. J Neurosci 31(47):17134–17148
8. Priebe NJ, Mechler F, Carandini M, Ferster D (2004) The contribution of spike threshold to the dichotomy of cortical simple and complex cells. Nat Neurosci 7(10):1113–1122
9. Carandini M (2004) Amplification of trial-to-trial response variability by neurons in visual cortex. PLoS Biol 2(9):E264
10. Medini P (2011) Cell-type-specific sub- and suprathreshold receptive fields of layer 4 and layer 2/3 pyramids in rat primary visual cortex. Neuroscience 190:112–126
11. Smith SL, Smith IT, Branco T, Hausser M (2013) Dendritic spikes enhance stimulus selectivity in cortical neurons in vivo. Nature 503(7474):115–120
12. Brecht M, Sakmann B (2002) Dynamic representation of whisker deflection by synaptic potentials in spiny stellate and pyramidal cells in the barrels and septa of layer 4 rat somatosensory cortex. J Physiol 543(Pt 1):49–70
13. Anderson JS, Carandini M, Ferster D (2000) Orientation tuning of input conductance, excitation, and inhibition in cat primary visual cortex. J Neurophysiol 84(2):909–926
14. Borg-Graham L, Monier C, Fregnac Y (1996) Voltage-clamp measurement of visually-evoked conductances with whole-cell patch recordings in primary visual cortex. J Physiol Paris 90 (3–4):185–188
15. Monier C, Chavane F, Baudot P, Graham LJ, Fregnac Y (2003) Orientation and direction selectivity of synaptic inputs in visual cortical neurons: a diversity of combinations produces spike tuning. Neuron 37(4):663–680
16. Monier C, Fournier J, Fregnac Y (2008) In vitro and in vivo measures of evoked excitatory and inhibitory conductance dynamics in sensory cortices. J Neurosci Methods 169(2):323–365
17. Iurilli G, Olcese U, Medini P (2013) Preserved excitatory-inhibitory balance of cortical synaptic inputs following deprived eye stimulation after a saturating period of monocular deprivation in rats. PLoS One 8(12):e82044
18. Iurilli G, Ghezzi D, Olcese U, Lassi G, Nazzaro C, Tonini R, Tucci V, Benfenati F, Medini P (2012) Sound-driven synaptic inhibition in primary visual cortex. Neuron 73 (4):814–828
19. Priebe NJ, Ferster D (2005) Direction selectivity of excitation and inhibition in simple cells of the cat primary visual cortex. Neuron 45 (1):133–145
20. Wehr M, Zador AM (2003) Balanced inhibition underlies tuning and sharpens spike timing in auditory cortex. Nature 426(6965):442–446
21. Ma WP, Li YT, Tao HW (2013) Downregulation of cortical inhibition mediates ocular dominance plasticity during the critical period. J Neurosci 33(27):11276–11280
22. Gonchar Y, Wang Q, Burkhalter A (2007) Multiple distinct subtypes of GABAergic neurons in mouse visual cortex identified by triple immunostaining. Front Neuroanat 1:3
23. Gonchar Y, Burkhalter A (1997) Three distinct families of GABAergic neurons in rat visual cortex. Cereb Cortex 7(4):347–358
24. Margrie TW, Meyer AH, Caputi A, Monyer H, Hasan MT, Schaefer AT, Denk W, Brecht M (2003) Targeted whole-cell recordings in the mammalian brain in vivo. Neuron 39 (6):911–918
25. Gentet LJ, Avermann M, Matyas F, Staiger JF, Petersen CC (2010) Membrane potential dynamics of GABAergic neurons in the barrel cortex of behaving mice. Neuron 65 (3):422–435
26. Gentet LJ, Kremer Y, Taniguchi H, Huang ZJ, Staiger JF, Petersen CC (2012) Unique functional properties of somatostatin-expressing GABAergic neurons in mouse barrel cortex. Nat Neurosci 15(4):607–612
27. Runyan CA, Schummers J, Van Wart A, Kuhlman SJ, Wilson NR, Huang ZJ, Sur M (2010) Response features of parvalbumin-expressing interneurons suggest precise roles for subtypes of inhibition in visual cortex. Neuron 67 (5):847–857
28. Kuhlman SJ, Tring E, Trachtenberg JT (2011) Fast-spiking interneurons have an initial orientation bias that is lost with vision. Nat Neurosci 14(9):1121–1123
29. Kerlin AM, Andermann ML, Berezovskii VK, Reid RC (2010) Broadly tuned response properties of diverse inhibitory neuron subtypes in mouse visual cortex. Neuron 67(5):858–871
30. Olcese U, Iurilli G, Medini P (2013) Cellular and synaptic architecture of multisensory integration in the mouse neocortex. Neuron 79 (3):579–593
31. Madisen L, Mao T, Koch H, Zhuo JM, Berenyi A, Fujisawa S, Hsu YW, Garcia AJ 3rd, Gu X, Zanella S, Kidney J, Gu H, Mao Y, Hooks BM, Boyden ES, Buzsaki G, Ramirez JM, Jones AR, Svoboda K, Han X, Turner EE, Zeng H (2012) A toolbox of Cre-dependent optogenetic

transgenic mice for light-induced activation and silencing. Nat Neurosci 15(5):793–802
32. Moore AK, Wehr M (2013) Parvalbumin-expressing inhibitory interneurons in auditory cortex are well-tuned for frequency. J Neurosci 33(34):13713–13723
33. Katz Y, Yizhar O, Staiger J, Lampl I (2013) Optopatcher – an electrode holder for simultaneous intracellular patch-clamp recording and optical manipulation. J Neurosci Methods 214 (1):113–117
34. Chen X, Leischner U, Rochefort NL, Nelken I, Konnerth A (2011) Functional mapping of single spines in cortical neurons in vivo. Nature 475(7357):501–505
35. Lavzin M, Rapoport S, Polsky A, Garion L, Schiller J (2012) Nonlinear dendritic processing determines angular tuning of barrel cortex neurons in vivo. Nature 490(7420):397–401
36. Rancz EA, Franks KM, Schwarz MK, Pichler B, Schaefer AT, Margrie TW (2011) Transfection via whole-cell recording in vivo: bridging single-cell physiology, genetics and connectomics. Nat Neurosci 14(4):527–532
37. Cohen L, Koffman N, Meiri H, Yarom Y, Lampl I, Mizrahi A (2013) Time-lapse electrical recordings of single neurons from the mouse neocortex. Proc Natl Acad Sci U S A 110(14):5665–5670
38. Petersen CC, Grinvald A, Sakmann B (2003) Spatiotemporal dynamics of sensory responses in layer 2/3 of rat barrel cortex measured in vivo by voltage-sensitive dye imaging combined with whole-cell voltage recordings and neuron reconstructions. J Neurosci 23 (4):1298–1309
39. Bruno RM, Sakmann B (2006) Cortex is driven by weak but synchronously active thalamocortical synapses. Science 312(5780):1622–1627
40. Constantinople CM, Bruno RM (2013) Deep cortical layers are activated directly by thalamus. Science 340(6140):1591–1594
41. Kodandaramaiah SB, Franzesi GT, Chow BY, Boyden ES, Forest CR (2012) Automated whole-cell patch-clamp electrophysiology of neurons in vivo. Nat Methods 9(6):585–587
42. Tan AY, Chen Y, Scholl B, Seidemann E, Priebe NJ (2014) Sensory stimulation shifts visual cortex from synchronous to asynchronous states. Nature 509(7499):226–229
43. Lee AK, Manns ID, Sakmann B, Brecht M (2006) Whole-cell recordings in freely moving rats. Neuron 51(4):399–407
44. Spira ME, Hai A (2013) Multi-electrode array technologies for neuroscience and cardiology. Nat Nanotechnol 8(2):83–94
45. Pinault D (1996) A novel single-cell staining procedure performed in vivo under electrophysiological control: morpho-functional features of juxtacellularly labeled thalamic cells and other central neurons with biocytin or neurobiotin. J Neurosci Methods 65(2): 113–136
46. Kitamura K, Judkewitz B, Kano M, Denk W, Hausser M (2008) Targeted patch-clamp recordings and single-cell electroporation of unlabeled neurons in vivo. Nat Methods 5(1):61–67

Chapter 2

Juxtasomal Loose-Patch Recordings in Awake, Head-Fixed Rats to Study the Link Between Structure and Function of Individual Neurons

Christiaan P.J. de Kock

Abstract

The loose-patch juxtasomal recording method can be applied to characterize action potential spiking from single units in the extracellular configuration and includes the attractive option of labeling the neuron for post hoc identification and reconstruction. This ensures "observing without disturbing" (Schubert, J Physiol 581(Pt 1):5, 2007) since the juxtasomal loose-patch recording does not involve breaking into the neuron and modifying its intracellular environment until after *all* physiological parameters have been obtained. The fundamental difference with extracellular recordings is therefore that juxtasomal recordings generate a direct link between physiological properties and cellular morphology. The necessary step for juxtasomal labeling involves physical interaction between the recording patch pipette and somatic membrane to create a loose-seal patch-clamp recording (hence: juxtasomal) and electroporation for label dialysis (Joshi and Hawken, J Neurosci Methods 156(1–2):37–49, 2006; Pinault, J Neurosci Methods 65(2):113–136, 1996). Next, post hoc histology is performed to reveal cell-type identity and optionally to digitally reconstruct the recorded neuron. In this chapter, I will describe the basic experimental procedures to obtain juxtasomal recordings in primary somatosensory cortex of awake, head-fixed rats and illustrate the information content of these experiments.

Key words Juxtasomal, Loose patch, In vivo, Action potential, Morphology, Histology, Single unit

1 Introduction

The revolutionary work of Santiago Ramón y Cajal and Camillo Golgi on the structure of the nervous system continues to inspire the neuroscience community [4, 5]. Compared to these pioneering studies, the contemporary billion dollar initiatives such as B.R.A.I.N. (Brain Research through Advancing Innovative Neurotechnologies) [6] and H.B.P. (Human Brain Project) [7, 8] are contrasting initiatives in economical and collaborative aspects. Additionally, the speed of scientific progress will be several orders of magnitude away from manual documentation of single brain areas. To understand the human brain or even local microcircuits however, there is still an

Alon Korngreen (ed.), *Advanced Patch-Clamp Analysis for Neuroscientists*, Neuromethods, vol. 113,
DOI 10.1007/978-1-4939-3411-9_2, © Springer Science+Business Media New York 2016

urgent need to carefully map the function and structure of individual neurons to be able to reliably ascend from subcellular level studies to meta-analysis and comprehensive models [9].

For a long time, the experimental standard to study physiological properties of neurons was to characterize single- or multi-unit activity using extracellular electrodes. Recording depth was the only parameter available to coarsely subclassify recorded neurons [10–12] and morphological reconstruction of recorded neurons was not performed. Probably driven by the question how physiology emerges from morphology, a few pioneering studies at that point started to identify recorded neurons and changed the frontier of cellular physiology [13–17]. The patch-clamp technique by Erwin Neher and Bert Sakmann [18] in combination with biocytin or neurobiotin loading for post hoc morphological reconstruction [19, 20] revolutionized studies linking physiological properties and morphology of individual neurons and local microcircuits at subcellular resolution [21–24].

With the patch-clamp technique and biocytin labeling available after the 1980s, a feasible experimental approach was suddenly at hand to study the structure and function of individual neurons within the same datasets and determine for the first time how function could emerge from structure. Perhaps one of the best examples of the strength of this approach has been the uncovering of the function of different types of neurons that together constitute the cortical column (for instance [25–32]). In retrospect, the introduction of both the patch-clamp technique and biocytin/neurobiotin labeling techniques boosted the number of studies showing cell-type-specific structure and function. At present, the neuroscience community seems to have realized that the new standard should be to determine the identity of recorded neurons, independent of brain area, species, slice preparation or in vivo.

One of the available techniques is the juxtasomal (or juxtacellular) loose-patch recording technique to obtain "morpho-functional features," first published by Didier Pinault in the *Journal of Neuroscience Methods* [3]. This technique has proven to be applicable across an impressive range of experimental settings including different species (rat, mouse, monkey, goldfish), brain areas (cerebral cortex, thalamus, striatum, ventral tegmental area, locus coeruleus, cerebellum), and perhaps most importantly, behavioral states (anesthetized, awake head-restrained and freely moving animals) [2, 24, 33–42]. The most obvious limitation of the technique is almost certainly the lack of information on subthreshold membrane dynamics and therefore only generates data on action potential spiking of the recorded neurons (Table 1). However, if one aims to understand the cellular basis of relatively simple behaviors, such as sensory-guided decision making [43], spatial navigation [39, 44, 45], or sensory detection [46, 47], action potential spiking of

Table 1
Characteristics of juxtasomal recordings

Advantages	Disadvantages
General: stable recordings, also in awake, head-fixed animals	
General: applicable across brain areas, species, behavioral state	
Morphology: dense labeling for morphology	No online control of labeling quality
Physiology: single-unit isolation	No information on population dynamics/synchrony
Physiology: action potential spiking without intracellular dialysis	No information on subthreshold membrane potential dynamics
Unbiased sampling possible, irrespective of action potential spiking frequencies	No optical control of unit selection
	Narrow bandwidth of current injections, optimal result requires experience

individual projection neurons is much more relevant for behavioral output compared to subthreshold voltage fluctuations.

In the protocol below, the methods of obtaining a juxtasomal recording is exemplified for an awake, head-restrained Wistar rat (P37, bodyweight 144 g, ♂). To obtain juxtasomal recordings from (urethane) anesthetized Wistar rats, only modifications to the surgical procedure are necessary and were described in detail previously [48].

2 Protocol: Juxtasomal Recordings in Somatosensory Cortex of Awake, Head-Fixed Wistar Rats

All experimental procedures are carried out in accordance with the Dutch law and after evaluation by a local ethical committee at the VU University Amsterdam, The Netherlands.

- For presurgical training, see Sect. 2.2.

2.1 Preparation of the Animal (Mounting Head-Post)

- Anesthetize a Wistar rat (P35-P45) with isoflurane (2–3 % in 0.4 l/min O_2, 0.7 l/min N_2O) and subsequently decrease isoflurane to 1.6 % to maintain stable anesthesia throughout the surgical procedure. Depth of anesthesia should be checked by monitoring pinch withdrawal, eyelid reflexes, and vibrissae movements.

  Note: without intubation, the isoflurane concentration is not calibrated and small differences between setups are likely to occur.

- Position the anesthetized rat in a stereotactic frame equipped with a heating pad, blunt ear bars, and a mouth clamp (e.g., RA-6N, Narishige, Japan). Insert the rectal temperature probe and maintain the rat's body temperature at 37.5 ± 0.5 °C using the heating pad.
- Trim the hair on the operational site using scissors.
- Inject 100 μl 1 % lidocaine (in 0.9 % NaCl) subcutaneously at the operational site for local anesthesia. After 3–5 min, make a 3 cm incision along the rostro-caudal axis and move the skin laterally using vascular clamps (standard micro-serrefines).
- Remove the periosteum and clean the exposed skull extensively with 0.9 % NaCl, 1 % H_2O_2, and finish with a few drops of 70 % ethanol to completely dry the skull.
- Add gel etchant (Kerr Corporation, Orange, USA) to the exposed and cleaned surface of the skull and wait 30 s.
- Clean the exposed skull extensively with 0.9 % NaCl, 1 % H_2O_2, and 70 % ethanol.
- Add OptiBond FL primer and adhesive (Kerr Corporation, Orange, USA) to establish a thin first layer of cement.
- Use a dental drill to scrape off the dental cement only at the site where the craniotomy is to be made. The advantage of this approach is that a maximal surface of the skull is used to establish adhesive contact between skull and head-post.
- Thin the skull at the site of the craniotomy and make a small (0.5 mm × 0.5 mm) craniotomy, avoiding damage to the dura mater and blood vessels. To target primary somatosensory cortex of adolescent Wistar rats, center the craniotomy at 2.5 mm posterior and 5.5 mm lateral with respect to Bregma.
- Position and fasten the small ring over the craniotomy with Tetric evo flow (Ivoclar Vivadent, Amherst, USA) to protect the craniotomy during the habituation training sessions yet leaving the craniotomy accessible for the recording day.
- Add Charisma dental cement (Kerr Corporation, Orange, USA) to establish a second layer of cement and carefully position the head-post on the unpolymerized Charisma.
- When head-post is positioned correctly, polymerize Charisma and finish with Tetric evo flow (Ivoclar Vivadent, Amherst, USA).
- Use superglue to glue the skin onto the last layer of Tetric evo flow and make sure that the skin tightly seals around the head-post.
- Extensively rinse the craniotomy with 0.9 % NaCl, leave the craniotomy moist, and seal the ring with the screw cap.

2.2 Animal Training

- Rats should be habituated to head restraining prior to the recording session to avoid stress-related effects on electrophysiological parameters.
- The habituation schedule involves pre- and postsurgical components. In the week before surgery (day −7 to −1 with respect to surgery on day 0), rats are handled twice a day (at fixed time points) to accustom the rat to interaction with the experimenter. Additionally, enriched housing (bedding, shelter, nesting material, wooden sticks) provides obvious welfare advantages, also after surgical preparation.
- Monitor bodyweight during the morning session and keep food and water ad libitum.
- After surgery (on day 0), habituate the rat to head restraining by head-fixing the rat twice per day (at fixed time points on day 1–3) using increasing duration of head fixation. Typically, the schedule of 5–10, 20–25, 30–40 min results in habituated rats allowing stable juxtasomal recordings on the experimental day (day 4). After each training session, rats are placed back in the enriched cage and receive a special food reward on top of their standard food pallets. For instance, the standard food pallet can be soaked in sugar water to produce an appealing food reward associated with the head-fixation procedure.
- Rats of P30-45 will show linear increase in body weight during the complete experimental paradigm (handling–surgery–habituation) except for a relatively stable body weight on the day of surgery. During habituation trainings, rats will gain body weight at a rate comparable to handling sessions. Rats that do not habituate to head fixation (reflected in reduced weight gain or even weight loss) in conjunction with signs of aberrant stress during head fixation (increased number of feces during head fixation, freezing, or bloodshot eyes) should be taken out of the experiment. In practice only a very small fraction of rats do not habituate (<1 %).
- At the end of the habituation sessions, whisking behavior during head fixation closely resembles normal exploratory behavior and allows studying sensory processing during free whisking or active object touch [28, 40, 49–51].

2.3 Juxtasomal Recordings and Biocytin Labeling

- As indicated previously, to target rat primary somatosensory (barrel) cortex, center stereotactic coordinates at 2.5 mm posterior, 5.5 mm lateral with respect to Bregma. Extracellular mapping techniques can be used to target individual barrel columns or alternatively, intrinsic optical imaging allows anatomical mapping at single barrel column resolution through the thinned skull [24].

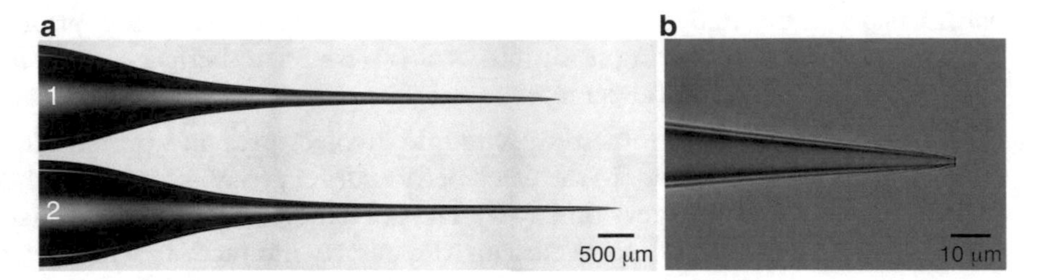

Fig. 1 *Electrode characteristics.* (**a**) The ideal electrode has a long, tapered shank whose length is adjusted to the recording depth. Electrode 1 aims at recording from supragranular layers in rat; electrode 2 has a longer shank and allows recording from granular and infragranular layers without damaging superficial layers at point of entry. The maximal diameter of the electrode inside the brain is ~75 μm. (**b**) Inner diameter of electrode tip is ~1 μm, resulting in electrode resistance of 3–6 MΩ

- Patch pipettes of borosilicate glass are optimal for single-unit isolation and biocytin labeling using the juxtasomal recording and labeling technique (Fig. 1). Patch pipettes are filled with Normal Rat Ringer (in mM: 135 NaCl, 5.4 KCl, 1.8 $CaCl_2$, 1 $MgCl_2$, and 5 HEPES, pH adjusted to 7.2 with NaOH, and 20 mg/ml biocytin) and yield electrodes with resistances of 3–6 MΩ. The ideal pipette morphology for the juxtasomal recording is a gradual slender taper, a low cone angle, and a tip with ~1 μm inner diameter (optional: check tip shape with 100× air objective from Olympus, MPLFLN 100×/0.90 M Plan Fluorite WD 1.0 mm).
- Depending on the recording depth (with respect to pial surface), the taper dimensions of the recording electrode are adjusted. The taper diameter should be <75 μm at point of entry in the brain to avoid mechanical damage or stress to the recording area. For recordings in supragranular layers of rat primary somatosensory cortex, a taper of 300–500 μm with an outer diameter of maximally 75 μm suffices (Fig. 1, electrode 1) whereas recordings from granular and infragranular layers require electrodes with a taper of 600–2000 μm, again with an outer diameter that does not exceed 75 μm at point of entry (thus: at 600–2000 μm from electrode tip).
- To target the D2 column of adolescent Wistar rat primary somatosensory cortex (P35-45), set the angle of the electrode to 34° with respect to the sagittal plane.
- Connect the head stage to an amplifier in bridge- or current-clamp mode.
- Position the electrode in close proximity of the craniotomy. Fill the recording chamber with 0.9 % NaCl and determine the electrode resistance by applying a square pulse of 1 nA positive current injection (200 ms on/off).

- Apply 100–150 mbar overpressure on the recording electrode and continuously monitor the electrode resistance. Advance with 1 μm steps until the resistance increases, which reflects contact with the dura mater. At this point, set the coordinates of the micromanipulator to "zero" to allow accurate depth measurement after single-unit isolation.
- Advance in 1 μm steps until the patch pipette penetrates the dura mater, which can be observed as a sudden drop in electrode resistance. Remove the holding pressure from the electrode.
- Search for single units while advancing in 1 μm steps and monitor the electrode resistance continuously using square pulse current injection. Proximity and physical contact of individual neurons lead to an increase in electrode resistance. Slowly advance the electrode until positive action potential waveforms of ~2 mV are recorded.

  Spiking frequencies obtained in rat somatosensory cortex across behavioral states are typically in the order of 0.1–5.0 Hz, characteristic of sparse coding [1, 40, 49, 52] although a subset of (inter)neurons have been recorded at higher spiking rates [28, 49, 53]. Regardless of spiking frequencies, the juxtasomal loose-patch recording ensures unprecedented single-unit isolation using conventional cluster cutting procedures adapted from extracellular recording methods. A signal to noise ratio of 4:1 (~2 mV spikes) allows reliable detection of spikes using MClust (David Redish, University of Minnesota, USA) based on either peak/valley or principle component analysis (PCI1 vs. PCI2). Additionally, interspike interval distributions can be plotted to confirm the presence of a refractory period of ~3 ms [54, 55], highly indicative of single-unit isolation. Single-unit isolation for electrophysiology, labeling, and reconstruction of only the recorded neuron is critical when studying the function of individual neurons in brain areas with intermingled cell types (e.g., cerebral cortex).
- To study action potential spiking characteristics of individual units in rat primary somatosensory cortex in awake, head-restrained rats, equip the setup with high-speed videography to monitor whisker position and movement. Free whisking involves stereotypic protraction and retraction of the whiskers at 4–12 Hz [51, 56–58] which can be captured sufficiently at ~100 Hz imaging resolution. To study active object touch, a higher temporal resolution (200–500 Hz) is typically used [28, 49, 53] (Fig. 2, 200 Hz).

  The combination of electrophysiology and high-speed videography captures action potential spiking synchronized to behavior. The electrophysiological data is analyzed relatively straightforward and spikes can be regarded as binary events as a first step (ignoring amplitude adaptation during bursts). The

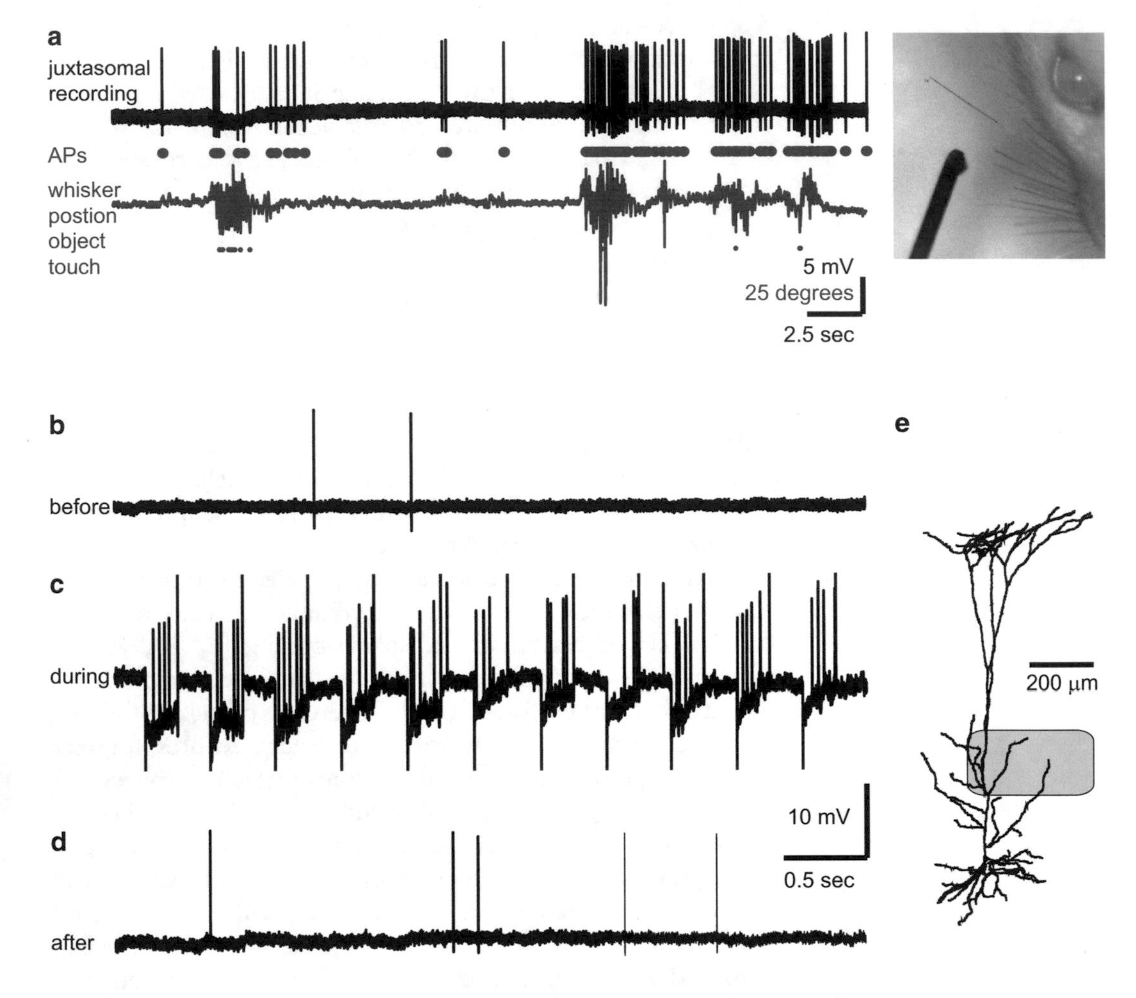

Fig. 2 *Juxtasomal recording of post hoc reconstructed L5B thick tufted pyramid in primary somatosensory cortex of awake, head-restrained rat.* (**a**) Example experiment combining juxtasomal recording with high-speed videography (@ 200 Hz) to obtain single-unit spiking properties during tactile exploration. Juxtasomal recording in *black*, spikes indicated as individual blue bullets, whisker position is tracked off-line (in *grey*) and in *red*, windows during which whisker was in contact with object. Note that spiking frequency was increased during free whisking and active object touch. (**b**) Spiking profile of same neuron at increased temporal resolution before biocytin labeling. (**c**) Spiking profile during biocytin loading. Note the increased action potential spiking frequency during on-phase of current injections (200 ms on/off). (**d**) Spiking profile after biocytin loading. The recording of spontaneous activity after labeling allows recovery from high action potential frequencies associated with breaking into the neuron and excessive inflow of extracellular ions. (**e**) Post hoc Neurolucida reconstruction to classify recorded neuron from panels **a**–**d** (L4 barrel contour in *grey*). Extensive apical tuft branching is characteristic of L5B thick tufted pyramidal neuron. Thus, the procedure allows a link between physiological properties of recorded neuron during somatosensory processing to single-cell morphological identity

behavioral data is much more complex and parameters on whisker use are tracked off-line [59] and can be represented as whisker position (degree), velocity (degrees/s), acceleration (degrees/s^2), whisker curvature, touch times, or a multidimensional combination to correlate single-unit spiking to sensory behavior.

- After obtaining all physiological parameters, the neuron can be biocytin labeled for post hoc identification and morphological reconstruction.
- For juxtasomal biocytin labeling, advance the electrode until the electrode resistance is 25–35 MΩ and spikes have amplitudes of 3–8 mV to obtain optimal conditions of juxtasomal filling. Start the juxtasomal filling by applying square pulses of positive current (1 nA, 200 ms on/off). Slowly and gradually increase the current by steps of 0.1 nA while closely monitoring the action potential waveform and frequency (Fig. 2).
- Monitor the membrane opening as a clear increase in action potential frequency during the on-phase of the block pulse (Fig. 2c). The spike waveform during filling shows an increased width and reduced after-hyperpolarization (Fig. 2c, d). Additional parameters include increased noise or a small (1–5 mV) negative DC shift [48].
- To maintain stable biocytin infusion after opening of the membrane (reflected by robust increase in action potential frequency during on-phase of the block pulse), the amplitude of current injections can typically be reduced (1–3 nA). Stop or even further reduce the current pulses upon sudden increase of the action potential frequency (also during off-phase) to avoid toxicity by excess influx of extracellular ions.
- Closely monitor the action potential spiking frequency after stopping the current injection. The spike waveform after a filling session is usually broadened and shows a strongly reduced after-hyperpolarization. Wait for recovery of the neuron, which is apparent when the spike waveform and action potential frequency return to its original properties (i.e., presence of normal after-hyperpolarization, Fig. 2).
- Repeat biocytin filling sessions after complete recovery of the neuron to increase biocytin load for improved staining quality.
- Retract the patch pipette in steps of 1 μm until the spike amplitude decreases to reduce any mechanical stress to the neuron. For cell-type identification and/or dendritic reconstructions, a typical diffusion time of 15–20 min is sufficient.
- After biocytin labeling, take the rat out of the head-fixation apparatus and anesthetize deeply with non-gaseous anesthesia (e.g., urethane or ketamine/xylazine) for transcardial perfusion with 0.9 % NaCl and subsequent fixation with 4 % paraformaldehyde (in 0.12 M phosphate buffer).

2.4 Perfusing the Animal and Removing the Brain

- Prepare the perfusion setup, rinse and pre-load the tubing with 0.9 % NaCl.
- Secure the rat on a surgical tray. Ensure sufficient depth of the anesthesia; foot pinch and eyelid reflexes should be absent.
- Make a medial to lateral incision through the abdominal wall just beneath the rib cage and proceed in posterior-anterior direction to expose the sternum. Pull the sternum in anterior direction, make a small incision in the diaphragm, cut through the lower ribs, and continue the incision along the entire length of the abdominal cavity to expose the heart.
- Remove the pericardium.
- Insert the needle into the left ventricle and make an incision in the right atrium. Perfuse with 0.9 % NaCl (~8 ml/min).
- Switch the infusion to 4 % paraformaldehyde (PFA) to fix the rat until stiffness of front paw and lower jaw is apparent.
- Decapitate the rat using a pair of scissors.
- Trim the remaining neck muscles and expose the skull completely.
- Position the scissors in the brain stem on the dorsal side and cut the bone carefully along the sagittal suture, maintaining the dorsal position.
- Remove the bones from both sides of the sagittal suture to expose the brain by using a forceps. Carefully remove the dura to avoid damage.
- Carefully insert a blunt spatula to the ventral side of the brain and remove the brain gently.
- Post-fix the entire brain overnight in 4 % PFA at 4 °C. Switch the brain to 0.05 M phosphate buffer (PB) and store at 4 °C.
- To slice the brain in 100 μm tangential sections, take the brain out of the 0.05 M PB and put it on a filter paper facing anterior. Use a sharp razor blade to cut off the cerebellum along the coronal plane and separate the hemispheres by cutting along the midsagittal plane.
- Apply superglue on the mounting platform and mount the left hemisphere on its sagittal plane with anterior facing right. Secure the mounting platform at an angle of 45° on a vibratome and submerge the brain in 0.05 M PB.
- Secure a razor blade on the vibratome and make sure that the first contact with the brain surface is in the middle of the anterior-posterior plane of the hemisphere. Cut 24 100 μm sections and collect them in a 24-well plate containing 0.05 M PB.

2.5 Histological Procedures

- Histological protocols for the cytochrome oxidase staining and the avidin-biotin-peroxidase method are performed according to previously described methods [19, 48, 60]. Optional: visualize biocytin using fluorescent avidin/streptavidin-Alexa conjugates. This additionally allows double staining with retrograde or anterograde tracing techniques.
- Wash sections 5 × 5 min with 0.05 M PB and prepare the 3,-3′-diaminobenzidine tetrahydrochloride (DAB)-containing solution (0.2 mg/ml CytC, 0.2 mg/ml Catalase, 0.5 mg/ml DAB in 0.05 PB) for the cytochrome oxidase staining to visualize barrels in layer 4 of primary somatosensory cortex. Incubate sections 6–12 from the pia in the preheated solution for 30–45 min at 37 °C.
- Rinse sections with 0.05 M PB for 6 × 5 min and quench endogenous peroxidase activity by incubating all sections in 3 % H_2O_2 in 0.05 M PB for 20 min at room temperature (RT).
- Rinse sections with 0.05 M PB for 5 × 10 min. Incubate sections in ABC solution overnight at 4 °C containing 0.05 M PB, 0.5 % Triton, 1 drop of components A and B/ 10 ml 0.05 PB (ABC Kit Vector Laboratories, Burlingame, USA).
- Rinse sections with 0.05 M PB for 5 × 10 min and prepare the DAB solution containing 0.05 M PB, 0.5 mg/ml DAB, 0.1 % H_2O_2 to visualize the biocytin-filled neuron. Incubate sections in filtered solution for 45–60 min at RT.
- Rinse sections with 0.05 M PB for 5 × 10 min Mount sections on microscope slides and cover slip with mowiol.
- Determine labeling quality using light microscopy (Fig. 3).

3 Outlook

In conclusion, the juxtasomal recording method generates data on action potential spiking of single, identified neurons in anesthetized or awake, behaving animals. This allows careful dissection of neuronal microcircuits consisting of a wide range of cell types, for instance the cortical column, and aims to address questions on cell-type-specific function during information processing and behavioral output. At present, not only patch-clamp techniques such as juxtasomal recordings (this chapter) or the tight-seal whole-cell recording method (for instance Chaps. 1, 6 and 9 [61]) are routinely applied to study related research problems but alternative methods exist such as 2-photon imaging alone or in combination with electron microscopy reconstruction [62, 63]. In general, these methods can be highly synergistic with juxtasomal recordings since they generate information on the level of network

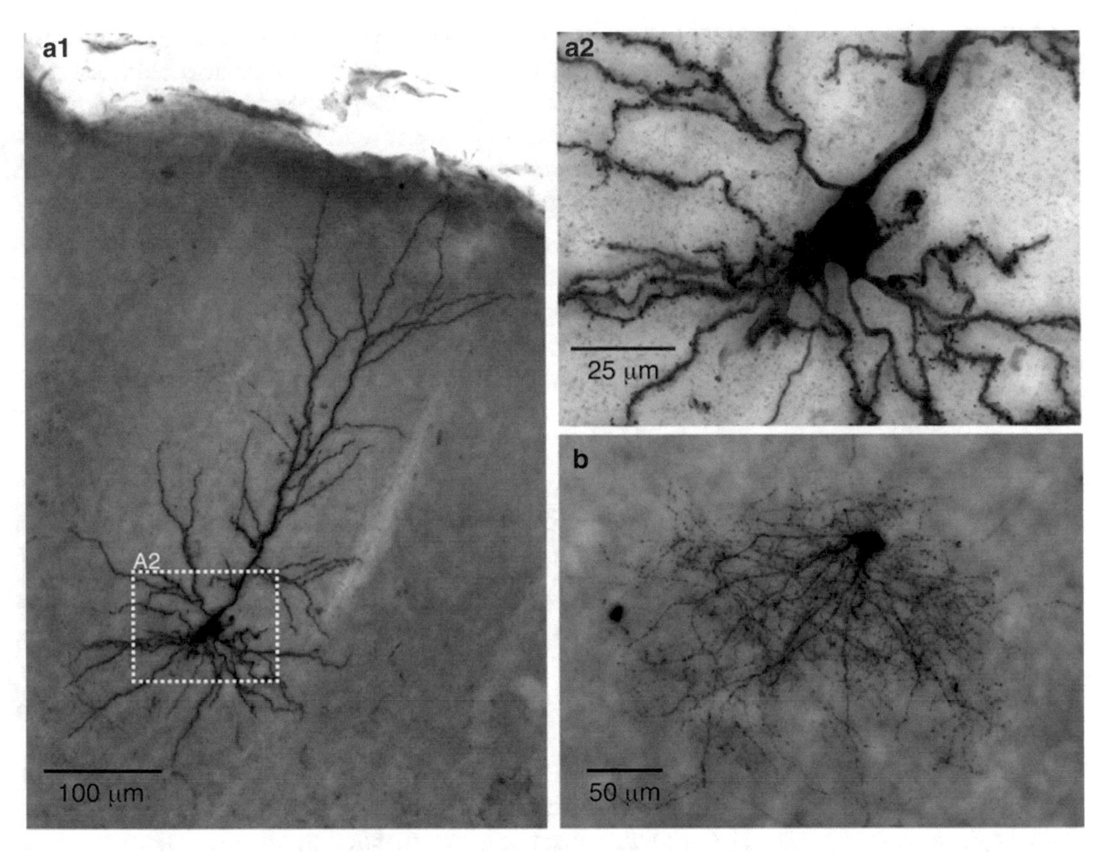

Fig. 3 *Photographs of biocytin-labeled neurons.* (**a1**) Coronal view of a rat layer 3 pyramidal neuron with 4× objective. (**a2**) Same neuron as (**a1**) but at high magnification (100× objective). (**b**) Tangential view of a mouse layer 4 spiny stellate (20× objective)

structure and function. Briefly, whole-cell patch-clamp recordings (unlike juxtasomal recordings) are highly suitable to study spontaneous or stimulus-evoked subthreshold membrane voltage dynamics. To record these membrane potential fluctuations however, it is necessary to break into the cell and dialyze the intracellular which inherently will affect electrolyte balance and action potential generation [64]. Including biocytin in the electrode solution permits post hoc reconstruction of cellular morphology at micrometer resolution [23, 30, 65]. The 3D volume that is occupied by distally projecting axons can frequently be up to several cubic millimeters and complete reconstruction of neurons from in vivo whole-cell recordings (as well as juxtasomal recordings) thus involves relatively large volumes [66–68]. In contrast, 2P imaging and dense EM reconstruction at nanometer resolution of the imaged network exclusively involves much smaller volumes (100 s of μm^3) [69]. The major advantage is obviously the possibility to study population activity and neuronal synchrony in addition to connectivity parameters at single-synapse resolution, but 2P imaging techniques are limited to optically accessible (hence superficial) brain areas and the dimensions of EM reconstructed brain tissue can never compete with reconstruction of relatively large volumes obtained with whole-cell or juxtasomal recordings. Ideally these different patch-clamp and imaging techniques are combined to reach a comprehensive understanding on the structure and function of individual neurons and/or networks. Eventually, merging data from different approaches will lead to comprehensive models on brain function and a full understanding of the cellular basis of simple behaviors [25, 43, 70–74].

Acknowledgements

I was introduced to the juxtasomal loose-patch technique by Randy Bruno (Columbia University, NY, USA) under the supervision of Prof. Dr. Bert Sakmann and thank both Randy and Prof. Sakmann for their continuous support, enthusiasm, and fruitful collaborations. Additionally, I'd like to thank Anton Pieneman for excellent technical support.

References

1. Schubert D (2007) Observing without disturbing: how different cortical neuron classes represent tactile stimuli. J Physiol 581(Pt 1):5
2. Joshi S, Hawken MJ (2006) Loose-patch-juxtacellular recording in vivo – a method for functional characterization and labeling of neurons in macaque V1. J Neurosci Methods 156 (1–2):37–49
3. Pinault D (1996) A novel single-cell staining procedure performed in vivo under electrophysiological control: morpho-functional features of juxtacellularly labeled thalamic cells and other central neurons with biocytin or neurobiotin. J Neurosci Methods 65 (2):113–136
4. DeFelipe J (2013) Cajal and the discovery of a new artistic world: the neuronal forest. Prog Brain Res 203:201–220
5. Sotelo C (2011) Camillo Golgi and Santiago Ramon y Cajal: the anatomical organization of the cortex of the cerebellum. Can the neuron doctrine still support our actual knowledge on the cerebellar structural arrangement? Brain Res Rev 66(1–2):16–34
6. Devor A et al (2013) The challenge of connecting the dots in the B.R.A.I.N. Neuron 80 (2):270–274
7. Markram H (2006) The blue brain project. Nat Rev Neurosci 7(2):153–160
8. Markram H (2012) The human brain project. Sci Am 306(6):50–55
9. Markram H (2013) Seven challenges for neuroscience. Funct Neurol 28(3):145–151
10. Armstrong-James M, Fox K, Das-Gupta A (1992) Flow of excitation within rat barrel cortex on striking a single vibrissa. J Neurophysiol 68(4):1345–1358
11. Mountcastle VB, Davies PW, Berman AL (1957) Response properties of neurons of cat's somatic sensory cortex to peripheral stimuli. J Neurophysiol 20(4):374–407
12. Simons DJ (1978) Response properties of vibrissa units in rat SI somatosensory neocortex. J Neurophysiol 41(3):798–820
13. Gilbert CD, Wiesel TN (1979) Morphology and intracortical projections of functionally characterised neurones in the cat visual cortex. Nature 280(5718):120–125
14. Mitani A et al (1985) Morphology and laminar organization of electrophysiologically identified neurons in the primary auditory cortex in the cat. J Comp Neurol 235(4):430–447
15. Larkman A, Mason A (1990) Correlations between morphology and electrophysiology of pyramidal neurons in slices of rat visual cortex. I. Establishment of cell classes. J Neurosci 10(5):1407–1414
16. Mason A, Larkman A (1990) Correlations between morphology and electrophysiology of pyramidal neurons in slices of rat visual cortex. II. Electrophysiology. J Neurosci 10 (5):1415–1428

17. Powell TP, Mountcastle VB (1959) Some aspects of the functional organization of the cortex of the postcentral gyrus of the monkey: a correlation of findings obtained in a single unit analysis with cytoarchitecture. Bull Johns Hopkins Hosp 105:133–162
18. Hamill OP et al (1981) Improved patch-clamp techniques for high-resolution current recording from cells and cell-free membrane patches. Pflugers Arch 391(2):85–100
19. Horikawa K, Armstrong WE (1988) A versatile means of intracellular labeling: injection of biocytin and its detection with avidin conjugates. J Neurosci Methods 25(1):1–11
20. Marx M et al (2012) Improved biocytin labeling and neuronal 3D reconstruction. Nat Protoc 7(2):394–407
21. Markram H et al (1997) Physiology and anatomy of synaptic connections between thick tufted pyramidal neurones in the developing rat neocortex. J Physiol 500(Pt 2):409–440
22. Feldmeyer D et al (1999) Reliable synaptic connections between pairs of excitatory layer 4 neurones within a single "barrel" of developing rat somatosensory cortex. J Physiol 521(Pt 1):169–190
23. Brecht M, Sakmann B (2002) Whisker maps of neuronal subclasses of the rat ventral posterior medial thalamus, identified by whole-cell voltage recording and morphological reconstruction. J Physiol 538(Pt 2):495–515
24. de Kock CP et al (2007) Layer and cell type specific suprathreshold stimulus representation in primary somatosensory cortex. J Physiol 581 (1):139–154
25. Feldmeyer D et al (2012) Barrel cortex function. Prog Neurobiol 2013 Apr;103:3–27. doi: 10.1016/j.pneurobio.2012.11.002. Epub 2012 Nov 27. Review.
26. Mountcastle VB (1997) The columnar organization of the neocortex. Brain 120(Pt 4):701–722
27. Oberlaender M et al (2012) Cell type-specific three-dimensional structure of thalamocortical circuits in a column of rat vibrissal cortex. Cereb Cortex 22(10):2375–2391
28. Gentet LJ et al (2012) Unique functional properties of somatostatin-expressing GABAergic neurons in mouse barrel cortex. Nat Neurosci 15(4):607–612
29. Schubert D, Kotter R, Staiger JF (2007) Mapping functional connectivity in barrel-related columns reveals layer- and cell type-specific microcircuits. Brain Struct Funct 212 (2):107–119
30. Brecht M, Roth A, Sakmann B (2003) Dynamic receptive fields of reconstructed pyramidal cells in layers 3 and 2 of rat somatosensory barrel cortex. J Physiol 553(Pt 1):243–265
31. Brecht M, Sakmann B (2002) Dynamic representation of whisker deflection by synaptic potentials in spiny stellate and pyramidal cells in the barrels and septa of layer 4 rat somatosensory cortex. J Physiol 543(Pt 1):49–70
32. Manns ID, Sakmann B, Brecht M (2004) Sub- and suprathreshold receptive field properties of pyramidal neurones in layers 5A and 5B of rat somatosensory barrel cortex. J Physiol 556(Pt 2):601–622
33. Klausberger T et al (2003) Brain-state- and cell-type-specific firing of hippocampal interneurons in vivo. Nature 421 (6925):844–848
34. Mileykovskiy BY, Kiyashchenko LI, Siegel JM (2005) Behavioral correlates of activity in identified hypocretin/orexin neurons. Neuron 46 (5):787–798
35. Bevan MD et al (1998) Selective innervation of neostriatal interneurons by a subclass of neuron in the globus pallidus of the rat. J Neurosci 18 (22):9438–9452
36. Varga C, Golshani P, Soltesz I (2012) Frequency-invariant temporal ordering of interneuronal discharges during hippocampal oscillations in awake mice. Proc Natl Acad Sci U S A 109(40):E2726–E2734
37. Voigt BC, Brecht M, Houweling AR (2008) Behavioral detectability of single-cell stimulation in the ventral posterior medial nucleus of the thalamus. J Neurosci 28(47):12362–12367
38. Aksay E et al (2000) Anatomy and discharge properties of pre-motor neurons in the goldfish medulla that have eye-position signals during fixations. J Neurophysiol 84 (2):1035–1049
39. Burgalossi A et al (2011) Microcircuits of functionally identified neurons in the rat medial entorhinal cortex. Neuron 70(4):773–786
40. de Kock CP, Sakmann B (2009) Spiking in primary somatosensory cortex during natural whisking in awake head-restrained rats is cell-type specific. Proc Natl Acad Sci U S A 106 (38):16446–16450
41. Jorntell H, Ekerot CF (2006) Properties of somatosensory synaptic integration in cerebellar granule cells in vivo. J Neurosci 26 (45):11786–11797
42. Boudewijns ZS et al (2013) Layer-specific high-frequency action potential spiking in the prefrontal cortex of awake rats. Front Cell Neurosci 7:99
43. Helmstaedter M et al (2007) Reconstruction of an average cortical column in silico. Brain Res Rev 55(2):193–203
44. Ray S et al (2014) Grid-layout and theta-modulation of layer 2 pyramidal neurons in medial entorhinal cortex. Science 343(6173): 891–896

45. Burgalossi A, Brecht M (2014) Cellular, columnar and modular organization of spatial representations in medial entorhinal cortex. Curr Opin Neurobiol 24(1):47–54
46. Doron G et al (2014) Spiking irregularity and frequency modulate the behavioral report of single-neuron stimulation. Neuron 81 (3):653–663
47. Houweling AR, Brecht M (2008) Behavioural report of single neuron stimulation in somatosensory cortex. Nature 451(7174):65–68
48. Narayanan RT et al (2014) Juxtasomal biocytin labeling to study the structure-function relationship of individual cortical neurons. J Vis Exp 84:e51359
49. O'Connor DH et al (2010) Neural activity in barrel cortex underlying vibrissa-based object localization in mice. Neuron 67(6): 1048–1061
50. Bagdasarian K et al (2013) Pre-neuronal morphological processing of object location by individual whiskers. Nat Neurosci 16 (5):622–631
51. Carvell GE, Simons DJ (1990) Biometric analyses of vibrissal tactile discrimination in the rat. J Neurosci 10(8):2638–2648
52. Barth AL, Poulet JF (2012) Experimental evidence for sparse firing in the neocortex. Trends Neurosci 35(6):345–355
53. Curtis JC, Kleinfeld D (2009) Phase-to-rate transformations encode touch in cortical neurons of a scanning sensorimotor system. Nat Neurosci 12(4):492–501
54. de Kock CP, Sakmann B (2008) High frequency action potential bursts (>or = 100 Hz) in L2/3 and L5B thick tufted neurons in anaesthetized and awake rat primary somatosensory cortex. J Physiol 586(14):3353–3364
55. Fee MS, Mitra PP, Kleinfeld D (1996) Variability of extracellular spike waveforms of cortical neurons. J Neurophysiol 76(6):3823–3833
56. Gao P, Bermejo R, Zeigler HP (2001) Whisker deafferentation and rodent whisking patterns: behavioral evidence for a central pattern generator. J Neurosci 21(14):5374–5380
57. Berg RW, Kleinfeld D (2003) Rhythmic whisking by rat: retraction as well as protraction of the vibrissae is under active muscular control. J Neurophysiol 89(1):104–117
58. Hill DN et al (2008) Biomechanics of the vibrissa motor plant in rat: rhythmic whisking consists of triphasic neuromuscular activity. J Neurosci 28(13):3438–3455
59. Knutsen PM, Derdikman D, Ahissar E (2005) Tracking whisker and head movements in unrestrained behaving rodents. J Neurophysiol 93 (4):2294–2301
60. Wong-Riley M (1979) Changes in the visual system of monocularly sutured or enucleated cats demonstrable with cytochrome oxidase histochemistry. Brain Res 171(1):11–28
61. Sakmann B, Neher E (1995) Single-channel recording. Plenum Press, New York
62. Bock DD et al (2011) Network anatomy and in vivo physiology of visual cortical neurons. Nature 471(7337):177–182
63. Briggman KL, Helmstaedter M, Denk W (2011) Wiring specificity in the direction-selectivity circuit of the retina. Nature 471 (7337):183–188
64. Margrie TW, Brecht M, Sakmann B (2002) In vivo, low-resistance, whole-cell recordings from neurons in the anaesthetized and awake mammalian brain. Pflugers Arch 444(4): 491–498
65. Oberlaender M, Ramirez A, Bruno RM (2012) Sensory experience restructures thalamocortical axons during adulthood. Neuron 74(4): 648–655
66. Bruno RM et al (2009) Sensory experience alters specific branches of individual corticocortical axons during development. J Neurosci 29 (10):3172–3181
67. Boudewijns ZS et al (2011) Semi-automated three-dimensional reconstructions of individual neurons reveal cell type-specific circuits in cortex. Commun Integr Biol 4(4):486–488
68. Oberlaender M et al (2011) Three-dimensional axon morphologies of individual layer 5 neurons indicate cell type-specific intracortical pathways for whisker motion and touch. Proc Natl Acad Sci U S A 108(10):4188–4193
69. Helmstaedter M (2013) Cellular-resolution connectomics: challenges of dense neural circuit reconstruction. Nat Methods 10(6): 501–507
70. Sarid L et al (2007) Modeling a layer 4-to-layer 2/3 module of a single column in rat neocortex: interweaving in vitro and in vivo experimental observations. Proc Natl Acad Sci U S A 104(41):16353–16358
71. Petreanu L et al (2009) The subcellular organization of neocortical excitatory connections. Nature 457(7233):1142–1145
72. Xu NL et al (2012) Nonlinear dendritic integration of sensory and motor input during an active sensing task. Nature 492(7428): 247–251
73. O'Connor DH, Huber D, Svoboda K (2009) Reverse engineering the mouse brain. Nature 461(7266):923–929
74. Kleinfeld D, Deschenes M (2011) Neuronal basis for object location in the vibrissa scanning sensorimotor system. Neuron 72(3):455–468

Chapter 3

Studying Sodium Channel Gating in Heterologous Expression Systems

Jannis E. Meents and Angelika Lampert

Abstract

Voltage-gated sodium channels (Na_vs) are essential for the initiation and propagation of action potentials in most excitable tissues, such as neurons or cardiac myocytes. Mutations in Na_vs are linked to several severe conditions, such as pain syndromes, epilepsy, and cardiac arrhythmias and these ion channels are therefore among the most promising drug targets. The development of Na_v modulators is complicated by the intricate gating mechanisms of these ion channels. They activate extremely quickly and subsequently inactivate equally fast. There are several additional gating modes that are physiologically relevant and that may be involved in the pathophysiology of numerous conditions, such as a variety of pain syndromes.

Whole-cell voltage clamp is a valuable technique to study the different gating modes of Na_vs and their possible physiological roles. It can be conducted in a variety of tissue preparations; however, for the basic investigation of Na_v activity, heterologous expression systems offer numerous advantages. The fast kinetics of Na_v activity make it difficult to accurately measure these events. The following chapter therefore aims to provide the necessary steps and protocols in order to study Na_v gating.

Key words Voltage-gated sodium channel, Nav, Patch-clamp electrophysiology, Voltage clamp, HEK cells

1 Introduction

Voltage-gated sodium channels (Na_vs) are a large family of ion channels that are selective for sodium ions and regulated by variations of the voltage across the cell membrane. Na_vs use these properties for the initiation and propagation of action potentials in excitable tissues, such as neurons, cardiac, or skeletal muscle. The channels are activated upon depolarization of the cell membrane, which leads to influx of sodium ions into the cytoplasm and thereby to a further depolarization, which may eventually generate an action potential. There are nine mammalian Na_v isoforms, Na_v 1.1 through Na_v 1.9, and each isoform is more or less restricted to certain tissues [1]. The majority of these isoforms has been found to be blocked by tetrodotoxin (TTX), a potent neurotoxin

Alon Korngreen (ed.), *Advanced Patch-Clamp Analysis for Neuroscientists*, Neuromethods, vol. 113,
DOI 10.1007/978-1-4939-3411-9_3, © Springer Science+Business Media New York 2016

produced by some marine animals, such as the pufferfish. The TTX-sensitive (TTXs) isoforms are Na_v 1.1–Na_v 1.4 as well as Na_v 1.6 and Na_v 1.7. The remaining three isoforms, Na_v 1.5, Na_v 1.8, and Na_v 1.9, are relatively resistant to TTX (TTXr) [1]. Na_vs consist of a large conducting α subunit, which can be associated with smaller β subunits. The latter are not required for channel function but influence expression, kinetics, and gating properties. The α subunit consists of four domains, DI through DIV, each of which contains six transmembrane segments, S1 through S6, and an additional pore loop between segment S5 and S6 that includes the selectivity filter. All 24 transmembrane segments are formed of a single protein of approximately 2000 amino acids in length and 260 kDa in size [1]. The S4 segment of each domain carries a certain amount of positively charged amino acids, and these positive charges function as the channel's voltage sensor. Na_vs have a range of different modes of activity, including activation, fast and slow inactivation, and deactivation. Some of these processes occur extremely quickly, within milliseconds or even microseconds, which often impedes accurate measurement of these events. The present chapter aims to provide the necessary tools and protocols for measuring Na_v activity in heterologous expression systems using whole-cell voltage-clamp electrophysiology. The following sections will give brief introductions into the different modes of activity and provide the appropriate protocols for their investigation. However, as whole-cell patch clamp on Na_vs is notoriously difficult and prone to errors, we will first attempt to describe the appropriate conditions that need to be met before reliable voltage clamp can be performed.

2 Requirements for Voltage Clamping of Sodium Channels

2.1 Cellular Requirements

Since the pioneering work of Hodgkin and Huxley on giant squid axons [2], a range of different neuronal cell lines as well as numerous heterologous expression systems have been established to study Na_vs. This chapter focuses on the use of heterologous expression systems and does not refer to experiments in neuronal cultures or tissue preparations. While Na_vs can be studied using different cell lines, we focus here on human embryonic kidney (HEK) 293 cells, as they provide several advantages and are very widely used. These cells endogenously express very few ion channels, mostly K^+ and Ca^{2+} channels [3, 4]. However, low endogenous expression of Na_v α and β subunits has been reported [5–7]. Luckily, these TTXs currents are usually small (<400 pA) [5, 6]. It is therefore important to achieve substantial current amplitudes in transfected cells and to exclude those cells from the analysis that carry only small amounts of current so as to be relatively certain that the recorded cell had been successfully transfected. An alternative approach involves mutagenesis of a TTXs channel, making it resistant to the neurotoxin [8, 9]. The mutated channel can then be investigated during bath application of TTX and its current analyzed in isolation.

Cell culture protocols for HEK293 cells are well established and the cells are easily transfected with most Na_{V}s either transiently or stably. If stably transfected cells are used, it is necessary to apply a selection marker, which is encoded on the vector DNA and provides resistance to antibiotics against eukaryotic cells, such as G418, zeocin, or puromycin. Adding the antibiotic to the culture medium promotes predominant growth of Na_{V}-expressing cells. Most isoforms are readily expressed in HEK293 cells. Especially expression of Na_{V} 1.2, Na_{V} 1.3, Na_{V} 1.4, Na_{V} 1.5, and Na_{V} 1.7 is comparatively uncomplicated in these cells. Na_{V} 1.8 and Na_{V} 1.6 reveal a much better expression when a neuronal cell line is chosen, such as ND7/23 or N1E cells. Heterologous expression of Na_{V} 1.9 has proven difficult in the past for reasons that are not understood. Recently, Vanoye et al. [10] have been able to develop a stably Na_{V} 1.9-expressing ND7/23 cell line by incubating the cells at 28 °C overnight. However, the authors were not able to generate the same result using HEK293 or Chinese hamster ovary cells [10].

To allow accurate recordings of sodium currents, it is imperative to record from small isolated cells that are not connected to other cells and that bear no or almost no cellular processes. Reasons for this will be discussed below (Sects. 2.4.1 and 4.3). These conditions are easily met when using HEK293 cells. These can be cultured and grown on plastic and glass surfaces in low confluency using the correct protocol, and their cell bodies are reasonably small, providing isolated cells with limited processes.

2.2 Bath and Pipette Solution

Composition of intracellular and extracellular solutions depends on the type and purpose of the recordings and varies between laboratories. If a physiological environment is desired, both solutions should contain ionic concentrations that mirror the composition of the native intra- and extracellular compartments of the cells. Here, we focus on recordings of isolated Na_{V} currents in heterologous expression systems and on the study of the biophysical properties of these Na_{V}s, and it is therefore advisable to substitute certain ions and to add blockers of endogenous channels, when present. The following extracellular (ECS) and intracellular (ICS) solutions are examples commonly used for the recording of Na_{V}s in transfected HEK293cells. ECS (concentrations in mM): NaCl 140, $MgCl_2$ 1, $CaCl_2$ 1, HEPES 10, Glucose 5, potentially sucrose/mannitol to adjust osmolarity (see below). ICS: CsF 140, NaCl 10, HEPES 10, EGTA 1. These solutions will produce a reversal potential (V_{rev}) for sodium of +67 mV, which is well outside the experimentally applied voltage range.

As Na_{V}s are sensitive to changes in pH, it is important to set the pH to consistent values throughout one experimental series.

Regardless of the cell type, the pH for the ECS should be set to the physiological pH of 7.4. The ICS is set slightly more acidic at pH 7.3. It is important to adjust the pH using an appropriate titration base/acid, which does not have a large impact on the ion concentration, e.g., CsOH for the ICS or NaOH for the ECS. However, if the total Na^+ concentration of the ECS has been reduced by substitution with, e.g., choline-chloride (see below), one can use CsOH for setting its pH, as Cs is a blocker of K_v channels and has no or only little impact on Na_v gating. The osmolarity of both solutions should always be in the range of 300–310 mOsm. To facilitate gigaohm (GΩ) seal formation and its stability, it is recommended to set the ICS to a slightly higher osmolarity (approximately 5–10 mOsm higher than ECS). This leads to a slight swelling of the cell after breaking of the membrane patch and thus stabilizes the seal. Small increases of the osmolarity can be achieved by adding glucose or preferentially an inert alternative, such as sucrose or mannitol.

Long-lasting experiments that require a high-quality seal are facilitated by adding fluoride (F^-) to the ICS, commonly in the form of CsF. It has been shown that the presence of F^- in the ICS greatly facilitates formation and stability of the GΩ seal and reduces space clamp artifacts [11]. It is not known exactly how this beneficial effect is achieved but the fact that F^- is likely to bind Ca^{2+} [12] suggests that at least the effect on seal stability might rely on complex formation between the two ions and subsequent "plugging" of potential leak sites. However, the experimenter should be aware that the presence of CsF also influences voltage-activated sodium currents. Both activation and inactivation kinetics of TTXr Na_vs are modified and the voltage dependence of activation as well as fast and slow inactivation are shifted towards more negative potentials [13, 14]. In addition, CsF leads to a considerable increase in current amplitude and current density [11, 13]. Against previous beliefs, it was shown that at least in Na_v 1.8, these effects are not mediated by binding of F^- to trace amounts of Aluminum and are independent of activation of G proteins, adenylate cyclase, and protein kinase A or C [13, 15]. It is a commonly observed phenomenon that the voltage dependence of Na_v activation shifts to more hyperpolarized potentials and current amplitude increases over time during a single experiment. These two phenomena are likely to be linked and are probably due to progressive dialysis of the intracellular medium with CsF [14, 16]. This has of course important ramifications for the design of experimental protocols, as will be discussed in Sect. 3.1.

The CsF-induced increase in current amplitude is beneficial for recordings of cells that express low amounts of Na_vs. However, in case of a very high expression, which often occurs with Na_v 1.5 or when recording Na_vs in native neurons, such as DRGs, it can lead to currents that are too large to be accurately measured (see

Sect. 2.4.3). In such a case, it may be necessary to reduce the extracellular Na^+ concentration. Part of the extracellular NaCl is then substituted with an equimolar concentration of, e.g., choline-Cl or NMDG. Depending on the size of the current, reduction of extracellular NaCl from 140 to 40 mM may be sufficient; however, in large DRG neurons, it may be necessary to reduce NaCl to 10 mM. It should be considered that reducing the extracellular Na^+ concentration will also reduce V_{rev} for Na^+, potentially bringing it close to the voltage range that is applied during experiments. This should be avoided if possible as it could interfere with data analysis as will be discussed in Sect. 4.1.

Finally, it is recommended to add a Ca^{2+} chelator and to exclude any Ca^{2+} and K^+ from the ICS. Ca^{2+} may have regulatory effects on Na_Vs [12, 17], and the addition of sufficient amounts of EGTA or BAPTA guarantees that even Ca^{2+} released from intracellular stores is sufficiently buffered. Exclusion of K^+ from the ICS and addition of Cs^+ avoids interference of outward currents through endogenous K^+ channels.

2.3 Patch Pipettes

The patch pipette forms the link between the head stage's electrode and the interior of the cell. As it is part of the electrical circuit, it influences the electrical signals measured by the amplifier and numerous advances have been made over the years to reduce this influence. Several properties of a patch pipette need to be considered before starting the experiment. The most important ones will be discussed in short here with regard to patch clamp in heterologous expression systems. The vast details of the properties of different pipette glass as well as construction and shape of the pipette have been discussed in detail elsewhere [18].

It is advisable to use glass capillaries containing a filament. The filament helps when filling the pipette with solution by increasing capillary action and thus avoiding the formation of bubbles. Generally, increasing the thickness of the capillary glass reduces the pipette capacitance that has to be canceled by the amplifier circuitry before the start of the recording. Because optimal series resistance (R_s) compensation is crucial for recording Na_Vs and because incomplete pipette capacitance cancelation can lead to poor R_s compensation, we suggest the use of thick-walled capillary glass. However, there are ways to reduce pipette capacitance regardless of the type of glass: reducing the bath solution level reduces the surface of the pipette that is exposed to liquid and that can act as a capacitor. This is certainly the easiest way to reduce pipette capacitance but might not always be possible. Alternatively, coating patch pipettes with a hydrophobic elastomer such as Sylgard® reduces the capacitance and prevents the bath solution from forming a liquid film along the surface of the pipette. The coating should reach to within ~100 μm of the pipette tip but must not flow over the tip as this will make seal formation impossible. The elastomer is applied to the pipette glass under constant visual control through a dissection microscope. After coating, the elastomer is dried by moving it into close proximity of a heated wire under visual control or by incubating the pipettes in a laboratory oven (e.g., 80 °C for 1 h).

The two aspects that influence the output of patch-clamp recordings the most and that depend mostly on the shape of the patch pipette are formation of a GΩ seal and a low access resistance (R_a). A stable GΩ seal guarantees that almost all current from the membrane patch flows into the pipette and is not lost to the bath as leak current. As the name suggests, R_a is a measure of quality for the connection between recording electrode and the cell's interior and thus determines how easily current can flow from one to the other. To guarantee accurate voltage clamp and to measure the fast inward current of Na_vs, R_a needs to be minimal. Unfortunately, while smaller pipette openings facilitate seal formation, they also increase R_a. The experimenter therefore needs to find a good balance between the two aspects. Fire polishing the pulled and potentially coated patch pipettes will increase the likelihood of seal formation while allowing for the opening to be rather large. It should be noted that during fire polishing, the glass of the pipette tip melts and this may lead to a shrinking of the opening. Thus, fire polishing can be a convenient way of fine-tuning the shape and size of the pipette opening. We and others have found that in order to achieve consistent GΩ seal formation, it is not always necessary to polish pipette tips, especially when using HEK293 cells. However, to reduce R_a enough to allow recording of Na_vs, pipette openings need to be fairly large, in which case fire polishing can be beneficial. In our experience it is possible to achieve R_a of <5 MΩ prior to R_s compensation using pipettes with an initial pipette resistance of 0.8–1.5 MΩ after fire polishing.

Polishing not only the pulled tip of the patch pipette but also the back opening is also advisable as this protects the washers inside the pipette holder as well as the electrode wire during insertion of the pipette. Fire polishing can be done using a variety of commercially available devices or even with a self-made apparatus, a description of which can be found, e.g., in [18].

2.4 Compensation and Voltage Errors

2.4.1 Space Clamp

Good voltage clamp requires recording from small isolated and round cells that ideally bear no cellular processes. The reason for this is that voltage clamp becomes very prone to error the further away the cell extends from the recording electrode. This phenomenon has been termed space clamp and it is a notorious problem in voltage-clamp recordings of cells that bear long processes, such as neurons. It is important to remember that the amplifier clamps the voltage of the tip of the recording electrode. In the case of long and thin processes, such as axons or dendrites however, the cytoplasm can act as a significant resistance that accumulates with length of the branch, and which prohibits true voltage clamp in distant parts of the cell. In addition, the increased membrane area of such cells leads to an increased cell capacitance that effectively delays any change in the clamped voltage. Such a delay can lead to a considerable voltage error. Because Na_vs have a steep voltage dependence, i. e., activation occurs over a narrow voltage range, they require accurate and reliable measurement of membrane voltage. What is more, Na_vs activate and inactivate very quickly; in case of a voltage error, it will be impossible to know at which precise voltage activation has occurred.

Unfortunately, space clamp problems cannot easily be avoided in single-electrode voltage clamp. One way is to increase the cell's impedance. This can be done by pharmacologically silencing ion channels that are not of interest (e.g., K^+ or Cl^- channels), a technique that is used in most selective recordings of sodium currents. An easier and fairly reliable way to reduce space clamp errors is to limit recordings to small and ideally round cells with no or almost no cellular processes.

2.4.2 Series Resistance Compensation

Proper R_s compensation is essential for reliable and successful voltage clamp. Generally, poor R_s compensation leads to a slower charging of the cell membrane and to potentially considerable voltage errors that can have significant effects on the obtained data (Fig. 1). Na_vs activate very quickly (mostly within 1 ms) and in addition inactivate only a few milliseconds after initial activation. Achievement of voltage clamp therefore needs to be as accurate and as fast as possible in order to allow resolution of these fast events. The time it takes for a voltage change to take effect is described by the time constant $\tau = R_s C_m$, with R_s being the series resistance and C_m being the membrane capacitance of the cell. Thus, assuming for example an uncompensated R_s of 10 MΩ and for simplicity a C_m of 100 pF, the charging of the membrane will be slowed by more than 1 ms, which is highly relevant in a recording where the event of interest only lasts a few milliseconds. While 100 pF may be very high and describes a large cell, the above example quickly illustrates the advantage of patching small cells (<20 pF) and of using large patch pipette openings with a smaller R_a (which is in series with R_s), as well as the importance of compensating R_s. The delay in the setting of the membrane voltage in an uncompensated cell leads to a corresponding voltage error. In other words, the experimenter can never be sure at which precise voltage a cellular response was obtained. This is of course highly problematic dealing with ion channels where 30 mV can make the difference between no response and a maximum response. What is more, such a voltage error in an uncompensated cell can lead to an artifactual shift of the current–voltage (*IV*) relationship and the corresponding activation curve (Fig. 1). Care should be taken not to mistake such a shift for a physiological effect. Another issue in the case of Na_vs arises from channel inactivation. The effective slowing of the voltage clamp in a

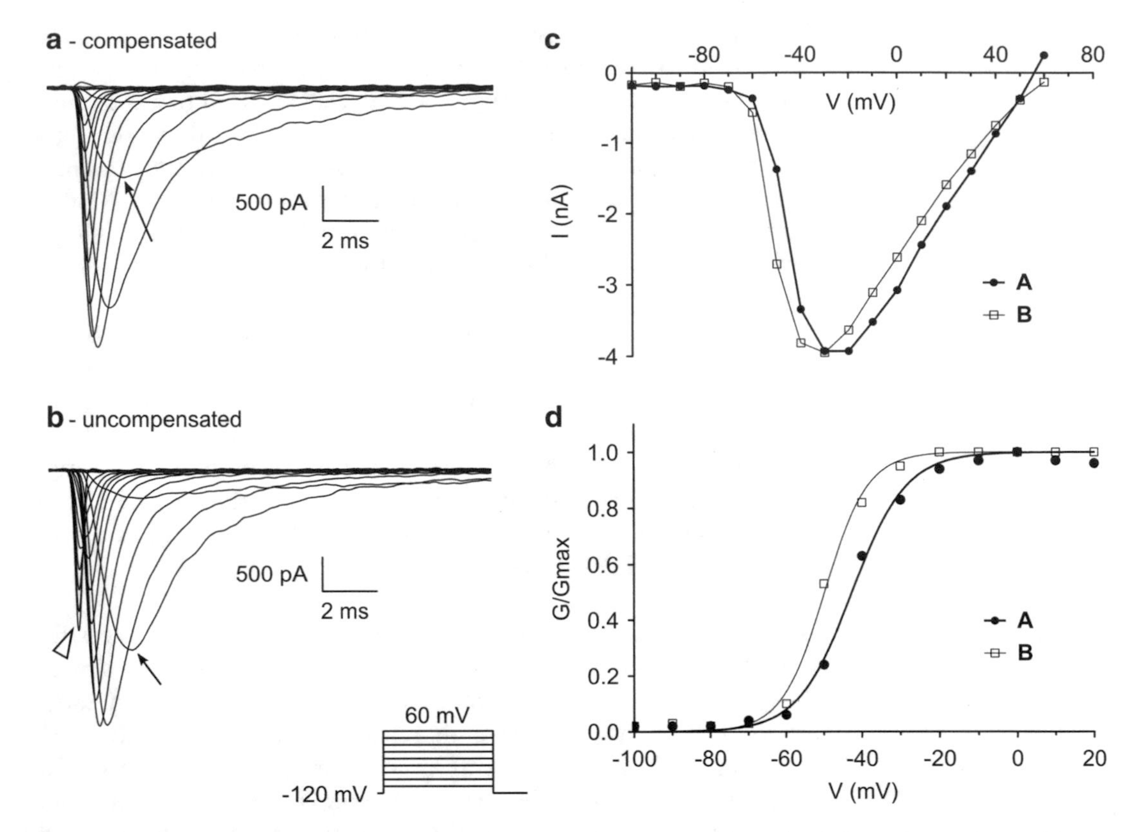

Fig. 1 Direct comparison of compensated and uncompensated recordings. (**a**, **b**) Current recordings obtained from the same HEK293 cell, stably expressing Na_V 1.5, using the indicated step protocol (see *inset* in **b**). The recording in (**a**) has been obtained under appropriate R_S compensation. The trace in (**b**) has been obtained without R_S compensation. 70 mM Na^+ had been replaced by 70 mM choline-chloride, reducing the final Na^+ concentration in the extracellular solution to 70 mM. The theoretical V_{rev} was +49 mV. *Arrows* point to the current trace obtained by a voltage step to −50 mV. Note the large uncompensated capacitance transients (*arrowhead* in **b**) that interfere with Na_V current. (**c**) *IV* relationships obtained from the recordings in (**a**) and (**b**). Note the negative shift due to missing R_S compensation. Also note the difference in peak currents between −60 and −30 mV between the two recordings. (**d**) Activation curves obtained from the recordings in (**a**) and (**b**). The negative shift in voltage-dependent activation is only due to missing R_S compensation and does not reflect a physiological change

poorly compensated cell can force an increased number of Na_vs into closed-state inactivation (see Sect. 3.4), which will prevent the channels from opening leading to a reduced or even no measurable current around the activation threshold. In summary, trustworthy recordings of Na_vs are only possible after appropriate R_s compensation. Because of the unreliability of the voltage clamp at high resistances, we advise that any cell with a compensated R_s of >5 M Ω at any time throughout a recording should be considered with care and ideally be excluded from the analysis.

R_s compensation is carried out after establishing the whole-cell configuration. It is vital that the fast capacitance transient of the patch pipette has been canceled effectively beforehand. The patch-clamp amplifier compensates for R_s by over-injecting into the cell the appropriate amount of current that is needed to overcome both R_s and C_m. The experimenter therefore needs to set the correct values for both parameters on the amplifier or within the amplifier software. R_s should ideally be compensated for by 80–85 %, but care must be taken not to drive the amplifier into oscillation when such high compensation is attempted. In our experience, using a compensation speed of 10–20 μs allows such high compensation in good cells.

Throughout the recording, it is important that R_a does not increase due to clogging of the electrode tip with cell membrane. Such a "re-sealing" of the membrane patch often occurs and can be reduced by applying gentle suction throughout the recording. It should be noted that applying suction during the recording is not recommended in cells that express mechano-sensitive ion channels, such as sensory neurons. During voltage clamp, R_a is in series with R_s and can therefore significantly contribute to the latter and influence R_s compensation. Throughout the experiment, it is therefore advisable to check R_s compensation as often as possible. If R_a has increased due to re-sealing of the membrane patch, R_s compensation will appear maladjusted and needs to be corrected.

2.4.3 Current Size

Some Na_V isoforms, such as Na_V 1.5, express very well in heterologous expression systems and display large inward currents of well above 10 nA in response to a standard activation protocol (Fig. 2b, c). Endogenous Na_Vs in large diameter neurons, e.g., from DRGs, are known to have a high current density. Such large currents pose a considerable problem in voltage-clamp recordings. First, during large inward current flux, the amplifier needs to inject a large amount of current into the cell to clamp it at the desired voltage. Currents of above 20 nA exceed the limit of some amplifiers so that currents cannot be measured exactly but will be clipped at 20 nA. Injecting large amounts of current can lead to delays in establishing the desired voltage, depending on the cell's capacitance and resistance. More importantly, however, large currents inevitably produce large voltage errors that prevent accurate voltage clamp (Fig. 2). For example, a 1 nA current flow across an R_s of 10 MΩ produces a voltage drop of 10 mV and thus a 10 % error. It should be noted that 10 mV shifts have been reported for disease causing mutations in Na_Vs [19]; thus a mistake of this size is relevant. With R_s compensation set to 80 %, the voltage drop becomes smaller (2 mV). However, considering a 10 nA current across the same R_s now produces a 100 mV drop! Even when compensated by 80 %, the voltage error is still 20 mV, which is obviously not acceptable for measuring channels with a steep voltage dependence, such as Na_Vs. In summary, large current flux will always lead to unreliable voltage clamp and it is

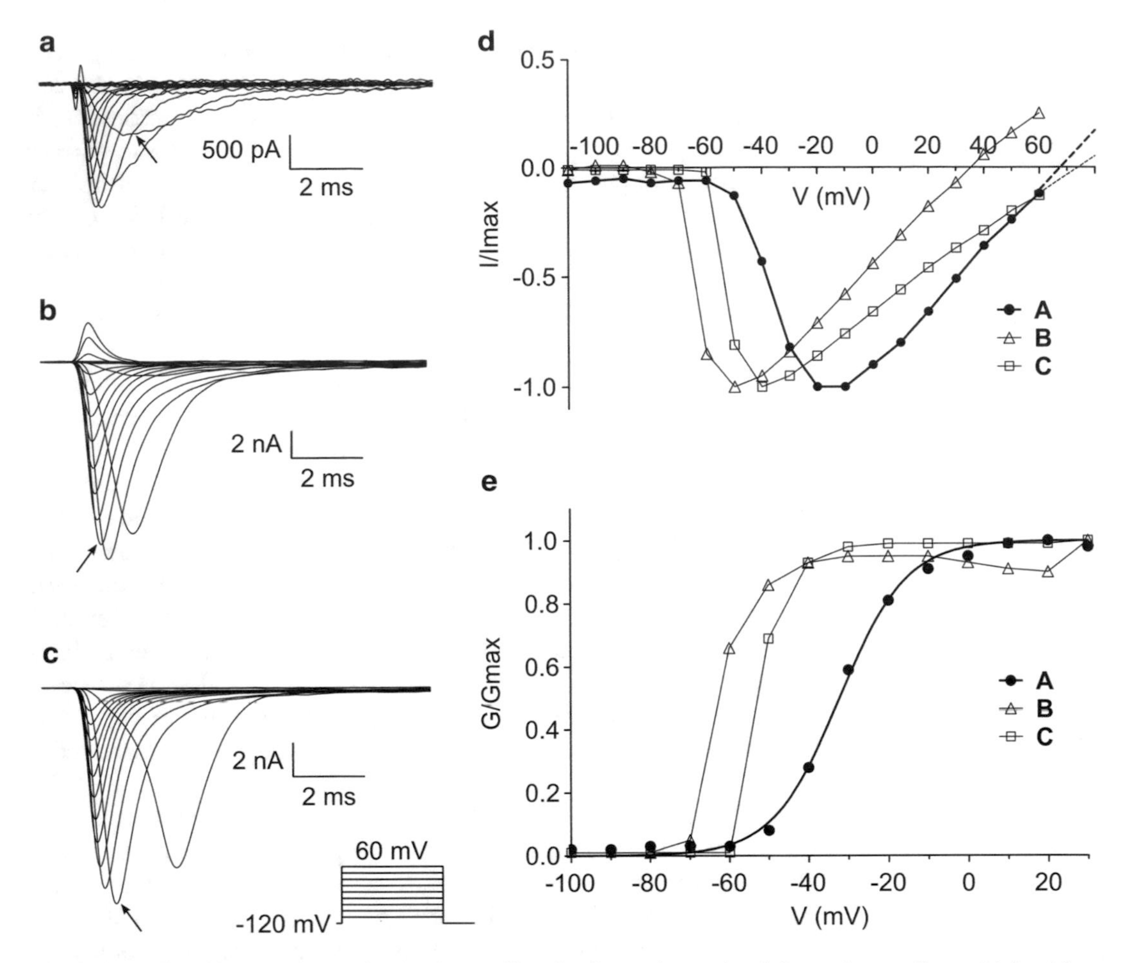

Fig. 2 *IV* relationships used to estimate the quality of voltage clamp. (**a**–**c**) Current recordings obtained from three different HEK293 cells, stably expressing Na_v 1.5, using the same step protocol (see *inset* in **c**). The composition of intra- and extracellular solutions is given in the main text. *Arrows* point to the current trace obtained by a voltage step to −40 mV. Note the very large current amplitudes in (**b**) and (**c**). (**d**) *IV* relationships obtained from the recordings in (**a**)–(**c**). The calculated V_{rev} for sodium was +67 mV. Only the recording in (**a**) has been obtained under appropriate voltage clamp. The extrapolated V_{rev} (*strong dotted line*; see main text) corresponds well with the theoretical value. In (**b**) and (**c**), currents at threshold are masked and currents activate only at near-maximal values. Both traces accordingly display a shift in the *IV* relationship, over- or underestimating V_{rev}. (**e**) Activation curves obtained from the recordings in (**a**)–(**c**). Only trace A displays an appropriate trend, showing clear voltage dependence, and could be well fitted by a Boltzmann function (*strong solid line*). Traces B and C (not fitted) display distortions leading to an artifactual negative shift in the activation curve

therefore imperative to reduce current amplitude to a level that decreases the voltage drop to an acceptable level. This is easily achieved by reducing the concentration gradient of the ion that carries the current, in this case Na^+. Replacing a part of the extracellular Na^+ with choline-Cl or NMDG effectively decreases current amplitude and voltage error (see also Sect. 2.2).

2.5 Filtering and Sampling

Filtering describes the process of improving the signal-to-noise ratio and performing data reduction that is carried out either by the amplifier during the recording or digitally by the analysis software after the recording. Before any filtering can be considered, it is vital to minimize both external noise that comes from outside the experimental setup and noise that is intrinsic to the recording system. Steps for external and intrinsic noise reduction are not covered here but are discussed in detail in the majority of basic patch-clamp literature [18, 20]. Once noise generation has been minimized, it is essential to filter the recorded data to optimize the signal-to-noise ratio and to isolate the signal of interest. The experimenter needs to be aware, however, that by definition, filtering leads to loss of information. Filter settings therefore need to be carefully considered. The fastest signals recorded under whole-cell conditions are in the range of 200–500 μs (2–5 kHz) [21], whereas noise has a much wider frequency range. The most commonly used filter type in whole-cell recordings is therefore a low-pass filter, which blocks out high-frequency noise. The majority of modern patch-clamp amplifiers have an integrated 4-pole Bessel filter. In some cases, it may be possible to select other filter types, such as Butterworth or Chebyshev, but Bessel filters are the most widely used and they have excellent characteristics for whole-cell recordings. The low-pass filter cut-off frequency needs to be set by the experimenter and greatly depends on the recorded current and on the sampling rate. Generally, the analogue signal is continuous and consists of an infinite number of data points. The analogue-to-digital converter samples this analogue signal at a certain time interval (the sampling rate) to produce a digital discrete (step-like) representation. This sampling rate has to be set by the experimenter and, again, leads to data reduction and can distort the signal and create artifacts (aliasing). Thus, it is important to select an appropriate filter and sampling rate to accurately record the signal of interest. According to the Nyquist-Shannon sampling theorem, the sampling rate needs to be at least twice the signal bandwidth in order to accurately represent an analogue waveform without losing information. Going further than that, it has become general practice to choose a sampling rate of at least five times the filter bandwidth for actual recordings. That means that at a low-pass filter cut-off frequency of 2 kHz, a minimum sampling rate of 10 kHz is recommended. Na_Vs display very fast kinetics that belong to the fastest signals recorded under whole-cell conditions. It is not unusual for peak inward sodium currents to be reached within 400 μs. Filtering with a cut-off frequency of 2 kHz, which corresponds to 500 μs, would completely cut off such a high-frequency signal! It is therefore essential to record sodium currents using a cut-off filter frequency of at least 10 kHz. Accordingly, we recommend a sampling rate of at least 50 kHz. In the case of measuring deactivation of Na_Vs (see Sect. 3.5), the activating stimulus only lasts a few hundred microseconds. Consequently, the time window to measure these currents before they deactivate is very narrow. It is therefore suggested to increase filter cut-off frequency and sampling rate in this protocol to 30 kHz and 100 kHz, respectively.

2.6 Leak Subtraction

Currents measured during whole-cell recordings not only consist of those flowing through ion channels but also contain capacitive currents that charge the membrane and leak currents. These passive components can be subtracted from the current of interest to improve the latter and to possibly reveal small current components that might otherwise be masked. The process of such leak subtraction can easily be achieved by a simple procedure, which is integrated into most patch-clamp software and is applied throughout the recording. This so-called P/N procedure consists of a simple voltage step protocol. Each test pulse in the voltage-clamp protocol is either preceded or followed by a series of N (usually four) voltage steps of $1/N$ or $-1/N$ amplitude of the test pulse [22] (Fig. 3). In a typical *IV* protocol that gradually steps the voltage from a holding potential of −120 to 60 mV in 10 mV increments, this would mean that the first test pulse (to −110 mV) is followed by four leak pulses of identical length from −120 mV to either −122.5 or −117.5 mV, depending on the selected polarity. The final test pulse (to 60 mV) would be followed by four leak pulses to −165 or −75 mV. It is important that leak pulses do not activate voltage-gated currents through ion channels, so that only capacitive and leak currents are elicited. Because leak pulses contain no active component, the four obtained leak current traces can be summed together and be subtracted from the current trace of interest. The above example illustrates that, depending on the test pulse amplitude, leak pulses can be quite strong depolarizations (e.g., to −75 mV) that may potentially induce Na_V activation or inactivation. In the case of heterologously expressed channels, it is therefore often advisable to apply leak pulses in a hyperpolarizing direction. However, in the above example, the maximum leak pulse would reach voltages of −165 mV, which most cells do not tolerate easily. In this case, it is advisable to reduce leak pulse amplitude to, e.g., 10 % of the test pulse (instead of 25 % in the above example). The generated smaller leak currents are then averaged and mathematically multiplied by the amplifier software and subtracted from the current of interest (Fig. 3).

It needs to be appreciated that it is dangerous to monitor only leak-subtracted traces during an experiment. As leak subtraction is performed by the recording software in the background, the experimenter might not be aware that he/she is dealing with a severely deteriorated cell. It is therefore strongly recommended to record both the leak-subtracted trace and the leak trace itself and to monitor leak traces together with traces of interest either during the recording or at least prior to data analysis.

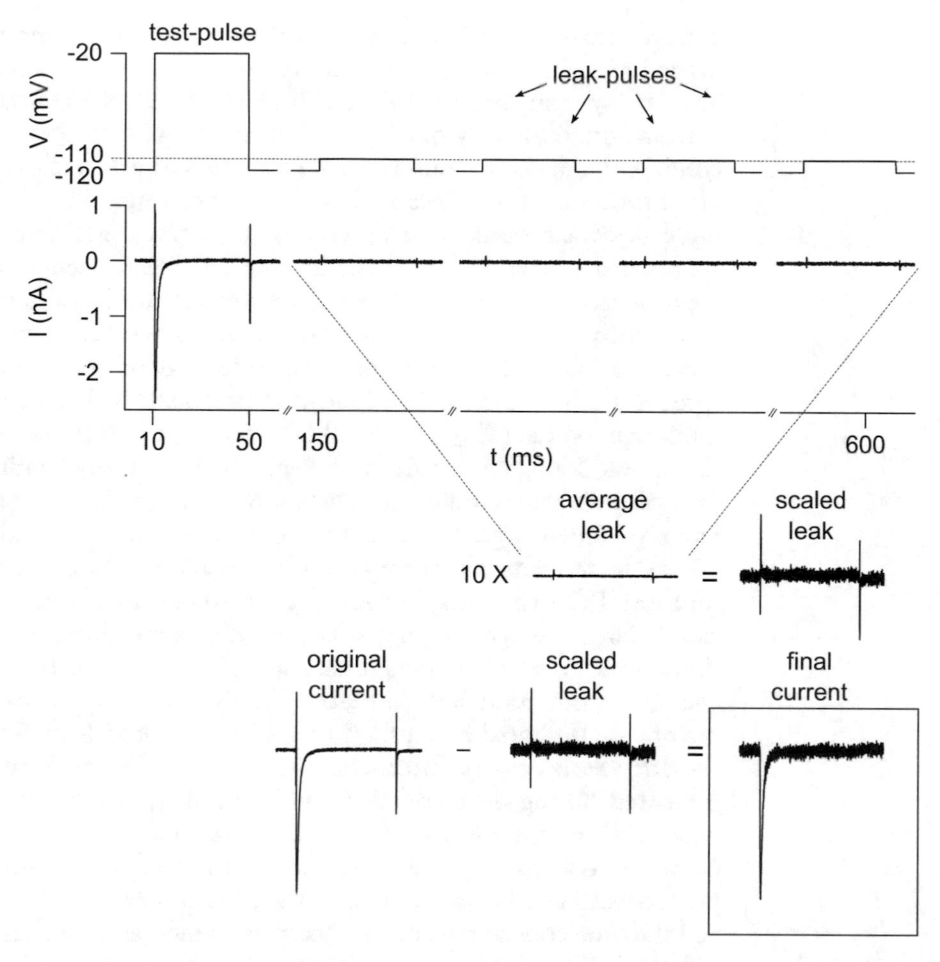

Fig. 3 Example of a leak subtraction protocol. Four current responses to a 10 mV pulse (10 % of test pulse) are averaged and then multiplied by 10 to represent the leak current that is expected during the test pulse to −20 mV. This scaled leak current is then subtracted from the original test pulse current to provide the final output. Note the increase in noise that is inherent in this form of leak subtraction

3 Voltage-Clamp Protocols

Proper channel function and action potential generation in excitable tissues rely on the very quick Na_v kinetics and their various modes of gating. The following sections are meant to describe the different modes of Na_v gating and to provide the proper voltage-clamp protocols that can be used to measure these processes.

3.1 Activation

Na_v activation occurs within a few milliseconds and leads to the conduction of a sodium current that is often of large amplitude. Activation takes place during depolarization of the cell's membrane potential and relies on the displacement of the channel's charged voltage sensor towards the extracellular side of the plasma membrane [23]. This movement induces a conformational change that leads to opening of the channel. Because these voltage-activated currents establish very quickly, within the range of a few milliseconds, voltage-clamp conditions need to be optimal to allow accurate measurement of these events. As has been mentioned above, under less-than-optimal clamp conditions, current activation may be masked by capacitive transients that have not been canceled appropriately (e.g., Fig. 1b). Assuming that clamp conditions and R_s compensation are satisfactory, the following voltage-clamp protocol can be used to measure voltage-dependent Na_v activation (protocol a in Table 1). The following protocols will all start at a holding potential (V_{hold}) of −120 mV. This hyperpolarized potential allows Na_v recovery from all types of inactivation while still providing stable recording conditions. Starting at V_{hold}, the membrane potential is gradually stepped in a depolarizing direction in 10 mV increments with a step duration of about 40 ms (sufficient for most TTXs channels). To measure channel activity over a sufficiently large voltage range, it is advisable to apply enough repetitions (sweeps) to reach voltages around 60–80 mV. To relieve the channels from both fast and slow inactivation (see below), the membrane potential is returned to −120 mV for at least 5 s after each depolarizing step (protocol a in Table 1). The peak currents measured during depolarization can be used to plot current–voltage relationships and to calculate channel conductance and half-activation voltage ($V_{1/2}$) as described in the analysis section. This protocol can also be used to obtain the time-to-peak (see Sect. 4.1) and the time constant of current decay of voltage-activated currents (see Sect. 4.2) as well as the amount of persistent current.

3.1.1 Notes

1. It is well known that the voltage activation curve and therefore the $V_{1/2}$ of Na_vs measured in an individual cell can shift during the process of a single voltage-clamp experiment [14]. It is believed that this is due to progressive dialysis of the intracellular medium of the cell and might rely on the effects of CsF contained in the ICS (see Sect. 2.2). The effect can indeed be reduced by replacing CsF with CsCl in the pipette solution [14]. It is therefore important to always begin the above protocol at the same time point after establishment of the whole-cell configuration.
2. Peak current amplitude in response to the same depolarizing voltage step can often be seen to increase during the first few minutes after establishing whole-cell configuration. This is certainly the case when measuring Na_v activity in heterologous expression systems, such as HEK293 cells [14, 16]. This phenomenon is linked to the effects of CsF in the intracellular solution (see Sects. 1 and 2.2) and could also be due to

Table 1
Summary of voltage-clamp protocols used to study Na_v activation and fast and slow inactivation

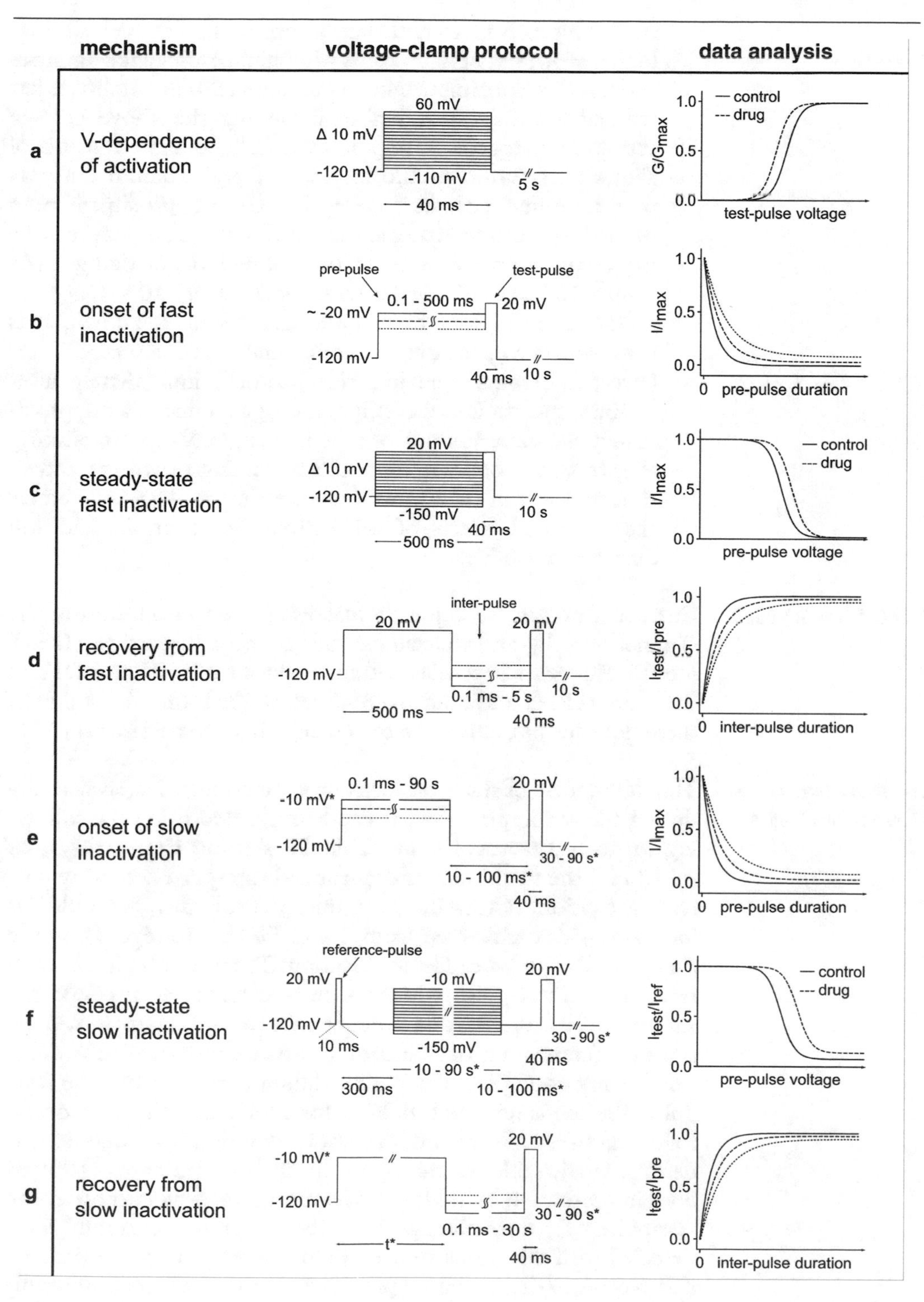

The data analysis column contains schematic plots that represent typical data fits, which can be obtained from the respective recordings.
*The exact value depends on measurements obtained with other protocols from this table. See main text for details.

trafficking of Na_vs to and their insertion into the cell's membrane. However, peak currents will stabilize after a few minutes and it is therefore important to wait until current amplitude has reached a steady state before beginning the above voltage activation protocol. Using a specifically designed protocol allows observation of the increase of peak sodium currents over time and permits estimation of the approximate time needed for currents to reach a steady state. This simple voltage-clamp protocol consists of repeated depolarizing steps from −120 to 0 mV for 100 ms, applied every 10 s. Once the average time needed to reach a steady state is known, it is possible to program the according amount of step repetitions (sweeps). Always applying this protocol immediately after establishing whole-cell configuration guarantees that approximately the same amount of time is spent between establishing the whole-cell configuration and beginning of voltage activation protocol in each individual experiment. This will reduce the undesired effects of cell dialysis (see Sect. 1) and will decrease variability.

3.2 Fast Inactivation

Currents through Na_vs quickly inactivate due to occlusion of the channel pore by an intracellular four-amino-acid-particle (IFMT motif). This particle resides within the linker of DIII and DIV of the channel itself and binds to residues on the intracellular side of these domains but only once the channel has been activated [23].

3.2.1 Time Course of Onset of Fast Inactivation

The time course of channel inactivation determines Na_v availability during the action potential. It can be measured using the following protocol (protocol b in Table 1): starting from a V_{hold} of −120 mV, the membrane is depolarized (pre-pulse), causing current activation. Crucially, the duration of this pre-pulse is increased with each sweep from 0.1 to 500 ms to investigate the time course of onset of fast inactivation. The pre-pulse is followed by a standard test-pulse, which we here define as a 40 ms depolarization to 20 mV. At this voltage near to all channels will be activated (compare with activation protocol above) while the duration is long enough to reach a maximum current. After the test-pulse, the cell is returned to V_{hold} for 10 s before the start of the following sweep. As the onset of fast inactivation is voltage-dependent, it is advisable to measure it at different potentials (dotted lines in protocol b in Table 1). Measuring the peak current at the test-pulse (I), normalizing it to the maximum current (I_{max}) detected with this protocol (i.e., following very short pre-pulses), and plotting I/I_{max} against pre-pulse duration allows a determination of the time course of onset of fast inactivation at the tested potential.

3.2.2 Voltage Dependence of Steady-State Fast Inactivation

Steady-state fast inactivation of Na_vs is measured as the fraction of channels that can be activated by a depolarizing step (test-pulse) after voltage-dependent inactivation has been induced by a range of depolarizing pre-pulses. In order to measure the amount of inactivation, a two-pulse protocol is typically applied (protocol c in Table 1): starting from a V_{hold} of -120 mV, the membrane potential is gradually stepped to a range of pre-pulse voltages. Because fast inactivation can often be observed at holding potential even though cells are held at relatively hyperpolarized voltages, it is advisable to start the pre-pulses at a very negative potential, such as -130 or -150 mV, increasing the value with each sweep in 10 mV increments up to a maximum of about 20 mV. The duration of the pre-pulse depends on the speed by which fast inactivation occurs at a given potential, and ideally this is determined beforehand (see Sect. 3.2.1). Often, 500 ms is chosen, as this is ample time for fast inactivation to occur and only little slow inactivation should have happened during this short interval. The pre-pulse is immediately followed by a standard test-pulse, after which the membrane voltage is returned to V_{hold} for 10 s before the next pre-pulse is applied. Peak currents elicited by the test-pulse can be used to calculate the fraction of available channels and the half-inactivation voltage as described in the analysis section.

3.2.3 Recovery from Fast Inactivation: Repriming

Repriming describes the recovery kinetics of Na_vs from steady-state fast inactivation. In this case, a three-pulse protocol is needed (protocol d in Table 1). Steady-state fast inactivation is induced by a standard pre-pulse to 20 mV for usually 500 ms (see Sect. 3.2.2). Alternatively, the pulse duration can be adjusted to what has been shown previously to induce maximum channel inactivation at 20 mV, using the protocol described above (Sect. 3.2.1). The pre-pulse is followed by a hyperpolarizing inter-pulse of varying duration, to measure the time dependence of recovery. Durations of 0.1 ms to 5 s should be applied to measure recovery over a sufficient time range. As the speed of recovery from inactivation depends on the amount of hyperpolarization, it is advisable to measure repriming over several inter-pulse voltages (dotted lines in protocol d in Table 1). The recovery-inducing inter-pulse is immediately followed by a standard test-pulse and a return to V_{hold} for 10 s before onset of the next sweep. For analysis, peak currents elicited by the test-pulse (I_{test}) are normalized to those measured during the pre-pulse (I_{test}/I_{pre}) and plotted against inter-pulse duration.

3.3 Slow Inactivation

In contrast to fast inactivation, which reduces the amount of excitable Na_vs very quickly, slow inactivation describes a process, during which the number of available channels decreases with a time course of seconds to minutes. This is a complex process and our understanding of the molecular underpinnings of this phenomenon is still limited. Several regions of the channel protein have been implicated in the development of slow inactivation, including the channel pore, the voltage-sensing domains, and the intracellular loops. For a more comprehensive review of the different mechanisms, the reader is referred to [24, 25]. To measure slow inactivation of $Na_{v}s$, it is important to first characterize the kinetics of fast inactivation (see above) so that the two mechanisms can be isolated and separated from each other.

3.3.1 Time Course of Onset of Slow Inactivation

The protocol to measure the onset of slow inactivation is similar to what has been described for fast inactivation (see Sect. 3.2.1). Pre-pulse voltages (dotted lines in protocol e in Table 1) can be selected based on the voltage dependence of steady-state slow inactivation as measured using the protocol described below (Sect. 3.3.2). The duration of the pre-pulse is changed between sweeps from 0.1 ms to 90 s. The pre-pulse is crucially followed by an inter-pulse to V_{hold}. This allows the channel to recover from fast inactivation and thus eliminates distortion by this separate phenomenon. Duration of the inter-pulse depends on the time course of recovery from fast inactivation (see Sect. 3.2.3). The inter-pulse is followed by a standard test-pulse and a return to V_{hold} for 90 s. The chosen duration for the return to V_{hold} before start of the following sweep depends on the time course of recovery from slow inactivation, which is measured using the protocol described below (see Sect. 3.3.3). This duration should be the same for all protocols involving slow inactivation (Sects. 3.3.1–3.3.3). During data analysis, normalized peak test-pulse currents (I/I_{max}) are plotted against pre-pulse duration.

3.3.2 Voltage Dependence of Steady-State Slow Inactivation

The protocol used to measure the voltage dependence of steady-state slow inactivation consists of a reference pulse, a pre-pulse to varying voltages, an inter-pulse to allow recovery from fast inactivation, and a standard test-pulse (protocol f in Table 1). These experiments are unusually long and it is very common to observe a considerable decrease in test-pulse peak current amplitude over time. In addition, if a compound is applied, time-dependent effects could influence test-pulse peak current amplitude. This means that test-pulse currents between sweeps cannot easily be compared. It is therefore necessary to include a reference-pulse at the beginning of each sweep. Each test-pulse current can then be normalized to its own reference-pulse current to account for current rundown and time-dependent changes. The reference-pulse is a 10 ms depolarization to 20 mV, followed by a return to V_{hold} for 300 ms to recover from fast inactivation. This is followed by a range of pre-pulse voltages that should ideally span from −150 to −10 mV, increased between sweeps in 10 mV increments. The duration of the pre-pulse depends on the Na_{v} isoform [24, 26, 27] and can be determined with the protocol described above (see Sect. 3.3.1). To be on the safe side while reducing experimental time, we suggest a pre-pulse duration of 30 s, which does not induce ultra-slow inactivation [27], but sufficient amount of slow inactivation. The pre-pulse is followed by an inter-pulse, a standard test-pulse and a return to V_{hold}, analogous to what was described directly above (Sect. 3.3.1). For current analysis, peak test-pulse currents are normalized to their respective reference-pulse current (I_{test}/I_{ref}) and plotted against pre-pulse voltage, providing a sigmoidal inactivation curve.

3.3.3 Recovery from Slow Inactivation

The protocol used to measure the time it takes for Na_vs to recover from steady-state slow inactivation is displayed in Table 1 (row g). First, a pre-pulse needs to be applied that induces slow inactivation. The duration and the voltage of this pre-pulse both depend on what has been measured for the onset and the voltage dependence of slow inactivation (Sects. 3.3.1 and 3.3.2). It is obvious that values should be chosen that guarantee maximum channel inactivation. The pre-pulse is followed by an inter-pulse of varying duration, to measure the time dependence of recovery. Duration of this inter-pulse depends on the channel isoform. In our experience it is advisable to use at least a range of 0.1 ms to 30 s, but longer recovery durations may be necessary. The short duration inter-pulses will reflect recovery from fast inactivation and the longer pulses will also release channels from slow inactivation. We suggest to measure recovery over several inter-pulse voltages (dotted lines in protocol g in Table 1), as recovery is a voltage-dependent process. However, care should be taken not to apply too depolarized voltages that might induce fast inactivation (see Sect. 3.2.2). The following test-pulse and return to V_{hold} are the same as described for the other slow inactivation protocols (Sects. 3.3.1 and 3.3.2). Peak test-pulse currents are normalized to peak pre-pulse currents and plotted against inter-pulse duration. It is likely that more than one kinetic will be visible (one for fast and at least one for slow inactivation).

3.3.4 Notes

Fast inactivation has been shown to interfere with slow inactivation. This can be seen in the fact that mutations that abolish fast inactivation by disrupting binding of the intracellular blocking particle (IFM-to-QQQ mutations) potentiate slow inactivation in some Na_v isoforms [28, 29]. To accurately measure recovery from slow inactivation without interference from fast inactivation, it is advisable to use a fast-inactivation deficient mutant, such as the IFM-to-QQQ or the CW-mutant isoform [28, 30]. However, expression may be an issue, and therefore elimination of fast inactivation by hyperpolarization as suggested here may be a more feasible alternative.

3.4 Closed-State Fast and Slow Inactivation

Closed-state inactivation describes a mechanism where small depolarizations that are not sufficient to activate Na_Vs induce inactivation of the channels, leading to decreased excitability to a following activating stimulus. It has been proposed that closed-state inactivation occurs if only the S4 segments of DIII and DIV are displaced by depolarization but not those of DI and DII. In this case, the activation gate does not open and no current flows, but the inactivation particle can already bind and occlude the channel pore [31].

The appropriate voltage-clamp protocol to investigate closed-state inactivation can be seen in Table 2 (row a). It begins with a pre-pulse of varying duration between 0.1 ms and 2 s. A pre-pulse

Table 2
Summary of voltage-clamp protocols used to study Na_v closed-state inactivation and deactivation as well as resurgent and ramp currents

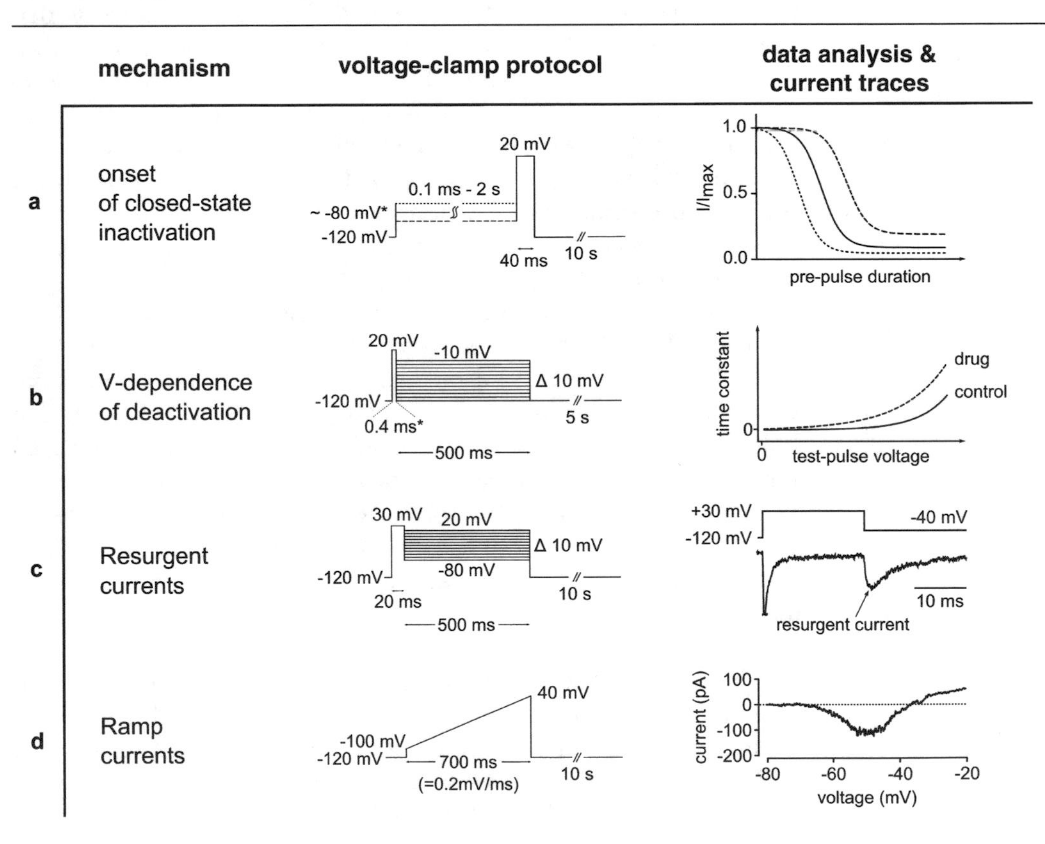

For onset of closed-state inactivation and voltage dependence of deactivation, the data analysis column contains schematic plots that represent typical data fits, which can be obtained from the respective recordings. For resurgent and ramp currents, example traces are shown in this column. In the case of resurgent currents, the displayed trace represents a zoomed in section. Please note the different scale bar for the time axis as compared to the voltage-clamp protocol column.

*The exact value depends on measurements obtained with protocols from Table 1. See main text for details.

voltage should be chosen that has previously been shown to induce steady-state fast and slow inactivation without causing activation of the channels using the protocols described in Sects. 3.1 (activation), 3.2.2 (fast inactivation), and 3.3.2 (slow inactivation). Several different pre-pulse voltages (dotted lines in protocol a in Table 2) can be applied to compare the time courses of closed-state inactivation at different potentials. The pre-pulse is immediately followed by a standard test-pulse causing activation of the cell. Closed-state inactivation will be measurable as a reduction in test-pulse current amplitude, indicative of a reduction in the fraction of available channels. Thus plotting normalized peak test-pulse currents (I/I_{max}) against pre-pulse duration provides a sigmoidal inactivation curve.

3.5 Deactivation

Deactivation of Na_vs describes the process during which the activation gate closes due to the voltage-sensing domains of the channel returning to their original pre-activation position. This process must not be confused with channel inactivation. While the latter leads to a termination of current flux due to a collapse or blockade of the channel pore, deactivation leads to closure of the channel's activation gate and its return to an excitable state. As both inactivation and deactivation require repolarization of the cell membrane after activation, it is imperative to use an experimental protocol that distinguishes between the two processes. Protocol b in Table 2 allows such a discrimination: Na_v activation is achieved by a very brief and strong depolarizing pre-pulse to 20 mV. The duration of this pre-pulse must be carefully chosen. It is important that the pre-pulse is short enough to minimize channel inactivation, while it lasts long enough to activate as many channels as possible. Typically, this will be after about a millisecond, depending on the channel isoform. At the same time, pre-pulse duration should be long enough to allow for maximum current development. Some Na_vs open faster than others; it is therefore important to determine the time-to-peak of inward current using the activation protocol in Sect. 3.1 (Table 1a). Thus, an ideal pre-pulse duration that allows for maximum current development while avoiding fast inactivation can be established. The brief activating pre-pulse is then followed by a 500 ms hyperpolarized pulse, to induce tail currents. Potentials are gradually stepped in a depolarizing direction starting from V_{hold}, in 10 mV increments up to −10 mV (test-pulse). After each sweep the channels are allowed to recover for 5 s. The brief activating pre-pulse induces a short inward current followed by the tail current that establishes due to the increase in driving force while the potential is switched to negative values. The rate of decay of this current, measured at each test potential, describes the time course of channel deactivation. This current decay can be fitted to obtain the time constant as described in Sect. 4.2. At more positive potentials, inactivation will become a relevant factor in slowing the current decline. Therefore the time constants measured at negative potentials are most appropriate for the interpretation of the deactivation time course.

3.5.1 Notes

1. Due to the very short duration of the activating pre-pulse, it is important that voltage-clamp conditions and, especially, R_s compensation are optimal (see Sect. 2.4.2). If these conditions are below optimal, current development might be delayed due to a voltage error and capacitance artifacts might occur, which take too long to subside and are likely to distort current recording and interfere with accurate measurement.
2. It is crucial to adjust the filter settings so that the very fast tail currents, activated by the hyperpolarizing voltage steps (hundreds of microseconds!), can be accurately resolved (see Sect. 2.5). We suggest low-pass filter settings of 30 kHz and a sampling rate of 100 kHz.

3.6 Resurgent Currents

Under certain circumstances, especially during short and strong depolarizations, Na_vs can undergo open channel block by an endogenous blocking particle that differs from the aforementioned inactivation particle (IFMT motif, see Sect. 3.2). During this open channel block, the inactivation particle is prohibited from binding and no fast inactivation occurs [32]. Upon slight repolarization, however, the blocking particle is expelled due to its positive charge, leading to the occurrence of resurgent currents (see current trace in row c of Table 2). The most likely candidate for the blocking particle that causes open channel block is the β4 subunit and, more precisely, its cytoplasmic "KKLITFILKKTREK" sequence. A synthetic peptide corresponding to this sequence has been shown to induce resurgent currents [33–35]. As they typically take place in the decaying phase of an action potential, resurgent currents have been proposed to contribute to increased excitability of a neuron [36].

Resurgent currents cannot normally be measured in heterologous expression systems and interestingly, co-transfection of the β4 subunit is not sufficient to restore resurgent currents. However, addition of the above blocking peptide "KKLITFILKKTREK" at 100 μM [33, 37] to the recording pipette allows detection of resurgent currents in cell lines. Why this peptide, which is a sequence from the β4 subunit, induces resurgent currents but expression of the entire β4 subunit does not, has not been fully elucidated. It is possible that the cytoplasmic part of the β4 subunit needs to be processed by enzymatic processes only found in some neurons [38].

If the above blocking peptide is present in the intracellular solution, resurgent currents can be measured in some heterologously expressed Na_vs using protocol c shown in Table 2. A short

20 ms depolarization to 30 mV will drive the blocking peptide into the channel pore, leading to open channel block. After this pre-pulse, the membrane potential is gradually stepped in a hyperpolarizing direction from 20 to −80 mV with increments of 10 mV per sweep (test-pulse). This will lead to voltage-dependent expulsion of the blocking peptide and occurrence of resurgent current. A very short and quickly deactivating tail current can appear during test-pulse repolarization, which indicates that some channels have not undergone open channel block. During repolarization, the driving force for the persistent current through these unblocked channels is increased, leading to a tail current. Care should be taken not to confuse these tail currents with resurgent currents. Luckily however, tail currents quickly deactivate via closing of the activation gate (see Sect. 3.5). Resurgent currents on the other hand have a slow activation kinetic and may clearly be distinguished from tail currents due to their much slower decay. The peak of the resurgent current is then typically plotted against test-pulse voltage, and its decay may be fitted.

3.7 Ramp Currents

Applying a slow voltage ramp protocol to heterologously expressed Na_vs can lead to the manifestation of ramp currents in some isoforms [8, 37]. Ramp currents are small inward currents that are evoked by slow depolarizations. Because such slow depolarizations can be thought of as the equivalent of naturally occurring subthreshold stimuli, any channel that responds with conduction of inward current might play an important role in boosting such weak stimuli and thus contribute to neuronal excitability. Ramp currents are impeded by closed-state inactivation, which is likely to be induced by slow depolarizing stimuli (see Sect. 3.4) before any activation and inward current can take place. Accordingly, any channel that does display ramp currents must have slow closed-state inactivation kinetics [39]. As a consequence, ramp current investigation can be used as an indirect measure of closed-state inactivation.

Testing for ramp currents is conducted using a very simple voltage ramp protocol, such as protocol d in Table 2. The ramp starts at −100 mV and increases the potential slowly by 0.2 mV/ms up to a maximum of 40 mV. Such a ramp takes 700 ms to complete, after which the cell membrane is returned to V_{hold} of −120 mV. If a ramp current establishes, it will typically do so with a threshold around −70 to −60 mV, depending of the Na_v subtype's activation voltage. Ramp current peak amplitude can be normalized to transient peak amplitude measured beforehand using a regular activation protocol.

4 Data Analysis

The types of software packages used for data acquisition and analysis normally depend on the type and brand of amplifier that is used. After data acquisition, the majority of data analysis, fitting of data, statistical analysis, and graphical output can be conducted using a range of commercially available software packages.

It has already been mentioned that in order to exclude the possibility of measuring currents mediated by endogenous Na_vs (see Sect. 2.1), only cells that display substantial current amplitudes (we recommend >400 pA) should enter the data analysis. The use of CsF in the ICS helps to establish large current amplitudes (see Sect. 2.2).

4.1 Voltage Dependent Activation of Sodium Channels

Data obtained by the protocol described in Sect. 3.1 can be used to derive the time-to-peak, current amplitude, current density, current–voltage (*IV*) relationship, conductance, and half-activation voltage ($V_{1/2}$). The time-to-peak of voltage-activated currents is determined by measuring the time from pulse onset until the peak of the sodium current. Current amplitude describes the transient peak inward current at each test potential and is easily determined by the analysis software. The maximum induced inward current can be used to derive current density by dividing it by the cell's capacitance, as provided by the amplifier after R_s compensation at the start of the experiment. Plotting current amplitude measured in each sweep against the corresponding test voltage provides a simple *IV* relationship (Figs. 1c and 2d), which allows some conclusions about recording quality, voltage-clamp conditions, and R_s compensation. Typical indications for less-than-optimal experimental conditions will be discussed below (Sect. 4.3). Current amplitude can be used to measure the fraction of available channels at each test voltage and to calculate $V_{1/2}$. However, because each current was elicited by a different voltage and because the difference in current amplitude at each voltage is not only due to changes in ion channel activity but also due to changes in the driving force, it is necessary to transfer current amplitude into membrane conductance. Conductance (G) is calculated from measured current (I) as follows, taking the driving force ($V - V_{rev}$) into account:

$$G = \frac{I}{(V - V_{rev})}$$

where V is the voltage applied at current I and V_{rev} is the reversal potential of the conducting ion, in this case Na^+. While V_{rev} can be calculated using the Nernst equation, a sufficiently accurate measure can be obtained by extrapolation from the plotted *IV* relationship. Once Na_vs are fully open, current through the channel should behave in an ohmic way, i.e., decrease in a linear fashion as the potential approaches V_{rev}. Often this is not exactly the case and the decay of the *IV* relationship may display a nonlinearity, especially when close to V_{rev}. It is therefore advisable to use recording conditions (composition of recording solutions, see Sect. 2.2) that produce a V_{rev}, which lies in the potential range where channels are already fully open. By fitting a linear regression to those points of the decaying *IV* relationship that behave as close to linear as possible, one can obtain a reading for the slope and the intercept of the regression line. Dividing the intercept by the slope provides a confident measure of V_{rev} (e.g., trace A in Fig. 2d). This way, conductance can easily be calculated. Normalizing G to G_{max} and plotting G/G_{max} against voltage provides a standard *GV* relationship that will show a sigmoidal shape (Figs. 1d and 2e). Calculated conductances around V_{rev} are notoriously unstable, which may lead to *GV* relationships that reach a maximum at one data point and then fall back below 1 or to *GV* relationships that never reach the maximum until the final data point. If the voltage-clamp protocol that has been described here has been used in combination with full sodium solutions (140 mM external, 10 mM internal; for protocol see Sect. 3.1, Table 1a), the final data point does not reflect the only point of maximal channel opening as it is far too depolarized. In this case the final data point can confidently be regarded as a calculation artifact and be eliminated from the *GV* relationship, giving the latter a sigmoidal shape that plateaus at a maximum of 1. By fitting the following Boltzmann equation to the *GV* relationship, one can obtain readings for the slope and the midpoint of the fitted curve, the latter being the value of activation $V_{1/2}$:

$$G = \frac{G_{max}}{1 + e^{\frac{(Vm - V1/2)}{k}}}$$

where G_{max} is the maximum sodium conductance, V_m is the membrane potential, $V_{1/2}$ is the half-maximal activation voltage, and k is the slope factor.

A similar procedure as for the fitting of the activation *GV* relationship can be conducted to analyze the voltage dependence of Na_v fast, slow, and closed-state inactivation. In these cases, however, it is not necessary to calculate conductance as the current of interest is always measured at the same potential (the test-pulse) and therefore the driving force for current flow is always the same. Here, each current is normalized to the maximum elicited current (I/I_{max}) and plotted against pre-pulse voltage. Fitting the sigmoidal curves with a Boltzmann equation (see above) provides a readout for slope and $V_{1/2}$ of channel inactivation.

4.2 Fitting of Time Constants

Fitting of exponential functions to either a raw current trace or to a plotted data curve can provide valuable information on current behavior or shape of the data curve. The main value of interest when plotting exponential functions is the time constant tau (τ), which is defined as the point at which a given curve has reached ~63 % of its final value. Similar to $V_{1/2}$ in the case of sigmoidal curves, τ allows quantification of the shape of an exponential curve and comparison between curves or traces obtained under different conditions. Exponential fitting of raw current traces is used when analyzing current decay, recorded using an activation protocol (see Sect. 3.1), or when analyzing current deactivation as described in Sect. 3.5. This can be done using most statistics and analysis tools, usually including most amplifier's recording software. The following mono-exponential function is fitted to the decaying part of the current trace roughly from the turning point of the trace to the steady-state level:

$$y(x) = \text{Amp0} + \text{Amp1}\left(1 - e^{\frac{-x}{\tau}}\right)$$

where x is time and Amp0 and Amp1 are the amplitude at time 0 and at steady state, respectively. Each τ measured using different test-pulses as described in Sect. 3.5 is plotted against the test-pulse voltage to obtain a deactivation curve.

It is also possible to fit plotted curves with exponential functions using the above formula. While this is not strictly necessary, it can help in quantifying differences between data curves obtained under different conditions (e.g., wildtype vs. mutant isoform or control vs. drug treatment). This can be done with the above deactivation curve as well as to curves describing onset and recovery from fast or slow inactivation. In these latter cases it is customary to plot normalized test-pulse peak current against pre-pulse or interpulse duration and to fit the obtained curves with a mono-exponential function.

4.3 Troubleshooting and Determining the Clamp Quality from Analyzed Data

As has been discussed in Sect. 2.4, a good quality of the voltage clamp is imperative for stable and representative recordings and the technical requirements for good quality voltage clamp have been given above. Recorded data obtained via a simple activation protocol (see Sect. 3.1, Table 1a) can be used to determine recording quality, clamp quality, and accuracy of R_s compensation. Thus, the raw current trace already provides valuable information on the recording conditions and whether or not it is advisable to include the trace in the analysis: a current trace that displays the appearance of multiple peaks in each sweep, one of which is heavily delayed, typically stems from a Na_v-expressing cell that is connected to a second responding cell. The current response of the second unwanted cell can be seen with a clear delay and reduced peak amplitude. Similarly, a current trace of a single cell that displays its current maxima at different time points may be suggestive of space clamp artifacts (see Sect. 2.4.1). The delayed opening of channels in distant cell compartments can cause current responses to appear less sharp and with a clear lag. However, one needs to be careful with this interpretation, as with stronger depolarization, channels will open more quickly even in a good recording. The time difference between peaks in a good recording will be on a smaller scale than that from a recording with space clamp artifacts.

When it comes to the quality of the voltage clamp, several aspects have to be considered, some of which have already been mentioned. First, R_a needs to be minimal. If R_a is too large or increases throughout the recording, e.g., by clogging of the electrode with cell membrane, current flow between the cell and the electrode is hampered. This leads to a delay in measuring the current that flows into the cell through Na_vs, i.e., the time-to-peak will be delayed. Furthermore, a voltage error is created that consequently leads to a delay and a slowing in effective clamping of the membrane potential.

Typical problems in clamp quality, including R_s compensation, are usually recognizable in the activation *IV* relationship (Fig. 2). The first indication is provided by the location of V_{rev}. As discussed, the theoretical V_{rev} for Na^+ can easily be calculated based on the ionic concentrations, using the Nernst equation [40]. Calculation tools are available online. A large (>10 mV) deviation of the measured (or extrapolated, see above) V_{rev} from the theoretical value can be an indication of a poorly clamped cell (e.g., Fig. 2d). In addition, a cell that has not been voltage-clamped appropriately will often display a very steep increase of the *IV* relationship. In such a case, small currents around the activation threshold will escape detection completely only to suddenly reach the peak of the *IV* relationship with the next test-pulse. In other words, current activation from 0 to ~100 % takes place over only 10–20 mV, instead of about 40 mV in a good recording (Fig. 2d). The reason for this behavior probably lies in the fast inactivation of Na_vs. In a poorly voltage-clamped cell, e.g. because of inappropriate R_s compensation, establishing the desired membrane potential takes more time, leading to a rather slow depolarization, instead of a step-like voltage change. Such slow depolarization can lead to the majority of Na_vs undergoing closed-state inactivation. Close to the activation threshold, this means that channels do not open and no current will be recorded. Only at stronger depolarizations (i.e., later voltage steps) is the activation threshold reached quickly enough for channels to still be excitable. At that point, however, the stimulus will be large enough to elicit maximum or near-maximum inward current. The same steep increase that is thus noticeable in the *IV* relationship will also be visible in the activation *GV* relationship (Fig. 2e). As mentioned earlier, these curves have a sigmoidal shape and can be used to calculate $V_{1/2}$. Generally, a poorly voltage-clamped cell will display a *GV* relationship that is shifted towards more negative potentials (Fig. 2e). Accordingly, care must be taken not to misjudge a poorly voltage-clamped cell for a real shift of activation $V_{1/2}$. The indications that have been given above and that can be noticed in current trace and *IV* relationship can be used to avoid such a misjudgment.

Because the *IV* relationship is a good indication of clamp quality, it may be advisable to begin each recording with a shortened activation protocol, similar to that described in Sect. 3.1 (Table 1a) but with less voltage steps (e.g., from −75 to 30 mV with 15 mV increments). Using the recording software's quick analysis tools, one can derive a quick *IV* relationship from such a test protocol. Close inspection of the raw currents as well as shape and V_{rev} of such a test-*IV* can provide useful information on the clamp quality and can save the experimenter valuable time.

References

1. Catterall WA, Goldin AL, Waxman SG (2005) International union of pharmacology. XLVII. Nomenclature and structure-function relationships of voltage-gated sodium channels. Pharmacol Rev 57:397–409. doi:10.1124/pr.57.4.4
2. Hodgkin AL, Huxley AF (1952) A quantitative description of membrane current and its application to conduction and excitation in nerve. J Physiol 117:500–544
3. Jiang B et al (2002) Endogenous KV channels in human embryonic kidney (HEK-293) cells. Mol Cell Biochem 238:69–79. doi:10.1023/A:1019907104763
4. Berjukow S et al (1996) Endogenous calcium channels in human embryonic kidney (HEK293) cells. Br J Pharmacol 118:748–754
5. Cummins TR et al (1993) Functional consequences of a Na + channel mutation causing hyperkalemic periodic paralysis. Neuron 10:667–678. doi:10.1016/0896-6273(93)90168-Q
6. He B, Soderlund DM (2010) Human embryonic kidney (HEK293) cells express endogenous voltage-gated sodium currents and Nav1.7 sodium channels. Neurosci Lett 469:268. doi:10.1016/j.neulet.2009.12.012
7. Moran O, Nizzari M, Conti F (2000) Endogenous expression of the beta1A sodium channel subunit in HEK-293 cells. FEBS Lett 473:132–134
8. Cummins TR et al (2001) Nav1.3 sodium channels: rapid repriming and slow closed-state inactivation display quantitative differences after expression in a mammalian cell line and in spinal sensory neurons. J Neurosci 21:5952–5961
9. Leffler A et al (2005) Pharmacological properties of neuronal TTX-resistant sodium channels and the role of a critical serine pore residue. Pflugers Arch 451:454–463. doi:10.1007/s00424-005-1463-x
10. Vanoye CG et al (2013) Mechanism of sodium channel NaV1.9 potentiation by G-protein signaling. J Gen Physiol 141:193–202. doi:10.1085/jgp.201210919
11. Qu W et al (2000) Very negative potential for half-inactivation of, and effects of anions on, voltage-dependent sodium currents in acutely isolated rat olfactory receptor neurons. J Membr Biol 175:123–138. doi:10.1007/s002320001061
12. Van Petegem F, Lobo PA, Ahern CA (2012) Seeing the forest through the trees: towards a unified view on physiological calcium regulation of voltage-gated sodium channels. Biophys J 103:2243–2251. doi:10.1016/j.bpj.2012.10.020
13. Saab CY, Cummins TR, Waxman SG (2003) GTPγS increases Nav1.8 current in small-diameter dorsal root ganglia neurons. Exp Brain Res 152:415–419. doi:10.1007/s00221-003-1565-7
14. Coste B et al (2004) Gating and modulation of presumptive NaV1.9 channels in enteric and spinal sensory neurons. Mol Cell Neurosci 26:123–134. doi:10.1016/j.mcn.2004.01.015
15. Sternweis PC, Gilman AG (1982) Aluminum: a requirement for activation of the regulatory

component of adenylate cyclase by fluoride. Proc Natl Acad Sci U S A 79:4888–4891

16. Todt H et al (1999) Ultra-slow inactivation in mu1 Na + channels is produced by a structural rearrangement of the outer vestibule. Biophys J 76:1335–1345
17. Ben-Johny M et al (2014) Conservation of Ca2+/ calmodulin regulation across Na and Ca2+ channels. Cell 157:1657–1670. doi:10.1016/j.cell.2014.04.035
18. Levis RA, Rae JL (2007) Technology of patch-clamp electrodes. In: Walz W (ed) Patch-clamp analysis: advanced techniques, 2nd edn. Humana Press Inc., Totowa, NJ, pp 1–34
19. Lampert A et al (2010) Sodium channelopathies and pain. Pflugers Arch 460:249–263. doi:10.1007/s00424-009-0779-3
20. Molleman A (2003) Patch clamping: an introductory guide to patch clamp electrophysiology, 1st edn. John Wiley & Sons Ltd, Chichester, UK
21. Sontheimer H, Olsen ML (2007) Whole-cell patch-clamp recordings. In: Walz W (ed) Patch-clamp analysis: advanced techniques, 2nd edn. Humana Press Inc., Totowa, NJ, pp 35–68
22. Armstrong CM, Bezanilla F (1975) Currents associated with the ionic gating structures in nerve membrane. Ann N Y Acad Sci 264:265–277. doi:10.1111/j.1749-6632.1975.tb31488.x
23. Peters CH, Ruben PC (2014) Introduction to sodium channels. In: Ruben PC (ed) Volt. Gated sodium channels. Springer, Berlin, pp 1–6
24. Silva J (2014) Slow inactivation of Na + channels. In: Ruben PC (ed) Volt. Gated sodium channels. Springer, Berlin, pp 33–49
25. Goldin AL (2003) Mechanisms of sodium channel inactivation. Curr Opin Neurobiol 13:284–290. doi:10.1016/S0959-4388(03)00065-5
26. O'Reilly JP et al (1999) Comparison of slow inactivation in human heart and rat skeletal muscle Na + channel chimaeras. J Physiol 515:61–73. doi:10.1111/j.1469-7793.1999.061ad.x
27. Hilber K et al (2001) The selectivity filter of the voltage-gated sodium channel is involved in channel activation. J Biol Chem 276:27831–27839. doi:10.1074/jbc.M1019 33200
28. Featherstone DE, Richmond JE, Ruben PC (1996) Interaction between fast and slow inactivation in Skm1 sodium channels. Biophys J 71:3098–3109
29. Richmond JE et al (1998) Slow inactivation in human cardiac sodium channels. Biophys J 74:2945–2952. doi:10.1016/S0006-3495(98)78001-4
30. Wang S-Y et al (2003) Tryptophan scanning of D1S6 and D4S6 C-termini in voltage-gated sodium channels. Biophys J 85:911–920. doi:10.1016/S0006-3495(03)74530-5
31. Armstrong CM (2006) Na channel inactivation from open and closed states. Proc Natl Acad Sci 103:17991–17996. doi:10.1073/pnas.0607603103
32. Lampert A, Eberhardt M, Waxman SG (2014) Altered sodium channel gating as molecular basis for pain: contribution of activation, inactivation, and resurgent currents. In: Ruben PC (ed) Volt. Gated sodium channels. Springer, Berlin, pp 91–110
33. Grieco TM et al (2005) Open-channel block by the cytoplasmic tail of sodium channel β4 as a mechanism for resurgent sodium current. Neuron 45:233–244. doi:10.1016/j.neuron.2004.12.035
34. Wang GK, Edrich T, Wang S-Y (2006) Time-dependent block and resurgent tail currents induced by mouse β4154–167 peptide in cardiac Na + channels. J Gen Physiol 127:277–289. doi:10.1085/jgp.200509399
35. Lewis AH, Raman IM (2011) Cross-species conservation of open-channel block by Na channel β4 peptides reveals structural features required for resurgent Na current. J Neurosci 31:11527–11536. doi:10.1523/JNEUROSCI.1428-11.2011
36. Raman IM, Bean BP (1997) Resurgent sodium current and action potential formation in dissociated cerebellar purkinje neurons. J Neurosci 17:4517–4526
37. Eberhardt M et al (2014) Inherited pain: sodium channel Nav1.7 A1632T mutation causes erythromelalgia due to a shift of fast inactivation. J Biol Chem 289:1971–1980. doi:10.1074/jbc.M113.502211
38. Huth T et al (2011) β-Site APP-cleaving enzyme 1 (BACE1) cleaves cerebellar Na + channel β4-subunit and promotes Purkinje cell firing by slowing the decay of resurgent Na + current. Pflugers Arch 461:355–371. doi:10.1007/s00424-010-0913-2
39. Cummins TR, Howe JR, Waxman SG (1998) Slow closed-state inactivation: a novel mechanism underlying ramp currents in cells expressing the hNE/PN1 sodium channel. J Neurosci 18:9607–9619
40. Hille B (2001) Ion channels of excitable membranes, 3rd edn. Sinauer Associates, Sunderland, MA

Chapter 4

Elucidating the Link Between Structure and Function of Ion Channels and Transporters with Voltage-Clamp and Patch-Clamp Fluorometry

Giovanni Zifarelli and Jana Kusch

Abstract

Ion channels and transporters are membrane proteins whose functions are driven by conformational changes. This implies that to gain a deep understanding of their dynamic behavior, structural and functional information need to be integrated. Classical biophysical techniques provide insight into either the structure or the function of these proteins, but their correlation in time remains a challenging task. In this chapter, we illustrate how two related techniques, voltage-clamp fluorometry (VCF) and patch-clamp fluorometry (PCF), provide such a type of integrated information. They combine electrophysiological techniques, two-electrode voltage-clamp (VCF) and patch-clamp (PCF), with spectroscopic approaches to simultaneously detect conformational changes and ionic currents mediated by ion channels and transporters in a native membrane environment. The optical part is based on the environmental sensitivity of the fluorescence emission of probes attached at specific sites of ion channels and transporters. This allows the correlation between structural conformation and defined functional states. VCF and PCF have been applied to a variety of ion channel and transporter families to investigate several biophysical problems ranging from structural changes linked to activation by various stimuli to the analysis of the process of inactivation and deactivation. Additionally, these techniques allowed for reading out gating-dependent ligand binding and protein mobility. In this chapter, illustrating some typical examples of the application of VCF and PCF, we try to show their potential and flexibility and to highlight some of the technical caveats.

Key words Ion channels, Transporters, Voltage-clamp fluorometry, Patch-clamp fluorometry, VCF, PCF, Gating, Conformational changes, Ligand binding, Transport

1 Introduction

The inner workings of proteins can be condensed in the following fundamental question: What is the relation between structural changes and functional transitions? The case of ion channels and transporters is a paradigmatic example. Several decades of electrophysiological and biochemical investigations unveiled a variety of sophisticated mechanisms by which ion channels open and close the permeation pathway in response to physiological stimuli, defined as

Alon Korngreen (ed.), *Advanced Patch-Clamp Analysis for Neuroscientists*, Neuromethods, vol. 113,
DOI 10.1007/978-1-4939-3411-9_4, © Springer Science+Business Media New York 2016

gating mechanisms. Similar considerations apply to transporters that function with an elaborate sequence of conformational changes to transport substrates across membranes.

The intrinsically dynamic nature of these transport mechanisms (and proteins) can be inspected with high time resolution by electrophysiological measurements. However, this purely functional readout does not provide structural insights into the rearrangements associated with those processes. Structural methods, like X-ray crystallography, on the other hand, produce a wealth of information, down to the atomic level, on specific protein conformations. However, these structures provide only static pictures of those highly dynamic processes, and in most cases the functional state associated with a given structure is not fully defined. Moreover, an important limitation is that structures of ion channels and transporters are not obtained in the native membrane environment, which is not just an inert landscape that hosts those proteins but actively shapes ion channels and transporters' structure and function.

To address the fundamental question that opened this chapter, the conformations of ion channels and transporters associated with defined functional states need to be determined in real time in a native membrane environment. But how to obtain simultaneous structural and functional information? Fluorescence spectroscopy provides powerful and versatile tools that, combined with electrophysiological techniques, assisted in this task. In general, biophysical techniques based on fluorescence can track processes on time scales that span several orders of magnitude and down to the sub-nanosecond level with very high sensitivity allowing also single-molecule detection [1]. Moreover, the different fluorophore's features such as the spectral properties (absorption and emission), the fluorescence intensity ("brightness"), the quantum yield, the fluorescence lifetime, and anisotropy probe different parameters of the protein to which they are attached such as chemical environment, mobility, proximity to other proteins or fluorophores, and size. Because of these properties, fluorescence techniques have been successfully applied to countless biological problems, like conformational rearrangements of proteins, protein mobility, or ligand binding processes (Fig. 1).

However, in the context of ion channel and transporter research, it was the environmental sensitivity of fluorescent dyes (Fig. 1a, *left panel*) that has been the key to the use of fluorescence techniques in combination with electrophysiological measurements to obtain simultaneous structural and functional information in real time. This chapter will focus on two versions of this combined approach that proved particularly successful in investigating ion channels and transporters: voltage-clamp fluorometry (VCF) and patch-clamp fluorometry (PCF) (Figs. 2 and 3).

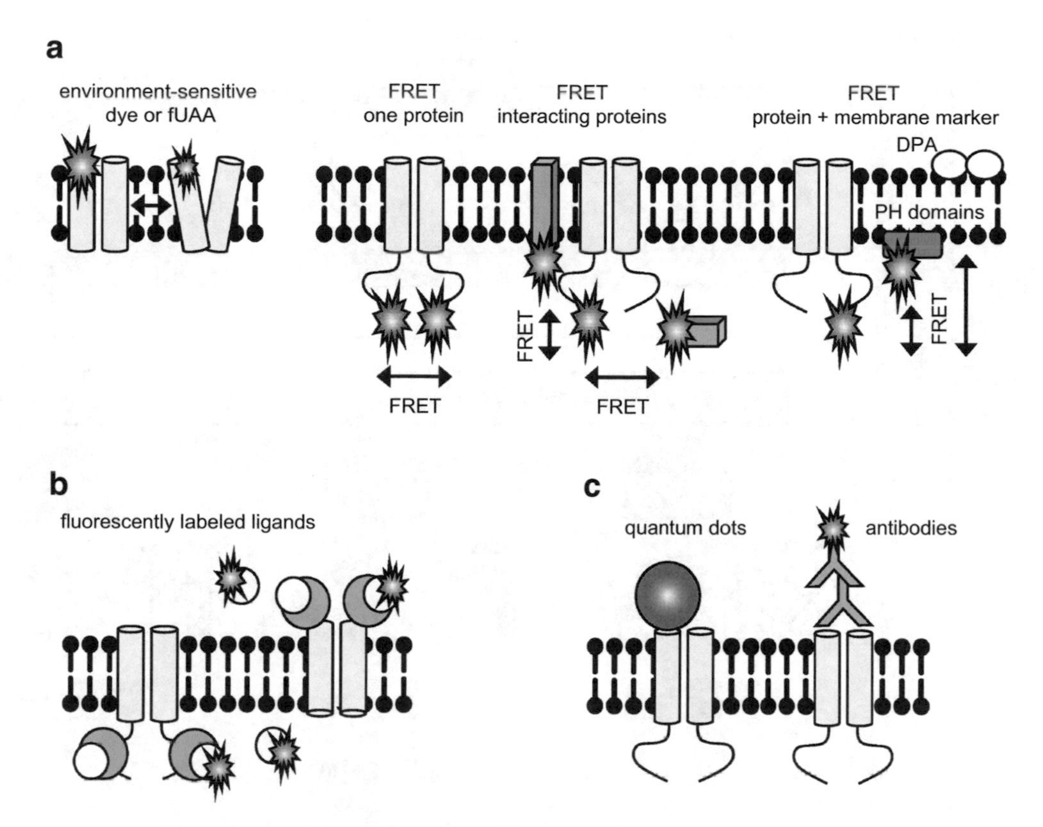

Fig. 1 Tagging strategies to explore different dynamic parameters. (**a**) Conformational rearrangements. *Left panel*: Environment-sensitive dyes or fluorescent unnatural amino acids (fUAAs) (*green symbol*) are added to the protein at the region of interest. Gating-dependent changes might cause changes in the emitted fluorescence signal due to effects on different photo-physical features of the fluorophore. *Right panel*: Measuring molecular distances between two fluorophores (donor D and acceptor A) or between a fluorescent donor and a nonfluorescent acceptor using FRET. FRET causes a decrease in D fluorescence (*blue symbol*) and an increase in A fluorescence (*yellow symbol*), and can therefore be quantified either by donor dequenching in which an increase in D fluorescence is measured after complete photobleaching of A, or by enhanced acceptor emission in which the recorded parameter is the increase in fluorescence intensity of A due to FRET with D. Alternatively, FRET quantification can be performed by fluorescence lifetime measurements of D and A. (**b**) Ligand binding. Ligand molecules can be tagged by covalent binding of fluorophore molecules. The tagged fluorophores as well as connecting linkers have to be chosen with great care to preserve the physiological function of the tagged ligand. (**c**) Protein mobility. Tagging proteins with quantum dots or fluorescently labeled antibodies can be used to monitor protein mobility in native and artificial membranes

2 Procedures and Techniques

2.1 Voltage-Clamp Fluorometry (VCF)

2.1.1 Principle

VCF is based on the concept that combined electrophysiological recordings (providing information on the functional states of ion channels, receptors, or transporters) and fluorescence detection (providing a readout of structural changes in the vicinity of the fluorophore) can lead to a better understanding of the interplay

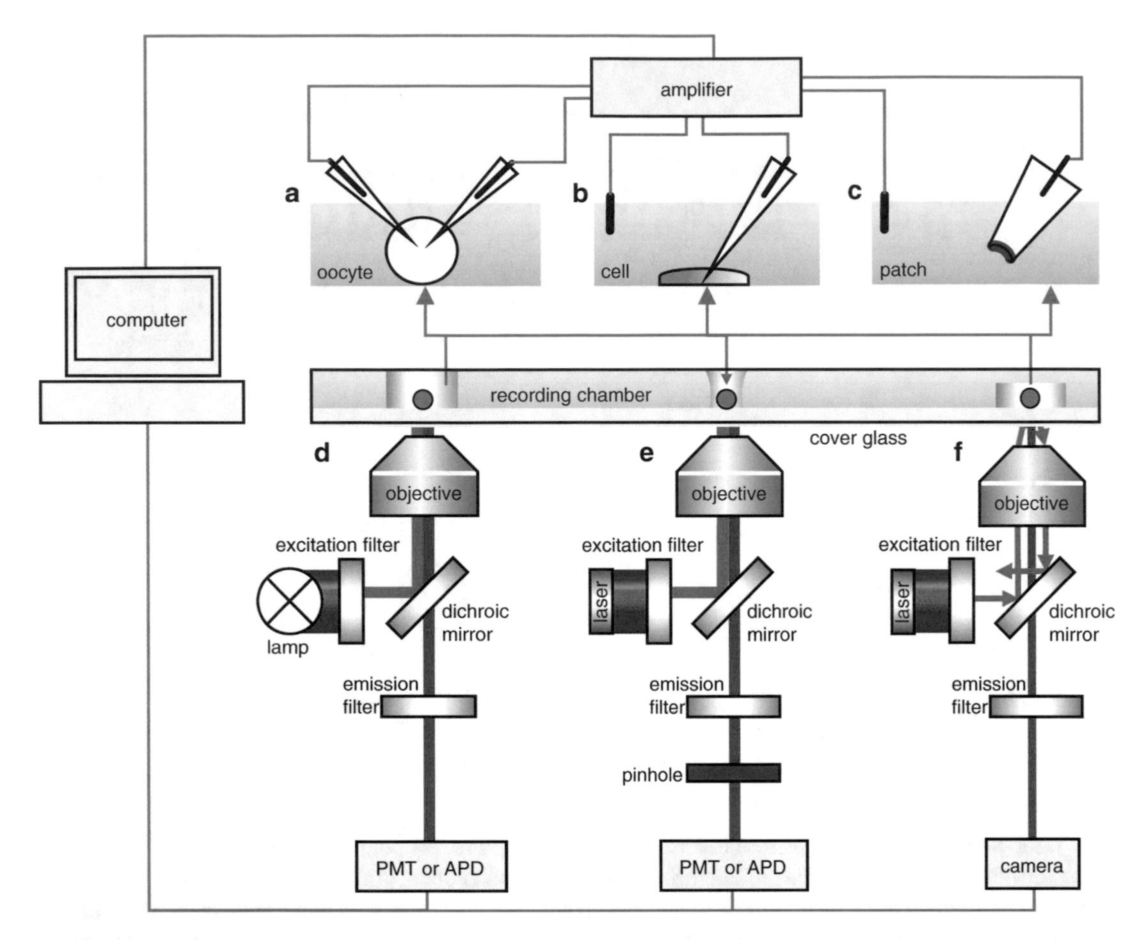

Fig. 2 Electrophysiological and optical components for PCF and VCF setups. The electrophysiological components (**a**: two-electrode voltage-clamp or cut-open oocyte technique, **b**: whole-cell and cell-attached configuration, **c**: excised patch configuration) are generally similar to traditional electrophysiological setups. The bottom of the recording chamber harvesting the cells should be a glass coverslip of 0.17 mm (#1.5) thickness to achieve bright images and to minimize spherical aberration. Each of the components **a–c** can be combined with one of the three microscopic components (**d**: wide-field microscopy; **e**: confocal microscopy; **f**: TIRF microscopy). Normally, inverted microscopes are preferred because they provide more room above the recording chamber for technical components like headstages, manipulators, or application systems. (**d**) In wide-field microscopy the entire specimen in the optical path is illuminated by light from a mercury or xenon lamp. Excitation light (*green line*) goes through the excitation filter and is reflected by the dichroic mirror. The light is focused on the specimen through the objective lens. All of the fluorescence emitted by the specimen (*orange line*) is transmitted by the dichroic mirror and reaches the detector, usually an avalanche photodiode (APD) or a photomultiplier (PMT). Backward reflected excitation light is removed by the emission filter. There can be a considerable background signal because of high out-of-focus fluorescence. (**e**) In confocal microscopy the light source is a laser. Two technical variations compared to **d** reduce the out-of-focus signal: (1) point-illumination (illuminating a single point at one time with a focused beam) (*green line*), (2) adding a variable pinhole aperture in front of the detector to block light emitted away from the point being illuminated. The smaller the diameter of this pinhole, the less light is detected from areas outside the focus. The detector is a photomultiplier (PMT) or an avalanche photodiode (APD). Because only a fraction of the total emitted fluorescence is collected, the excitation energy must be relatively high, increasing the risk of rapid photobleaching. Photobleaching can be minimized by using stable fluorophores, high detector sensitivity, and

between structure and function of these proteins [2, 3]. The electrophysiological part of VCF is a traditional two-electrode voltage clamp (TEVC) setup or a cut-open oocyte setup (Fig. 2a). The fluorescence application is based on site-directed labeling, the ability to place a fluorophore at a specific location in a protein, and on the dependence of the quantum yield and spectral properties of the dye on its proximal environment. Fluorescence emission is typically detected through a wide-field fluorescence microscope and a photodiode or photomultiplier (Fig. 2d). In a typical example that will be illustrated in detail, voltage-dependent changes in the fluorescence intensity emitted by a fluorophore, which is attached to a specific site, can be interpreted as environmental changes caused by voltage-dependent conformational rearrangements around this fluorophore. The functional state of the channel associated with those conformations is simultaneously monitored by electrophysiology (Fig. 3a) [2, 3].

Site-directed labeling in VCF is typically carried out by introducing a cysteine residue at a specific site in the channel via mutagenesis. The channels are heterologously expressed in the plasma membrane of oocytes. If the introduced cysteine is accessible from the extracellular side, it can be labeled by covalent bonding between its sulfhydryl group and thiol groups, which are attached to the fluorophore, typically a maleimide moiety [2, 3]. Labeling specificity can be very high if the native cysteines of the channel are removed. When the expression level of the protein under study is high, the background produced by labeling cysteines of endogenous proteins is not a major concern but can be reduced by incubating the oocytes in cysteine-reactive reagents, like *N*-ethyl-maleimide (NEM), before heterologous expression. This approach of site-directed fluorescence labeling is based on the accessibility of the introduced cysteines for the extracellular solution, which poses strict limitations on the ion channel region that can be investigated using this technique. Nonetheless, it has been successfully applied to different types of membrane proteins, as voltage-gated ion channels and various types of transporters.

To overcome these limitations, Kalstrup and Blunck very recently performed VCF with environmental-sensitive fluorescent unnatural amino acids (fUAAs) [4] (Fig. 1a, *left panel*). fUAAs are intrinsically fluorescent and can be incorporated into the protein

Fig. 2 (continued) maximum objective numerical aperture. (**f**) In total internal reflection fluorescence microscopy the excitation light originates from a laser as light source. When the excitation light (*green line, arrow up*) strikes the glass/specimen interface, a small portion of the reflected light penetrates through the interface and propagates parallel to the surface (evanescent field). Only fluorophores in this field (100 nm penetration depth) are excited. The dichroic filter efficiently reflects light totally reflected at the glass/specimen interface (*green line, arrow down*), but passes light emitted by the specimen (*orange line*). The emitted light is finally collected by a CCD (charged coupled device) camera

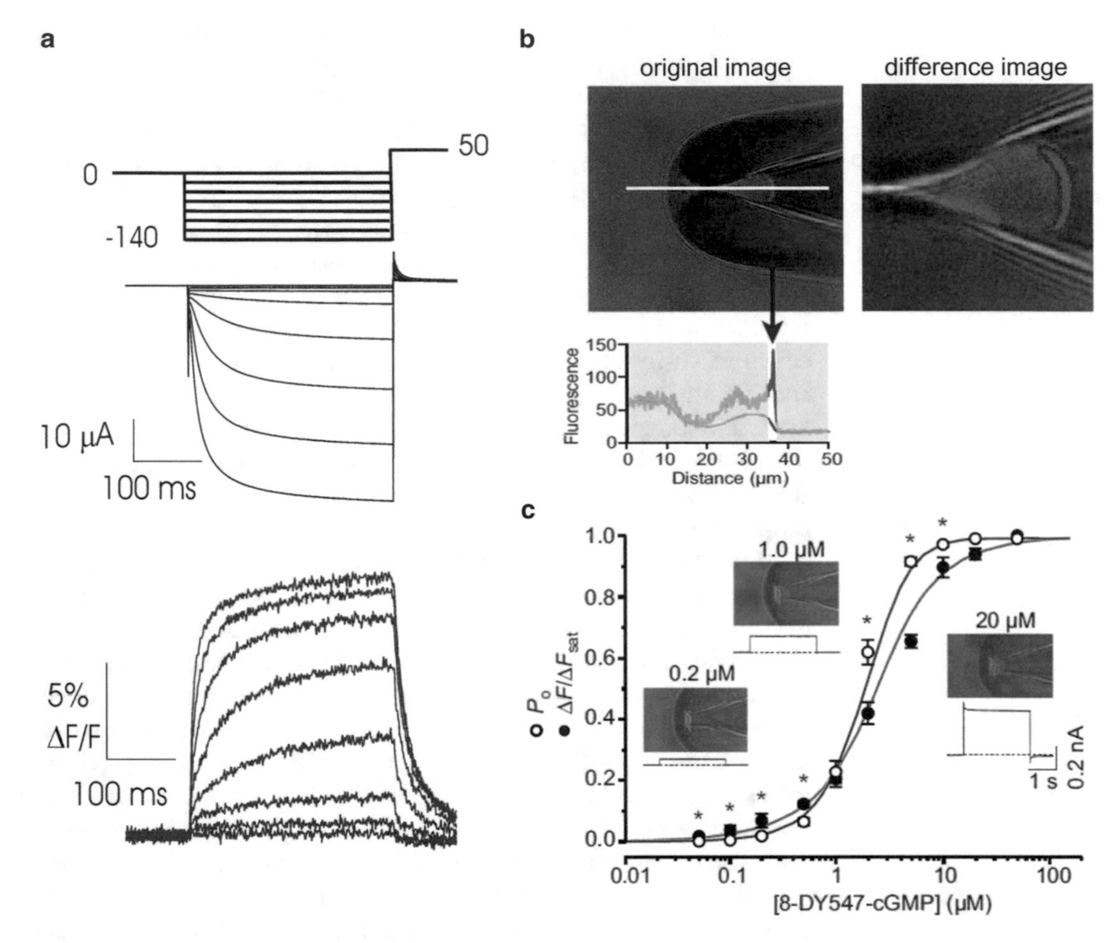

Fig. 3 Examples of VCF and PCF recordings. (**a**) A representative VCF experiments on HCN channels labeled with Alexa-488-maleimide at the cysteine residue substituting R322 in the S4 segment. Channels were held at 0 mV and then stepped to negative potentials (from 0 to −140 mV, ΔV = 20 mV), followed by a step to +50 mV. Channel opening elicited current and fluorescence changes. Image adapted from Bruening-Wright et al. (2007) with permission (**b**, **c**) A representative PCF experiment on CNG channels to quantify state-dependent ligand binding. The confocal images in (**b**) show the tip of a patch pipette containing an excised inside-out membrane patch. In the *left image*, the *green* fluorescence arose from 10 μM fcGMP (8-DY547-AET-cGMP), the *red* fluorescence from 1 μM of the reference dye DY647. DY647 was used to subtract the background fcGMP signal. The scaling procedure underlying the subtraction procedure is illustrated in the graph, showing both scaled fluorescence signals along the *white line*. The *right image* is the resulting difference image. The *red mask* indicates the region of interest used for fluorescence quantification. The plot in (**c**) shows the dose-binding curve (*filled circles*) and dose–response curve (P_o; *open circles*) as a function of the fcGMP concentration. Image modified from Biskup et al., 2007

during synthesis using an orthogonal tRNA synthetase pair. By introducing fUAAs in different regions of the Shaker K^+ channel, the authors were able to investigate voltage-dependent conformational changes for residues that are not accessible to labeling from the extracellular side [4].

Very interesting is also the recent combination of VCF experiments with fluorescent agonists [5] (Fig. 1b), a powerful approach that so far was restricted to the PCF technique (see Sect. 2.2.2). Bhargava et al. investigated ligand binding to the ATP-gated P2X1 receptor with the fluorescent agonist Alexa-647-ATP gaining insight into the interplay between binding and unbinding and entry into as well as recovery from the desensitized state [5].

VFC has been applied to an appalling variety of ion channels and transporters to address several biophysical problems which is impossible to cover in a comprehensive manner. Instead, in the next sections, we will try to describe in detail some of the applications that highlight general aspects of the technique. VCF has been illustrated in several other reviews [6–10].

2.1.2 Voltage Sensor Movement

The VCF technique was pioneered more or less at the same time in the laboratories of Isacoff and Bezanilla to gain insight into the molecular rearrangements accompanying voltage gating in K^+ channels of the Shaker family [2, 3]. These channels have a tetrameric architecture symmetrically arranged around a central pore that opens in response to changes in membrane potential [11]. Each monomeric component is composed of six transmembrane helices, S1 to S6. Helices S5, S6 and the pore loop (P-loop) between them form the central pore domain which contains the channel's selective pathway and the gates [12]. Voltage-gated channels also possess a specialized functional unit, the voltage sensor, that detects changes in the membrane potential and transmits this information to the pore. This is at the basis of the mechanistic and kinetic distinction between gating and permeation. From sequence analysis it was realized early on that the S4 transmembrane segment contained an unusual cluster of positively charged amino acids and would be ideally suited to act as the voltage sensor [13]. This subject has been intensively investigated for several decades. Sophisticated electrophysiological techniques allowed even to directly reveal and study the small currents, called "gating currents," that precede channel opening and are due to the movements of the charged amino acids in S4 determining channel's activation [14, 15]. However, it proved very difficult to unambiguously determine the role of S4 as the voltage sensor and to understand the conformational changes accompanying voltage gating. Using cysteine-scanning mutagenesis it was suggested that residues in S4 change their accessibility to extracellular sulfhydryl-reactive reagents providing a first clue of an S4 movement [16], but without explaining how or during which gating transition (activation, opening, or inactivation). VCF was devised to address these problems. Mannuzzu et al. adopted a VCF configuration in which electrophysiological measurements were performed with the two-electrode voltage-clamp technique (Fig. 2a, d) [2]. In their first study, they analyzed the fluorescence changes of TMRM (tetramethylrhodamine maleimide) attached to several positions in the extracellular loop S3-S4 and in S4 in the Shaker mutant W434F [2], which is nonconducting but preserves an intact activation mechanism, allowing reliable measurements of gating currents [17]. Most but not all of the cysteines introduced could be labeled and in some cases the extent of labeling could be influenced by incubating the cells in different solutions in order to depolarize or hyperpolarize the channel during labeling. S346C with the target residue in the S3-S4 loop furthest apart from the S4 segment was equally labeled in both conditions whereas A359C and R365C showed state-dependent labeling. This is consistent with the notion that these residues become more exposed upon depolarization. The mutation L366C, the furthest position towards the COOH-terminal end of S4 that was probed, could not be labeled, suggesting that this position was not accessible to TMRM [2].

Activation of the channel by positive potentials decreased the fluorescence intensity of several labeled cysteines on the extracellular end of S4 in a voltage-dependent manner, but not the one of the residue S346C, confirming that TMRM at this position did not change environment upon depolarization. M356C, A359C, and R365C showed fluorescence changes that closely correlated with gating currents both in terms of kinetics and steady-state voltage dependence suggesting that fluorescence reported on movement of the S4 related to channel's activation [2]. The decrease of the fluorescence indicated that the dye experienced a less hydrophobic environment upon activation [2].

To check the hypothesis that fluorophores' accessibility changes upon gating, quenchers can be added to the extracellular solution to enhance the fluorescence decrease. For this reason, Mannuzzu et al. applied the anion iodide, a membrane-impermanent collisional quencher. Iodide enhanced the fluorescence decrease upon depolarization at position A359C suggesting that indeed this residue became more accessible to the extracellular solution upon depolarization. However, there was almost no effect of iodide on other positions indicating that individual residues experienced a specific change in environment and exposure to the extracellular space [2].

Instead of using TEVC, Cha and Bezanilla applied the cut-open oocyte technique as the electrophysiological component of VCF (Fig. 2a, d) [3], which offers a superior time resolution compared to TEVC [18]. They applied this approach to explore the conformational changes of S4 and of other Shaker channel regions. A detailed kinetic analysis showed pronounced differences in the charge-voltage (Q–V) and fluorescence-voltage (F–V) curves at extreme potentials that did not emerge in the study of Mannuzzu et al. [2]. Importantly, while the close agreement of these curves directly indicate a correlation between gating charge movement and fluorescence, there is no inherent reason why the two should be identical. The gating charge is due to the sum of several contributions of individual charge movements and electrophysiologically these movements show several kinetic components [19]. On the other hand, the *F–V* curve measures the relative changes in fluorescence of a probe attached to one residue, dictated by the local environment of the probe [3]. In particular, Cha and Bezanilla could resolve two different components of the gating charge movement and observed that fluorescence changes reported from TMRM at position A359C correlated with the slower component [3]. They also investigated residues in the S2 segment because it was reported to contain a negative charge that contributes to gating [20]. They found that the S2 fluorescence signals had faster kinetics compared to the S4, resembling the fast component of the gating charge [21]. Cha et al. also analyzed fluorescence changes from a residue in the pore region, T449C [3]. Fluorescence changes from this site have also two components: one correlated with the voltage dependence of ionic current suggesting a link to the mechanism leading to channel opening and another one with the same time course of C-type inactivation (see below). Thus, this was the first work showing that the properties of the fluorescence changes from dyes at different positions correlated with different functional transitions of the channel [3].

An important aspect of this work was the comparison between the fluorescence signals elicited by different fluorophores. Fluorescein maleimide (FM) and Oregon Green (OrGrM) yielded very similar fluorescence signals at position M356C but in contrast to TMRM the fluorescence intensity increased with depolarization. In addition, the *F–V* curve in OrGrM was not monotonic as with TMRM suggesting that over the voltage range tested the probes experienced different microenvironments [3]. TMRM has a net charge of 0 and contains a delocalized positive charge at its tricyclic end whereas OrGrM has a net charge of −2 and a negative charge at its tricyclic end. The authors suggested that because of these differences the probes might dwell in electrostatically different environments with different quenching properties [3].

To gain a deeper insight into the origin of the fluorescence changes observed with TMRM, the same group additionally analyzed the fluorescence spectra of the dye attached to M356C and A359C [22]. Spectral analysis and anisotropy provide complementary information about the origin of the fluorescence changes compared to the fluorescence intensity. An important parameter in this respect is the so-called Stokes shift, defined as the difference between the wavelength of the excitation's peak and the emission's peak, which is highly dependent on the polarity of the environment to which the probe is exposed. The Stokes shift observed for TMRM attached to M356C and A359C at different potentials was close to the one in water and did not change significantly in spite of a substantial change in quantum yield. This suggests the idea that these residues are exposed to an aqueous environment which does not change significantly at the different voltages as confirmed also by quenching experiments with iodide [2, 3] and heavy water (deuterium oxide, D_2O) [22]. In addition, the quantum yield of TMRM does not depend on pH but the fluorescence of TMRM attached at positions M356C and A359C was pH-dependent, suggesting that S4 might interact with titratable residues in other regions of the channel [22].

In conclusion, the detailed analysis of several spectroscopic properties of the dyes attached at different positions in S4 indicated that the fluorescence changes upon activation are not due to the movement of the residues from a buried position in a protein hydrophobic surrounding to an aqueous, fully exposed environment. The data supported the notion that S4 moves in a narrow vestibule with water exposed crevices and the fluorescence changes are due to subtle environmental changes related to differential collisional quenching by water or other parts of the protein depending on the voltage and these environmental changes are strongly site-specific [22].

To estimate the extent of S4 movement, VCF was combined with Förster resonance energy transfer (FRET) performed with different approaches (Fig. 2a, *right panel*) to obtain a more quantitative estimate of S4 movement upon activation [23–25]. FRET is the transfer of the excited state energy from an initially excited fluorescent donor molecule to an acceptor molecule that does not necessarily need to be a fluorophore as well [1]. Employing two types of molecules, functioning as efficient FRET partners, this physical phenomenon has been widely used in elegant approaches for monitoring conformational changes and measuring molecular distances (see also Sect. 2.2). The efficiency of FRET depends on the spectral overlap of the donor emission spectrum and the acceptor absorption spectrum, the distance between the two molecules, and the relative orientation of the donor emission dipole moment and the acceptor absorption dipole moment. FRET typically occurs over distances comparable to the dimensions of macromolecules, and therefore has been used intensively as a type of spectroscopic ruler in studying conformational changes in membrane proteins.

Glauner et al. used fluorescein and TMRM as FRET pairs attached to cysteines placed within or just outside S4 [24]. Cha et al. and Posson et al. used a variation of FRET called luminescence or lanthanide-based resonance energy transfer (LRET) in which the donor is not a fluorescent molecule but a luminescent lanthanide chelate whereas the acceptor is a conventional fluorescent dye [23, 26]. LRET offers several advantages over classical FRET, like minimal spectral overlap of donor and acceptor and zero intrinsic anisotropy of the donor. In detail, Posson et al. used a terbium-chelate attached to introduced cysteines in the voltage sensor and several fluorescent dyes attached to the Shaker pore blocker charybdotoxin as acceptors. This arrangement, in principle, facilitates measurements of transmembrane movements [23] compared to classical FRET experiments performed with site-directed labeling [24, 26]. However, these different studies did not converge on the extent of S4 movement. The interpretation of these results is complicated because translational movements can be accompanied by possible tilting and rotations of the S4 helix and by focusing of the transmembrane voltage due to water-filled crevices around the S4. This means that the FRET changes observed can be accounted for by different models combining those ingredients to various degree [27]. Structural evidence from Mackinnon's group led to the proposal of a drastically new model to explain the S4 function and movement, the paddle model [28–30]. However, the detailed mechanism of S4 movement is still hotly debated [31].

2.1.3 Investigation of Nonconductive States of Ion Channels

Ion channels perform their physiological tasks switching from conductive to nonconductive states due to the mechanisms of inactivation, deactivation, or desensitization that shut off the permeation pathway either in the presence (inactivated and desensitized state) or in the absence of activating stimuli (resting, deactivated state). VCF is a valuable tool to investigate the conformational changes related to transitions to and between these nonconducting states which are electrically silent or, in any case, difficult to track with electrophysiological techniques.

Shaker channels are characterized by two mechanisms of inactivation: fast N-type inactivation which is well understood [32] and C-type inactivation, which is in general slower and mechanistically still not understood [33]. VCF allowed to directly link conformational changes in the extracellular mouth of the channel pore to slow inactivation processes [3, 34, 35] confirming previous electrophysiological evidence [33]. The combination of mutations in the pore region involved in slow inactivation and detection of fluorescence changes on the extracellular end of S4 indicated also that a biphasic interaction between these two regions takes place during C-type inactivation [34, 35]. This interaction provided a structural correlate for the electrophysiological observation of gating charge immobilization. Gating charge immobilization means that upon prolonged depolarization the channel enters into a stabilized inactivated state in which the gating charges are stabilized in an "active" state. From this state recovery is slower and require more hyperpolarized potentials compared to fast inactivation [36]. Additionally, VCF allowed monitoring the rearrangements leading to the recovery from C-type inactivation [34]. The ability to identify transitions between inactivated states that are electrically silent shows once again the power of the VCF approach [3, 34, 35]. This is further illustrated by the direct demonstration of closed-state inactivation by VCF [37].

VCF has been applied to investigate many other ion channel families like HERG [38–40], KCNQ [41, 42], BK [43], plant KAT [44], hyperpolarization-activated cyclic-nucleotide-gated (HCN) [45], sodium [25, 46], proton [47, 48], and ASIC [49, 50] channels. It has been also applied to several receptors to investigate conformational changes linked to transport or agonist binding; for example in the nicotinic acetylcholine [51, 52], GABA [53–56], glycine [57, 58], and P2X receptors [5, 59].

2.1.4 VCF Applied to Transporters

All the basic principles of VCF mentioned in relation to ion channels extend also to transporters. They mediate ion transport by intrinsically different thermodynamic mechanisms compared to ion channels but their function is also based on a concerted sequence of conformational changes associated with ion movement. A meaningful example of the power of the VCF approach in the field of transporters is its application to the gastric H,K-ATPase, the main transporter responsible for acid secretion in the stomach [60]. It belongs to the family of P-type ATPases and is closely related to the Na,K-ATPase which exchanges three Na^+ ions for two K^+ ions in each transport cycle [61]. In contrast, H,K-ATPase transport is electroneutral with a 2:2 (or 1:1) H^+/K^+ exchange stoichiometry [60].

In spite of this overall electroneutrality, individual steps in the transport cycle have been found to be voltage-sensitive, hence electrogenic [62]. Building on previous VCF studies on the Na, K-ATPase [7, 63, 64], Durr et al. investigated the fluorescence changes produced by TMRM labeling of the mutant S806C of the H,K-ATPase [65]. These changes are due to the transition between inward and outward facing states of the transporter [64]. In this manner, it was possible to directly relate the effect of several physiological parameters like voltage, intra- and extracellular pH, and extracellular cations with the conformational equilibrium between inward and outward facing states [65].

VCF has been applied to many other transporters to investigate different properties like substrate binding, conformational changes related to activation or transport, and effect of β subunits. Some examples are the glutamate transporters of the EAAT family [66, 67], the organic cation transporter rOct1 [68], the Na^+/phosphate cotransporter [69, 70], the Na^+/glucose cotransporter [71, 72], the gamma-aminobutyric acid transporter GAT1 [73], and a voltage-sensitive phosphatase (VSP) structurally related to the voltage sensor of cation channels [74].

2.2 Patch-Clamp Fluorometry (PCF)

The term "patch-clamp fluorometry" has been used in two different ways: on one hand to describe the combination of fluorescence microscopy and patch-clamp techniques in general, but on the other hand to designate the specific case of combining fluorescence microscopy with excised patches in the inside-out configuration. To avoid confusion, in a previous review we introduced for the latter case the term "classical PCF" [75], because it was the first PCF approach published [76]. Since then, PCF has been shown to be a very powerful tool, also when combined with other patch-clamp configurations. In the following, we will give an overview of how PCF has been applied to study different ion channel families. Even if theoretically possible, to our knowledge, up to now PCF has not been used to study electrogenic transporters.

2.2.1 PCF Using Whole-Cell Configuration

The whole-cell configuration is the most common patch-clamp configuration used in electrophysiological experiments and allows for monitoring the current flow through a large channel population at the same time. In the 2000s, many researches started to combine optical approaches with the whole-cell configuration employing immortalized mammalian cell lines (Fig. 2b, d–f). For some proteins, such cell lines like COS1, CHO or HEK293 are advantageous over non-mammalian cells like Xenopus laevis oocytes in achieving adequate expression levels and providing a suitable environment to maintain physiological protein function. For whole-cell PCF different cell types, protein labeling strategies (Fig. 1), and microscopic techniques (Fig. 2d–f) have been applied. The required components to perform the electrophysiological part of this approach are similar to traditional whole-cell recordings and will not be mentioned herein in detail.

Monitoring Gating-Dependent Conformational Changes of Proteins Using FRET

As mentioned earlier, a very elegant approach for monitoring conformational changes and measuring molecular distances is to take advantage of the physical phenomenon of Förster resonance energy transfer (FRET) between two fluorophores. Thus, in principle, FRET works as a spectroscopic ruler. Generally, this ruler can be used for exploring (1) intra- as well as intersubunit rearrangements of the same protein, (2) rearrangements of two different proteins relative to each other, and (3) rearrangements of protein parts relative to the plasma membrane (Fig. 1a, *right panel*).

A very efficient and often employed FRET pair is formed by the fluorescent proteins (FP) CFP (cyan) and YFP (yellow) [77], two mutants of GFP (green), a 27 kDa protein from the jellyfish *Aequorea victoria*. In general, they can be attached to the N- and C-termini without major perturbations of proteins' function. However, due to their quite large size relative to the size of channel proteins, the position of protein tagging different from the termini has to be chosen with care. Even in cases where the native protein structure is known, prediction of suitable insertion sites for fluorescent proteins is very difficult and generating individual fusion proteins using conventional molecular biology techniques is very time-consuming. Therefore Giraldez et al. proposed to use transposable elements, DNA sequences that can change its position within the genome, to randomly insert YFP and CFP into different channel domains to find appropriate tagging positions within the protein instead of just tagging the final termini. Doing so for large-conductance voltage- and calcium-dependent potassium (BK) channels, they succeeded in finding 19 out of 55 constructs to be expressed at the plasma membrane showing a normal Ca^{2+}- and voltage-dependent gating behavior [78].

Gating-dependent intra- and intersubunit movements (Fig. 1a, *right panel*) in CFP/YFP-tagged voltage-gated ion channels have been investigated intensively by Kobrinski and coworkers after heterologous expression in mammalian COS1 cells [79, 80]. To study conformational mobility of the N- and C-terminal tails of human $Ca_v1.2$ channel α_{1C} pore-forming subunits, they employed double-tagged subunits for intrasubunit movements and single-tagged subunits for intersubunit movements. They reported a state dependence of FRET associated with conformational refolding of the termini in the resting, conducting, and inactivated states when double-tagging but not single-tagging the subunits [79]. In contrast, in rat Kv1.2 channels relative movement between N- and C-termini occurred both within subunits and between subunits [80]. The relative changes of FRET found in $Ca_v1.2$ channels are ~30 times greater than in Kv2.1 channels, suggesting that gating-induced movements are translated into much larger rearrangements [80].

The FRET pair CFP/YFP has also been employed successfully to detect interactions between pore-forming and accessory subunits (Fig. 1a, *right panel*). This strategy has been used to directly compare the interaction between $Ca_v1.2$ channel α_{1C} subunits and different β-subunits in different functional states of the channel, and to determine molecular motions of the β-subunits relative to α_{1C} subunits [81, 82]. Fisher and coworkers tagged the C-terminus of ATP-gated P2X2 receptor subunits with either CFP or YFP and succeeded in providing a time-resolved measure of state-specific gating motions. They could show that movements of the cytosolic portion of the channel correspond to permeability changes of the channel [83]. In contrast to the studies described so far which all used epifluorescence microscopy under steady-state conditions of channel activation [79–82], Fisher and coworkers combined whole-cell patch clamp with TIRF microscopy at nonsteady state [83] (Fig. 2f). TIRF microscopy (total internal reflection fluorescence microscopy) [84, 85] allows visualizing single fluorophores in the membrane by reducing the background noise. This is realized by illuminating the specimen and exciting the fluorophores using an evanescent field, i.e., an area beyond a boundary surface, which is generated by total internal reflection of light at that surface.

When describing gating-dependent rearrangements of cytosolic parts of ion channels, also movements relative to the plasma membrane are of high interest (Fig. 1a, *right panel*). The position of the membrane can be marked by using CFP/YFP-labeled pleckstrin homology (PH) domains of phospholipase Cδ1 [86]. These membrane markers localize via PIP2 to the inner leaflet of the plasma membrane and distribute in a random and homogeneous fashion. Kobrinsky and coworkers used this approach to study relative movements of the cytoplasmatic tails of α_{1C} subunits and β-subunits of calcium channels. CFP-labeled N-termini of α_{1C} subunits and accessory β-subunits but not C-termini of α_{1C} subunits showed FRET with the YFP-labeled PH domain. Even if voltage-dependent inactivation generates strong FRET between α_{1C} and β_{1a} N-termini suggesting mutual reorientation of these tails, their distance relative to the plasma membrane is not appreciably changed [81].

Alternatively to CFP/YFP, other FRET pairs can be employed. Yang and coworkers used fluorescein maleimide (FM) and tetramethylrhodamine maleimide (TMRM), which are sulfhydryl-reactive fluorophores extracellularly applied to label native cysteines in TRPV1 channels [87]. A FRET analysis was performed to explore the structural changes during temperature-dependent activation. The authors suggested that the observed movements are specifically associated with temperature-dependent activation, indicating that this activation pathway is distinct from those for ligand- and voltage-dependent activation [87].

Monitoring Protein Mobility by Single-Particle Tracking

Neurotransmitter receptors are known to show different types of lateral mobility in the plasma membrane, besides Brownian movements, which have been proposed to play pivotal roles in signaling [88]. Single-particle tracking (SPT) of membrane proteins is an elegant approach to study this mobility. SPT uses video microscopy combined with digital computer processing to monitor the motion of single molecules of interest, which are often labeled with sub-micrometer fluorophores, such as quantum dots (Fig. 1c). Quantum dots are tiny particles or nanocrystals of a semiconducting material with diameters typically ranging from 2 to 10 nm, providing unique fluorescent properties [89]. Notably, the definite size of these particles is slightly extended in most biological applications due to surface attachment groups, which are grafted to mediate different functionalities [90]. These particles are orders of magnitude more photostable than organic dyes and fluorescent proteins, making them attractive candidates for long-term SPT experiments in live cells, obtaining long tracking trajectories.

Richler and coworkers used single-molecule imaging with simultaneous whole-cell recordings to track quantum dot-labeled

P2X2 receptors in the dendrites of cultured rat hippocampal neurons to explore P2X2 receptor mobility and its regulation. They found that lateral mobility of these receptors in the plasma membrane of dendrites is heterogeneous but mostly Brownian in nature, consisting of mobile and slowly mobile receptor pools. This mobility is subunit and cell type specific and additionally activation dependent. The lateral mobility is doubled, when the channels were opened and when calcium ions permeate [91].

2.2.2 PCF Using Excised Patch Configuration

In the early 2000s, Zheng and Zagotta established a very successful PCF approach, which combines fluorescence microscopy with the inside-out configuration of patch clamp [76, 92] (Fig. 2c). Compared to approaches employing intact cells, this so-called classical PCF allows direct and fast access to the intracellular side of ion channels providing the possibility to label intracellular amino acids and to apply intracellular ligands and channel modulators. It has been shown to be a suitable approach for studying both voltage- and ligand-gated ion channels, using different labeling strategies (see below) (Fig. 1a, c).

So far, oocytes of the South African claw frog *Xenopus laevis* have been used as heterologous expression system in classical PCF. With this approach the fluorescence signal arises from the same population of channels, which cause the electrical signal. Considering that around 1000 fluorophore molecules per membrane patch are sufficient to perform a reliable experiment [93] and that typically each channel subunit is tagged by one fluorophore molecule, a few hundred channels per patch are an appropriate number. However, the exact number of channels required is limited by the signal-to-background ratio: the lower the background signal, the lower the acceptable number of channels in the patch.

While for traditional patch-clamp experiments various pipettes shapes and sizes have been shown to be applicable, for classical PCF relatively thick-walled and fire-polished patch pipettes with large openings seem to be advantageous. They allow to obtain bright fluorescence signals [93, 94].

Monitoring Conformational Changes of Channel Proteins

The pioneering work of Zheng and Zagotta in the field of PCF with isolated membranes aimed to shed light on the gating-dependent rearrangements within the intracellular channel portion of cyclic-nucleotide-gated (CNG) channels [76]. While structurally similar to voltage-gated cation channels, CNG channels are only weakly voltage-dependent and are activated by the binding of cyclic nucleotides to an intracellular C-terminal binding domain (CNBD) present in each of the four subunits. The CNBD is linked to the pore region by a complex α-helical structure, the so-called C-linker [95]. In their approach, Zheng and Zagotta labeled channels in inside-out patches with the maleimide conjugated dye

AlexaFluor 488, which is able to covalently bind to the sulfhydryl group of cysteine residues and followed fluorescence quenching caused by the anionic quencher iodide and the cationic quencher thallium. Both ions lead to collisional quenching, a decrease in fluorescence, by providing a nonradiative route for loss of the excited state energy of the fluorophore upon direct contact [96]. After application of the ionic quencher to the excised patch, iodide had a higher quenching efficiency in the channel's closed state whereas thallium had a higher quenching efficiency in the open state. This state and charge dependence of quenching suggested gating-related movements of charged or dipolar residues near the fluorophore residing in the C-linker [76].

As described earlier, FRET occurs not only between two fluorophores but also between a fluorescent molecule and a nonfluorescent acceptor (see Sect. 3.1.1). For the fluorophores GFP, AlexaFluor 488 and AlexaFluor 568, dipicrylamine (DPA) functions as a suitable acceptor. DPA is a small, negatively charged membrane probe that moves either into the inner or outer leaflet of the membrane in a voltage-dependent manner [97]. Taraska and Zagotta labeled cysteine residues inserted at different sites in the cytoplasmatic portion of rod CNG channels with the mentioned fluorophores as donors and used FRET to measure the distances between the fluorophores and DPA residing in the plasma membrane. Comparison of FRET data in the presence and absence of the agonist cGMP showed that, upon channel activation, none of the C-terminal subdomains moved perpendicular to the membrane. There was only a minor lateral movement, indicating that subtle conformational changes are sufficient to drive channel activation [98].

FRET between the two fluorescent proteins CFP and YFP, as described above for whole-cell PCF (see Sect. 3.1.1), could also be shown to be a successful tool in inside-out patches of *Xenopus laevis* oocytes. Miranda and coworkers labeled large-conductance voltage- and calcium-dependent potassium (BK) channels at either of three sites at the so-called gating ring, a large intracellular channel portion, where Ca^{2+} binding promotes channel opening. FRET signals indicated larger rearrangements than predicted by existing X-ray structures of the isolated gating ring. Surprisingly, the described rotation-like rearrangements are not simply related to the open-closed transitions of the channel, which is in contrast to the currently accepted model of Ca^{2+}- and voltage-dependent gating in BK channels [99].

Monitoring State-Dependent Ligand Binding

Investigating binding of ligands or modulators to intracellular binding sites of membrane proteins has been realized by three approaches: (1) monitoring binding-induced changes in FRET signals between two fluorophores tagged to the protein,

(2) monitoring FRET signals caused by fluorophore-tagged ligands approaching fluorophore-tagged proteins, (3) monitoring fluorescence increase caused by the binding of tagged ligands to untagged proteins.

The first approach was applied by Zheng and coworkers to study Ca/CaM binding to olfactory cyclic-nucleotide-gated (CNG) channels. FRET signals were generated by fusing CFP and YFP to the N- and C-terminus, respectively, in different subunits. Upon application of Ca/CaM to the intracellular side of an excised patch, the FRET signal decreased with the same time course as the ionic current. This demonstrated not only that Ca/CaM binding occurred but also that binding was followed by a separation or reorientation of N- and C-termini, disturbing their interaction and thereby leading to a channel inhibition [100].

Applying the second option, Trudeau and Zagotta investigated the mechanism of Ca/CaM binding in rod CNG channels perfusing the intracellular side of excised patches with CaM molecules conjugated with AlexaFluor 488 (Ca/CaM-488) and monitoring FRET between Ca/CaM-488 and channel-attached CFP [101]. Additionally, they just followed the binding/unbinding time courses of Ca/CaM-488 to untagged CNG channels. Together, the results showed a direct association between CaM and the N-terminus of CNGB1 subunits and a rotation or separation of the N- and C-termini in the presence of Ca/CaM [101].

Fluorescence signals caused by tagged ligands to untagged proteins (Fig. 1b) were intensively applied to study cyclic nucleotide binding to CNG channels and to their close relatives HCN (hyperpolarization-activated and cyclic-nucleotide-modulated) channels. A fluorescent cGMP analog (8-DY547-AET-cGMP, fcGMP), synthesized by coupling the dye DY547 to cGMP, has been shown to be suitable for quantifying-ligand binding in homotetrameric CNGA2 channels, as it activated the channels in a similar manner as the unmodified ligand (Fig. 3b, c) [94]. To subtract the background signal caused by fcGMP in the bath solution, simultaneous application of a second dye (DY647) was required, which labeled the bath solution but did not bind to the membrane or to the channel protein [94]. Simultaneously measured concentration-activation and concentration-binding curves [94] as well as activation/deactivation and binding/unbinding time courses [102] have been analyzed to establish Markovian state models describing CNG channel gating. Furthermore, fcGMP has been used as a tool for comparing the ligand binding behavior of modulatory subunits and principle subunits in olfactory heterotetrameric CNG channels [103].

To investigate the relation between ligand binding and activation in HCN channels, the same dye DY547 has been coupled to cAMP (8-DY547-AET-cAMP, fcAMP) [104]. The tagged cAMP activated homomeric HCN2 channels in a similar manner as untagged cAMP [104, 105]. Again, the background was subtracted by application of a reference dye (DY647). The simultaneously obtained activation and binding data proved the idea that ligand binding to ion channels is state-dependent [104, 106]. A Markovian state model based on activation/deactivation and binding/unbinding data could be established [105]. Alternatively, a NBD (7-Nitrobenz-2-Oxa-1,3-Diazole)-coupled cAMP has been used, which is strongly fluorescent in a hydrophobic environment, such as in the cyclic-nucleotide-binding pocket, but weakly fluorescent in an aqueous environment [107–110]. This makes the application of a reference dye unneeded, but increases the risk of unspecific binding to the patch membrane. However, regardless of which analog has been chosen, all these studies agreed in showing that the interaction between cAMP and the HCN channel protein is state-dependent.

2.2.3 PCF Using Cell-Attached Configuration

The cell-attached configuration is the method of choice, if the current through a few channels or even only through one single channel is intended to be measured, while the cell is wished to stay intact (Fig. 2b). When studying voltage-gated ion channels, applied voltages can be changed easily. However, varying application of substances like ligands or modulators can be realized only to the extracellular but not to the intracellular side of the membrane. Because the latter has to be done via the patch pipette, only one concentration per experiment is possible. That might be one of the main reasons why, at least to our knowledge, this configuration has not been used in PCF approaches very often.

Nevertheless, Schmauder and coworkers showed that especially for experiments at the single-channel level, cell-attached patches can be advantageous. They used Cy5-coupled epibatidine [111], a full agonist for muscle-type nicotinic acetylcholine receptors, and confocal fluorescence microscopy to correlate optical and electrical single-channel events. HEK293 cells, transiently expressing nAChR, were probed by a patch pipette filled with a 150 nM solution of the fluorescent agonist. After formation of a stable giga-seal, the excitation laser beam was focused on the membrane patch within the glass pipette. Fluorescence and electrical recordings were initiated simultaneously. The authors suggested that the acetylcholine receptor, after binding an agonist, first explores a broad landscape of closed but liganded channel states of similar energy by conformational diffusion, followed by a fast gating step towards an ion-conductive channel [112].

2.2.4 PCF Using Artificial Bilayers

The planar lipid bilayer (PLB) technique, developed in the mid-1970s, allows to study ion channels and transporters electrophysiologically after reconstituting membrane proteins into an artificial membrane [113]. The main advantage over other electrophysiological approaches is the possibility to precisely define the lipid and electrolyte conditions throughout the experiment.

Conventionally, planar bilayers are formed vertically in aqueous environments. The electrolyte solutions at both sides of this membrane are connected to a patch-clamp amplifier through Ag/AgCl electrodes. To monitor current and fluorescence signals from dye-labeled membrane proteins at the same time, Ide and coworkers devised a setup with a membrane created in horizontal orientation [114, 115]. They painted a lipid solution (DΦPC/ml *n*-decane) over a 200 μm aperture to create the bilayer. As a model system, which promises large single-channel conductance, BK channels natively expressed in bovine tracheal smooth muscles were incorporated by allowing sarcolemmal vesicles to fuse with the lipid bilayer. Because the channel density in this tissue was very low, each vesicle contained no or only one channel. Before vesicle fusion, the channels were labeled by incubating the vesicles with Cy5-conjugated anti-BK channel antibodies (Fig. 1c). The fluorescence recording was realized using TIRF microscopy [84, 85] (Fig. 2c). The authors succeeded in demonstrating that there is a strong correlation between the appearance of single current events and fluorescence signals arising from the membrane. A similar approach was proposed by Borisenko and coworkers to study the correlation between gramicidin heterodimerization and channel activity. They also used a horizontal PLB, but labeled the channel protein by covalent binding of Cy3 and Cy5 and applied wide-field illumination and single-pair FRET (spFRET = FRET between a single donor and a single acceptor molecule) instead of TIRF [116].

Alternatively, an artificial membrane can be established at the tip of a patch pipette instead of being painted upon a hole [117]. Harms and coworkers used this approach to study gramicidin channel dynamics [118]. They formed the lipid bilayer at the pipette tip by the apposition of two monolayers using 4:1 diphytanoylphosphatidylethanolamine (dPhPE):diphytanoylphosphatidylcholine (dPhPC) [117]. Gramicidin monomers dissolved in ethanol were incorporated after being added to the aqueous subphases at both sides of the membrane. Protein labeling was realized by covalent binding of TMR (tetramethylrhodamine) or Cy5 *N*-hydroxysuccinimide (NHS) esters to a lysine residue at the C-terminus. The pipette tip was imaged using wide-field single-molecule fluorescence microscopy. Electrodes in the pipette and in the bottom chamber enabled simultaneous current recordings. The authors used spFRET, as well as fluorescence self-quenching, which is caused by short distances of two identical fluorophores within a gramicidin dimer. The optical together with the electrophysiological data suggested the occurrence of multiple intermediate states underlying the gramicidin channel dynamics.

3 Notes

In the following we will mention some of the possible technical difficulties in performing VCF and PCF experiments.

3.1 Protein Labeling

3.1.1 Labeling Membrane-Trafficked Proteins

It is generally desirable that the entire channel/transporter population in a cell or in a membrane patch is labeled and that all the labeled proteins contribute to the current response to guarantee that the correlated fluorescence and current signals arise from the same population. Nevertheless, in practice there might be unlabeled proteins which contribute to the current, or labeled proteins which are electrically silent. The experimental conditions should be chosen in such a way, that these cases are very rare. In this context it is important to assure that the entire membrane area which is selected for quantifying the fluorescence intensity was voltage clamped. This is particularly important for approaches using excised membranes, in which very often fluorescent membrane fragments stick to the pipette glass likely causing distorted current flow, but should kept in mind also in VCF experiments in which voltage-clamp homogeneity over the entire oocyte's membrane might be an issue.

In some instance, labeling of specific residues might be possible only in a specific state of the protein because of a specific conformational arrangement. In these cases, labeling conditions should be selected which maximize the probability of residing in that respective state. Provided that two residues to be tagged are available in different conformational states, this state-dependent availability can be advantageous for tagging these respective residues with two different fluorophores, for instance for FRET approaches.

3.1.2 Environmental-Sensitive Fluorophores

In VCF and PCF experiments the interpretation of fluorescence changes in terms of structural rearrangements around the probe requires great caution. First of all, the underlying mechanism for the fluorescence changes should be identified. The fluorescence intensity might change because of (1) reorientation of the fluorophore's transition dipole, leading to a change in absorption cross-section, (2) shifts in the excitation spectrum of the dye, for example due to fluorophore–fluorophore interactions which lead to changes in absorption at particular excitation wavelengths and a corresponding change in emission, (3) changes in the hydrophobicity of the fluorophore's environment which lead to changes in fluorophore quenching, and (4) interaction between fluorophore and nearby protein residues that also quenches the fluorophore [1, 119].

This implies that the structural insight provided by experiments employing environmental-sensitive fluorophores is relatively limited. They report on very local changes of the probe environment that might be due to movement (or reorientation) of the residue where the fluorophore is attached or to movements of the surrounding regions of the protein. Moreover, the environment probed by the fluorophore depends on its size and it might differ from the restricted surrounding of the site where it is attached to the protein. Finally, the labeling procedure (including the removal of the native cysteines and insertion of new cysteines for labeling with thiol-reactive dyes) might lead to a modification of the functional properties of the proteins.

Nevertheless, there are several strategies to address these limitations. Spectroscopic analysis can gauge the molecular mechanisms altering fluorophore's emission. Labeling of the same site with different types of fluorophores or labeling of several residues in the same or in different protein regions might give more global information about the nature of the conformational changes which are less dependent on the specific properties of a single dye.

However, the most important aspect, which is also the distinctive property of VCF and PCF, is that the comparison of the functional properties of the channel and of the fluorescence data for each construct and condition tested provides always an internal control to correlate functional and structural information.

3.2 Undesired Fluorescence Signals

One major source of undesired fluorescence signals is autofluorescence, the natural emission of light by biological components. In general, the risk of autofluorescence is higher in approaches using whole cells compared to approaches using excised membrane patches, which basically exclude autofluorescent intracellular constituents. Nevertheless, there are still sources of autofluorescence in the membrane, like tryptophan, tyrosine, or phenylalanine. These amino acids absorb at 260–295 nm and emit at 300–350 nm. Another source is NADH, absorbing at 340 nm maximum and emitting at 460 nm maximum. However, because the commonly used fluorophores need longer wavelengths for excitation, these components can be ignored in many cases.

If the experimental procedure includes application of fluorophores via the bath solution, free fluorophores can considerably contribute to the background signal. Thus, extensive washing prior to fluorescence recording is needed.

3.3 FRET

Even if FRET is used in several PCF and VCF studies to monitor conformational changes and relative movements, there are some limitations, particularly when using FPs such as CFP and YFP as FRET partners. In case of FPs one major problem is the large size of these proteins. They have a roughly cylinder-like shape that is 24 Å in diameter and 42 Å in length [120], large enough to potentially disturb channels' native behavior. Secondly, the distances, which usually can be reported by FRET, are in the order of the diameter of the channel protein itself (30–70 Å) [121], which is too large to detect subtle conformational changes. Thirdly, because FRET efficiency depends on the sixth power of the distance between the fluorophores, the distance dependence of the signal is extremely steep, making it sensitive to movements only in a very narrow range of distances, around R_0 [121].

To overcome these limitations, Islas and Zagotta used the phenomenon of bimane quenching by tryptophane [122]. Bimane covalently attached to a cysteine can be quenched by tryptophane either residing nearby in the protein or being applied via the bath solution. This allows for measuring structure and dynamics of short-range interactions. Bimane is a small, environmentally sensitive fluorophore, whose fluorescence is quenched due to photo-induced electron transfer from tryptophan to excited bimane. Fluorescence from bimane-modified CNGA1 channels could be reported in inside-out patches [121].

Another alternative to overcome the problems of traditional FRET approaches is to monitor FRET between a small fluorescent dye and a nickel ion bound to a dihistidine motif in the protein [123], the so-called transition metal ion FRET. This approach has the advantage of measuring the dynamics of close-range interactions, of using tiny probes with short linkers, of showing a low orientation dependence compared to traditional FRET and of allowing to flexibly add and remove acceptors [123]. As an example, the R_0 value for Ni^{2+} and fluorescein (12 Å) is four times shorter than standard FRET pairs such as CFP and YFP (50 Å) or fluorescein and rhodamine (54 Å) [1, 124].

One major concern about FRET is that the absence of a FRET signal does not indicate that two proteins are not interacting, but can also mean that the distance was too large to cause FRET, that the fluorophores were quenched, or that both fluorophores moved in the same way.

Recently, Jarecki et al. presented a new promising tool that can overcome FRET limitations in particular for the estimate of short distances lower than 10 Å [125]. It is based on new compounds, called "tethered quenchers." Tethered quenchers have at one extremity a functional group able to bind to a specific channel region and, at the other extremity, separated by linkers of different length, a quencher that is able to alter the fluorescence of a chromophore attached to another region of the channel. The quenching efficiency depends on the spatial proximity of the two molecules. Thus, linkers of different length can be used as calipers to estimate the distance between two sites in a channel [125]. The idea of tethered molecules used as distance rulers has been applied before in different manners [30, 126]. However, tethered quenchers are unique in that the functional readout is the modulation of a fluorescence signal [125]. Jarecki et al. applied the technique to obtain distance constraints of the closed state of Shaker channel using tethered molecules with a TEA moiety binding the extracellular mouth of the pore and a nitroxide quenching group at the other end. As readout they used the fluorescence quenching of TMRM attached at different sites in the channel in VCF experiments [125]. The resulting distances (with a resolution reported to be below 4 Å) are in agreement with a model of the closed state obtained for the Kv1.2/2.1 chimera by molecular dynamics (MD) simulations [127].

3.4 pH Dependence

Fluorescent proteins, as GFP, YFP, and CFP, are pH-sensitive. Therefore some fluorophores should be avoided for experiments in which acid quenching could produce artifacts [128]. However, newer fluorophores such as variants of CFP (mCerulean) and YFP (mVenus or mCitrine) are not only less sensitive to pH, but show also a photo-physical behavior which make them brighter and more resistant to bleaching [129], making them very satisfying alternatives for several experiments.

3.5 Free Radicals

Generally, fluorescence microscopy is an invasive approach, because the high-intensity light required to excite the fluorophores can have negative effects on the cell membrane or the whole cell. Toxic effects of light primarily arise due to generation of chemically reactive substances such as free radicals and specifically singlet and triplet forms of oxygen. The most effective and practical way to circumvent light-induced damages is to avoid any unnecessary exposure by establishing optimized setups and applying optimized experimental strategies [130].

References

1. Lakowicz JR (1999) Principles of fluorescence spectroscopy. Kluwer Academics/Plenum, New York, NY
2. Mannuzzu LM, Moronne MM, Isacoff EY (1996) Direct physical measure of conformational rearrangement underlying potassium channel gating. Science 271:213–216
3. Cha A, Bezanilla F (1997) Characterizing voltage-dependent conformational changes in the Shaker K+ channel with fluorescence. Neuron 19:1127–1140
4. Kalstrup T, Blunck R (2013) Dynamics of internal pore opening in K(V) channels probed by a fluorescent unnatural amino acid. Proc Natl Acad Sci U S A 110:8272–8277
5. Bhargava Y, Nicke A, Rettinger J (2013) Validation of Alexa-647-ATP as a powerful tool to study P2X receptor ligand binding and desensitization. Biochem Biophys Res Commun 438:295–300
6. Claydon TW, Fedida D (2007) Voltage clamp fluorimetry studies of mammalian voltage-gated K(+) channel gating. Biochem Soc Trans 35:1080–1082
7. Dempski RE, Friedrich T, Bamberg E (2009) Voltage clamp fluorometry: combining fluorescence and electrophysiological methods to examine the structure-function of the Na(+)/K(+)-ATPase. Biochim Biophys Acta 1787:714–720
8. Pless SA, Lynch JW (2008) Illuminating the structure and function of Cys-loop receptors. Clin Exp Pharmacol Physiol 35:1137–1142
9. Gandhi CS, Isacoff EY (2005) Shedding light on membrane proteins. Trends Neurosci 28:472–479

10. Gandhi CS, Olcese R (2008) The voltage-clamp fluorometry technique. Methods Mol Biol 491:213–231
11. Jiang Y, Lee A, Chen J et al (2003) X-ray structure of a voltage-dependent K+ channel. Nature 423:33–41
12. Doyle DA, Morais CJ, Pfuetzner RA et al (1998) The structure of the potassium channel: molecular basis of K+ conduction and selectivity. Science 280:69–77
13. Noda M, Shimizu S, Tanabe T et al (1984) Primary structure of Electrophorus electricus sodium channel deduced from cDNA sequence. Nature 312:121–127
14. Armstrong CM, Bezanilla F (1974) Charge movement associated with the opening and closing of the activation gates of the Na channels. J Gen Physiol 63:533–552
15. Bezanilla F (2008) How membrane proteins sense voltage. Nat Rev Mol Cell Biol 9: 323–332
16. Yang N, Horn R (1995) Evidence for voltage-dependent S4 movement in sodium channels. Neuron 15:213–218
17. Perozo E, MacKinnon R, Bezanilla F et al (1993) Gating currents from a nonconducting mutant reveal open-closed conformations in Shaker K+ channels. Neuron 11:353–358
18. Stefani E, Bezanilla F (1998) Cut-open oocyte voltage-clamp technique. Methods Enzymol 293:300–318
19. Bezanilla F (2000) The voltage sensor in voltage-dependent ion channels. Physiol Rev 80:555–592
20. Seoh SA, Sigg D, Papazian DM et al (1996) Voltage-sensing residues in the S2 and S4 segments of the Shaker K+ channel. Neuron 16:1159–1167
21. Bezanilla F, Perozo E, Stefani E (1994) Gating of Shaker K+ channels: II. The components of gating currents and a model of channel activation. Biophys J 66:1011–1021
22. Cha A, Bezanilla F (1998) Structural implications of fluorescence quenching in the Shaker K+ channel. J Gen Physiol 112:391–408
23. Posson DJ, Ge P, Miller C et al (2005) Small vertical movement of a K+ channel voltage sensor measured with luminescence energy transfer. Nature 436:848–851
24. Glauner KS, Mannuzzu LM, Gandhi CS et al (1999) Spectroscopic mapping of voltage sensor movement in the Shaker potassium channel. Nature 402:813–817
25. Cha A, Ruben PC, George AL Jr et al (1999) Voltage sensors in domains III and IV, but not I and II, are immobilized by Na + channel fast inactivation. Neuron 22:73–87
26. Cha A, Snyder GE, Selvin PR et al (1999) Atomic scale movement of the voltage-sensing region in a potassium channel measured via spectroscopy. Nature 402:809–813
27. Tombola F, Pathak MM, Isacoff EY (2006) How does voltage open an ion channel? Annu Rev Cell Dev Biol 22:23–52
28. Long SB, Campbell EB, Mackinnon R (2005) Crystal structure of a mammalian voltage-dependent Shaker family K+ channel. Science 309:897–903
29. Long SB, Campbell EB, Mackinnon R (2005) Voltage sensor of Kv1.2: structural basis of electromechanical coupling. Science 309: 903–908
30. Ruta V, Chen J, MacKinnon R (2005) Calibrated measurement of gating-charge arginine displacement in the KvAP voltage-dependent K+ channel. Cell 123:463–475
31. Borjesson SI, Elinder F (2008) Structure, function, and modification of the voltage sensor in voltage-gated ion channels. Cell Biochem Biophys 52:149–174
32. Hoshi T, Zagotta WN, Aldrich RW (1991) Two types of inactivation in Shaker K+ channels: effects of alterations in the carboxy-terminal region. Neuron 7:547–556
33. Hoshi T, Armstrong CM (2013) C-type inactivation of voltage-gated K+ channels: pore constriction or dilation? J Gen Physiol 141:151–160
34. Loots E, Isacoff EY (1998) Protein rearrangements underlying slow inactivation of the Shaker K+ channel. J Gen Physiol 112: 377–389
35. Loots E, Isacoff EY (2000) Molecular coupling of S4 to a K(+) channel's slow inactivation gate. J Gen Physiol 116:623–636
36. Olcese R, Latorre R, Toro L et al (1997) Correlation between charge movement and ionic current during slow inactivation in Shaker K+ channels. J Gen Physiol 110:579–589
37. Claydon TW, Vaid M, Rezazadeh S et al (2007) A direct demonstration of closed-state inactivation of K+ channels at low pH. J Gen Physiol 129:437–455
38. Smith PL, Yellen G (2002) Fast and slow voltage sensor movements in HERG potassium channels. J Gen Physiol 119:275–293
39. Tan PS, Perry MD, Ng CA et al (2012) Voltage-sensing domain mode shift is coupled to the activation gate by the N-terminal tail of hERG channels. J Gen Physiol 140:293–306

40. Wang Z, Dou Y, Goodchild SJ et al (2013) Components of gating charge movement and S4 voltage-sensor exposure during activation of hERG channels. J Gen Physiol 141: 431–443
41. Osteen JD, Gonzalez C, Sampson KJ et al (2010) KCNE1 alters the voltage sensor movements necessary to open the KCNQ1 channel gate. Proc Natl Acad Sci U S A 107:22710–22715
42. Li Y, Gao J, Lu Z et al (2013) Intracellular ATP binding is required to activate the slowly activating K+ channel I(Ks). Proc Natl Acad Sci U S A 110:18922–18927
43. Savalli N, Kondratiev A, Toro L et al (2006) Voltage-dependent conformational changes in human Ca(2+)- and voltage-activated K (+) channel, revealed by voltage-clamp fluorometry. Proc Natl Acad Sci U S A 103: 12619–12624
44. Latorre R, Olcese R, Basso C et al (2003) Molecular coupling between voltage sensor and pore opening in the Arabidopsis inward rectifier K+ channel KAT1. J Gen Physiol 122:459–469
45. Bruening-Wright A, Larsson HP (2007) Slow conformational changes of the voltage sensor during the mode shift in hyperpolarization-activated cyclic-nucleotide-gated channels. J Neurosci 27:270–278
46. Chanda B, Bezanilla F (2002) Tracking voltage-dependent conformational changes in skeletal muscle sodium channel during activation. J Gen Physiol 120:629–645
47. Qiu F, Rebolledo S, Gonzalez C et al (2013) Subunit interactions during cooperative opening of voltage-gated proton channels. Neuron 77:288–298
48. Tombola F, Ulbrich MH, Kohout SC et al (2010) The opening of the two pores of the Hv1 voltage-gated proton channel is tuned by cooperativity. Nat Struct Mol Biol 17:44–50
49. Passero CJ, Okumura S, Carattino MD (2009) Conformational changes associated with proton-dependent gating of ASIC1a. J Biol Chem 284:36473–36481
50. Bonifacio G, Lelli CI, Kellenberger S (2014) Protonation controls ASIC1a activity via coordinated movements in multiple domains. J Gen Physiol 143:105–118
51. Dahan DS, Dibas MI, Petersson EJ et al (2004) A fluorophore attached to nicotinic acetylcholine receptor beta M2 detects productive binding of agonist to the alpha delta site. Proc Natl Acad Sci U S A 101: 10195–10200
52. Lester HA, Dibas MI, Dahan DS et al (2004) Cys-loop receptors: new twists and turns. Trends Neurosci 27:329–336
53. Li P, Khatri A, Bracamontes J et al (2010) Site-specific fluorescence reveals distinct structural changes induced in the human rho 1 GABA receptor by inhibitory neurosteroids. Mol Pharmacol 77:539–546
54. Khatri A, Sedelnikova A, Weiss DS (2009) Structural rearrangements in loop F of the GABA receptor signal ligand binding, not channel activation. Biophys J 96:45–55
55. Muroi Y, Theusch CM, Czajkowski C et al (2009) Distinct structural changes in the GABAA receptor elicited by pentobarbital and GABA. Biophys J 96:499–509
56. Muroi Y, Czajkowski C, Jackson MB (2006) Local and global ligand-induced changes in the structure of the GABA(A) receptor. Biochemistry 45:7013–7022
57. Pless SA, Dibas MI, Lester HA et al (2007) Conformational variability of the glycine receptor M2 domain in response to activation by different agonists. J Biol Chem 282: 36057–36067
58. Han L, Talwar S, Lynch JW (2013) The relative orientation of the TM3 and TM4 domains varies between alpha1 and alpha3 glycine receptors. ACS Chem Neurosci 4: 248–254
59. Lorinczi E, Bhargava Y, Marino SF et al (2012) Involvement of the cysteine-rich head domain in activation and desensitization of the P2X1 receptor. Proc Natl Acad Sci U S A 109:11396–11401
60. Shin JM, Munson K, Sachs G (2011) Gastric H+, K + -ATPase. Compr Physiol 1: 2141–2153
61. Bublitz M, Poulsen H, Morth JP et al (2010) In and out of the cation pumps: p-type ATPase structure revisited. Curr Opin Struct Biol 20:431–439
62. van der Hijden HT, Grell E, de Pont JJ et al (1990) Demonstration of the electrogenicity of proton translocation during the phosphorylation step in gastric H + K(+)-ATPase. J Membr Biol 114:245–256
63. Geibel S, Kaplan JH, Bamberg E et al (2003) Conformational dynamics of the Na+/K + -ATPase probed by voltage clamp fluorometry. Proc Natl Acad Sci U S A 100:964–969
64. Geibel S, Zimmermann D, Zifarelli G et al (2003) Conformational dynamics of Na+/K + - and H+/K + -ATPase probed by voltage clamp fluorometry. Ann N Y Acad Sci 986:31–38

65. Durr KL, Tavraz NN, Friedrich T (2012) Control of gastric H, K-ATPase activity by cations, voltage and intracellular pH analyzed by voltage clamp fluorometry in Xenopus oocytes. PLoS One 7:e33645
66. Hotzy J, Schneider N, Kovermann P et al (2013) Mutating a conserved proline residue within the trimerization domain modifies Na + binding to excitatory amino acid transporters and associated conformational changes. J Biol Chem 288:36492–36501
67. Larsson HP, Tzingounis AV, Koch HP et al (2004) Fluorometric measurements of conformational changes in glutamate transporters. Proc Natl Acad Sci U S A 101:3951–3956
68. Egenberger B, Gorboulev V, Keller T et al (2012) A substrate binding hinge domain is critical for transport-related structural changes of organic cation transporter 1. J Biol Chem 287:31561–31573
69. Patti M, Forster IC (2014) Correlating charge movements with local conformational changes of a Na(+)-coupled cotransporter. Biophys J 106:1618–1629
70. Virkki LV, Murer H, Forster IC (2006) Voltage clamp fluorometric measurements on a type II Na + -coupled Pi cotransporter: shedding light on substrate binding order. J Gen Physiol 127:539–555
71. Gagnon DG, Frindel C, Lapointe JY (2007) Voltage-clamp fluorometry in the local environment of the C255-C511 disulfide bridge of the Na+/glucose cotransporter. Biophys J 92:2403–2411
72. Meinild AK, Hirayama BA, Wright EM et al (2002) Fluorescence studies of ligand-induced conformational changes of the Na (+)/glucose cotransporter. Biochemistry 41:1250–1258
73. Meinild AK, Loo DD, Skovstrup S et al (2009) Elucidating conformational changes in the gamma-aminobutyric acid transporter-1. J Biol Chem 284:16226–16235
74. Sakata S, Okamura Y (2014) Phosphatase activity of the voltage-sensing phosphatase, VSP, shows graded dependence on the extent of activation of the voltage sensor. J Physiol 592:899–914
75. Kusch J, Zifarelli G (2014) Patch-clamp fluorometry: electrophysiology meets fluorescence. Biophys J 106:1250–1257
76. Zheng J, Zagotta WN (2000) Gating rearrangements in cyclic nucleotide-gated channels revealed by patch-clamp fluorometry. Neuron 28:369–374
77. Kenworthy AK (2001) Imaging protein-protein interactions using fluorescence resonance energy transfer microscopy. Methods 24:289–296
78. Giraldez T, Hughes TE, Sigworth FJ (2005) Generation of functional fluorescent BK channels by random insertion of GFP variants. J Gen Physiol 126:429–438
79. Kobrinsky E, Schwartz E, Abernethy DR et al (2003) Voltage-gated mobility of the Ca2+ channel cytoplasmic tails and its regulatory role. J Biol Chem 278:5021–5028
80. Kobrinsky E, Stevens L, Kazmi Y et al (2006) Molecular rearrangements of the Kv2.1 potassium channel termini associated with voltage gating. J Biol Chem 281:19233–19240
81. Kobrinsky E, Kepplinger KJF, Yu A et al (2004) Voltage-gated rearrangements associated with differential beta-subunit modulation of the L-type Ca(2+) channel inactivation. Biophys J 87:844–857
82. Kobrinsky E, Tiwari S, Maltsev VA et al (2005) Differential role of the alpha1C subunit tails in regulation of the Cav1.2 channel by membrane potential, beta subunits, and Ca2+ ions. J Biol Chem 280:12474–12485
83. Fisher JA, Girdler G, Khakh BS (2004) Time-resolved measurement of state-specific P2X2 ion channel cytosolic gating motions. J Neurosci 24:10475–10487
84. Ambrose EJ (1956) A surface contact microscope for the study of cell movements. Nature 178:1194
85. Axelrod D (1981) Cell-substrate contacts illuminated by total internal reflection fluorescence. J Cell Biol 89:141–145
86. van der Wal J, Habets R, Várnai P et al (2001) Monitoring agonist-induced phospholipase C activation in live cells by fluorescence resonance energy transfer. J Biol Chem 276: 15337–15344
87. Yang F, Cui Y, Wang K et al (2010) Thermosensitive TRP channel pore turret is part of the temperature activation pathway. Proc Natl Acad Sci U S A 107:7083–7088
88. Triller A, Choquet D (2008) New concepts in synaptic biology derived from single-molecule imaging. Neuron 59:359–374
89. Walling MA, Novak JA, Shepard JR (2009) Quantum dots for live cell and in vivo imaging. Int J Mol Sci 10:441–491
90. Michalet X, Pinaud FF, Bentolila LA et al (2005) Quantum dots for live cells, in vivo imaging, and diagnostics. Science 307: 538–544
91. Richler E, Shigetomi E, Khakh BS (2011) Neuronal P2X2 receptors are mobile ATP sensors that explore the plasma membrane when activated. J Neurosci 31:16716–16730

92. Zheng J, Zagotta WN (2003) Patch-clamp fluorometry recording of conformational rearrangements of ion channels. Sci STKE 2003:PL7
93. Zheng J (2006) Patch fluorometry: shedding new light on ion channels. Physiology (Bethesda) 21:6–12
94. Biskup C, Kusch J, Schulz E et al (2007) Relating ligand binding to activation gating in CNGA2 channels. Nature 446:440–443
95. Craven KB, Zagotta WN (2006) CNG and HCN channels: two peas, one pod. Annu Rev Physiol 68:375–401
96. Lehrer SS (1971) Solute perturbation of protein fluorescence. The quenching of the tryptophyl fluorescence of model compounds and of lysozyme by iodide ion. Biochemistry 10: 3254–3263
97. Chanda B, Asamoah OK, Blunck R et al (2005) Gating charge displacement in voltage-gated ion channels involves limited transmembrane movement. Nature 436:852–856
98. Taraska JW, Zagotta WN (2007) Structural dynamics in the gating ring of cyclic nucleotide-gated ion channels. Nat Struct Mol Biol 14:854–860
99. Miranda P, Contreras JE, Plested AJ et al (2013) State-dependent FRET reports calcium- and voltage-dependent gating-ring motions in BK channels. Proc Natl Acad Sci U S A 110:5217–5222
100. Zheng J, Varnum MD, Zagotta WN (2003) Disruption of an intersubunit interaction underlies Ca2 + -calmodulin modulation of cyclic nucleotide-gated channels. J Neurosci 23:8167–8175
101. Trudeau MC, Zagotta WN (2004) Dynamics of Ca2 + -calmodulin-dependent inhibition of rod cyclic nucleotide-gated channels measured by patch-clamp fluorometry. J Gen Physiol 124:211–223
102. Nache V, Eick T, Schulz E et al (2013) Hysteresis of ligand binding in CNGA2 ion channels. Nat Commun 4:2864
103. Nache V, Zimmer T, Wongsamitkul N et al (2012) Differential regulation by cyclic nucleotides of the CNGA4 and CNGB1b subunits in olfactory cyclic nucleotide-gated channels. Sci Signal 5:ra48
104. Kusch J, Biskup C, Thon S et al (2010) Interdependence of receptor activation and ligand binding in HCN2 pacemaker channels. Neuron 67:75–85
105. Kusch J, Thon S, Schulz E et al (2012) How subunits cooperate in cAMP-induced activation of homotetrameric HCN2 channels. Nat Chem Biol 8:162–169
106. Colquhoun D (1998) Binding, gating, affinity and efficacy: the interpretation of structure-activity relationships for agonists and of the effects of mutating receptors. Br J Pharmacol 125:924–947
107. Marni F, Wu S, Shah GM et al (2012) Normal-mode-analysis-guided investigation of crucial intersubunit contacts in the cAMP-dependent gating in HCN channels. Biophys J 103:19–28
108. Wu S, Gao W, Xie C et al (2012) Inner activation gate in S6 contributes to the state-dependent binding of cAMP in full-length HCN2 channel. J Gen Physiol 140:29–39
109. Wu S, Vysotskaya ZV, Xu X et al (2011) State-dependent cAMP binding to functioning HCN channels studied by patch-clamp fluorometry. Biophys J 100:1226–1232
110. Xu X, Marni F, Wu S et al (2012) Local and global interpretations of a disease-causing mutation near the ligand entry path in hyperpolarization-activated cAMP-gated channel. Structure 20:2116–2123
111. Grandl J, Sakr E, Kotzyba-Hibert F et al (2007) Fluorescent epibatidine agonists for neuronal and muscle-type nicotinic acetylcholine receptors. Angew Chem Int Ed Engl 46:3505–3508
112. Schmauder R, Kosanic D, Hovius R et al (2011) Correlated optical and electrical single-molecule measurements reveal conformational diffusion from ligand binding to channel gating in the nicotinic acetylcholine receptor. Chembiochem 12:2431–2434
113. Miller C (1986) Ion channel reconstitution. Plenum Press, New York, NY
114. Ide T, Takeuchi Y, Aoki T et al (2002) Simultaneous optical and electrical recording of a single ion-channel. Jpn J Physiol 52:429–434
115. Ide T, Yanagida T (1999) An artificial lipid bilayer formed on an agarose-coated glass for simultaneous electrical and optical measurement of single ion channels. Biochem Biophys Res Commun 265:595–599
116. Borisenko V, Lougheed T, Hesse J et al (2003) Simultaneous optical and electrical recording of single gramicidin channels. Biophys J 84:612–622
117. Montal M, Mueller P (1972) Formation of bimolecular membranes from lipid monolayers and a study of their electrical properties. Proc Natl Acad Sci U S A 69:3561–3566
118. Harms GS, Orr G, Montal M et al (2003) Probing conformational changes of gramicidin ion channels by single-molecule patch-clamp fluorescence microscopy. Biophys J 85:1826–1838

119. Valeur B (2001) Molecular fluorescence: principles and applications. Wiley-VCH Verlag GmbH, Berlin
120. Ormo M, Cubitt AB, Kallio K et al (1996) Crystal structure of the Aequorea victoria green fluorescent protein. Science 273: 1392–1395
121. Islas LD, Zagotta WN (2006) Short-range molecular rearrangements in ion channels detected by tryptophan quenching of bimane fluorescence. J Gen Physiol 128:337–346
122. Mansoor SE, McHaourab HS, Farrens DL (2002) Mapping proximity within proteins using fluorescence spectroscopy. A study of T4 lysozyme showing that tryptophan residues quench bimane fluorescence. Biochemistry. 41:2475–2484
123. Taraska JW, Puljung MC, Olivier NB et al (2009) Mapping the structure and conformational movements of proteins with transition metal ion FRET. Nat Methods 6:532–537
124. Vogel SS, Thaler C, Koushik SV (2006) Fanciful FRET. Sci STKE 2006:re2
125. Jarecki BW, Zheng S, Zhang L et al (2013) Tethered spectroscopic probes estimate dynamic distances with subnanometer resolution in voltage-dependent potassium channels. Biophys J 105:2724–2732
126. Blaustein RO, Cole PA, Williams C et al (2000) Tethered blockers as molecular 'tape measures' for a voltage-gated K+ channel. Nat Struct Biol 7:309–311
127. Jensen MO, Jogini V, Borhani DW et al (2012) Mechanism of voltage gating in potassium channels. Science 336:229–233
128. Shaner NC, Steinbach PA, Tsien RY (2005) A guide to choosing fluorescent proteins. Nat Methods 2:905–909
129. Taraska JW, Zagotta WN (2010) Fluorescence applications in molecular neurobiology. Neuron 66:170–189
130. Magidson V, Khodjakov A (2013) Circumventing photodamage in live-cell microscopy. Methods Cell Biol 114:545–560

Chapter 5

Dendrites: Recording from Fine Neuronal Structures Using Patch-Clamp and Imaging Techniques

Sonia Gasparini and Lucy M. Palmer

Abstract

Dendrites are the principal site of synaptic input onto neurons but despite their importance in neuronal signaling, little is known about how they receive and transform this input. This is due largely to their typically submicron size, which has historically rendered them inaccessible for direct recording. However, the advent of electrophysiological patch-clamp and advanced imaging techniques over the past few decades has opened this field of research. Fuelled by Rall's theory of active dendritic integration, intracellular recording techniques proved that dendrites do indeed have active conductances which modify synaptic input and thereby alter neuronal output in response to certain patterns of information. Furthermore, advances in fluorescence imaging have highlighted the importance of dendritic activity during sensory processing and behavior. Here we summarize advances in experimental methods, namely electrophysiological and fluorescence imaging techniques, which have improved the accessibility of recording from fine dendritic structures.

Key words Dendrite, Imaging, Synapse, Dendritic channels, In vitro, Brain slice, Calcium indicators, GFP

1 Introduction

1.1 History of Dendritic Recording Techniques

The unique morphology of neurons stems from the elaborate branching of their dendritic tree. Despite their anatomical dominance, the functional importance of dendrites has been debated ever since they were first reported by Deiters in 1865 [1]. Systematic advances in recording techniques have revealed many important discoveries about dendritic function (Fig. 1). By interpreting the morphology of neurons using the Golgi staining technique, Ramon y Cajal generated a hypothesis over 100 years ago that electrical signals propagated from dendrites to the soma and axon [2]. Called the "neuron doctrine," this hypothesis was the first to place dendrites at the site of synaptic input and information transfer between cells and hence catapulted the role of dendrites to the forefront of understanding brain function as a whole. The foresight

Alon Korngreen (ed.), *Advanced Patch-Clamp Analysis for Neuroscientists*, Neuromethods, vol. 113,
DOI 10.1007/978-1-4939-3411-9_5, © Springer Science+Business Media New York 2016

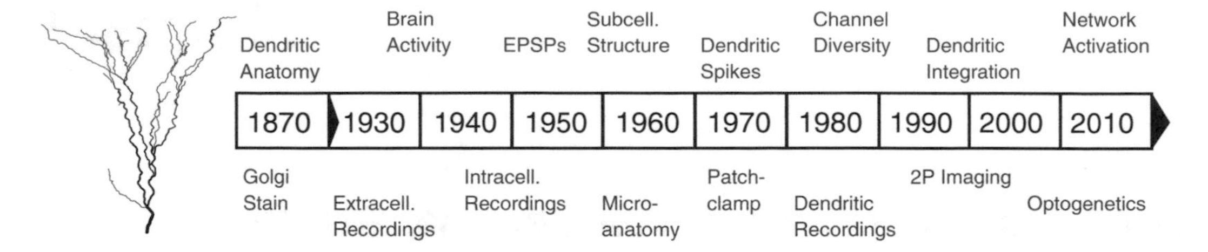

Fig. 1 Timeline for advances in neuronal imaging and recording techniques (*red*) and the associated discoveries (*blue*)

of Cajal in predicting the importance of dendrites, based purely on the morphology of neurons, is one of the great breakthroughs in neuroscience. The development of microelectrodes in the 1950s enabled intracellular postsynaptic potentials to be initially recorded in mammalian motoneurons [3], giant synapses of the squid stellate ganglion [4], and mammalian sympathetic ganglion [5]. These recordings along with the development of cable theory (for a review, see [6]) illustrated that dendrites play an important role in transferring information towards the soma in the form of excitatory postsynaptic potentials (EPSPs). These physiological events were correlated with structural features of dendrites using high-magnification imaging techniques (electron microscopy) to study the microanatomy of dendrites in the late 1950s [7, 8].

Despite realizing their importance in neuronal processing, dendrites were historically believed to purely serve as a passive cable to transfer information from the site of input to the site of action potential initiation in the axon. Following this, it was therefore generally accepted that dendritic integration was passive and action potentials were generated as a result of the graded summation of synaptic inputs in the soma/axon region. However, modeling work conducted by Rall (for a review, see [9]) raised the possibility that dendrites contain active conductances which can transform synaptic inputs. It was not until the development of the "patch-clamp technique" in the 1970s that neuronal properties were directly probed at a subcellular level [10, 11]. This led to Bert Sakmann and Erwin Neher winning the Nobel Prize in Physiology or Medicine in 1991 "for their discoveries concerning the function of single ion channels in cells." The patch-clamp technique offers a multitude of different technical approaches including "whole-cell," "cell-attached," "inside-out" and "nucleated," which are discussed in detail below with reference to dendritic recordings. By clamping onto the membrane of neurons/dendrites using the cell-attached patch-clamp technique, it was revealed that dendrites are endowed with a multitude of different voltage-sensitive ion channels. Combined with the use of infrared illumination for enhanced contrast [12], the density and distribution of these channels could be probed throughout the large apical dendritic tree. Numerous studies revealed that the main dendritic voltage-sensitive ion channel players are Na^+, K^+, Ca^{2+}, I_h; however, their distribution and density differ according to the cell type. For example, cortical layer 5 pyramidal neurons [13] and hippocampal CA1 neurons [14] have a uniform Na^+ channel distribution whereas Cerebellar Purkinje cells [15] have a rapid decrease in Na^+ channel density with increasing distance from the soma (for a review see [16]). These voltage-sensitive ion channels render dendrites "active"—that is, they can "actively" alter dendritic excitability. For example, Na^+ channels support "active" back-propagation of action potentials into the dendrites of numerous neuronal subtypes. These back-propagating action potentials have been shown to be extremely important in numerous cellular processing including spike-timing-dependent plasticity and local synaptic boosting [17]. Dendritic voltage-sensitive ion channels have also been shown to add to the complexity of cellular signaling by altering the "rules" of dendritic integration. That is, synaptic input does not necessarily sum linearly and if certain input conditions are met, input can sum sub- or supra-linearly through the activation of different dendritic ion channels (for a review, see [18]). Briefly, the dendrites of pyramidal neurons in both the cortex and hippocampus have been shown to support supralinear summation of synaptic input driven by the activation of Na^+ [19–22] and Ca^{2+} [23, 24] channels. Furthermore, supralinear dendritic integration can also be driven by NMDA glutamatergic receptors [25–27]. Active dendritic integration was first recorded in brain slices and it was not until the recent advances in fluorescence imaging that dendritic activity was recorded in vivo illustrating that dendritic spikes are important for encoding sensory-based behavior [28–30]. These findings shed new light on the computational capabilities of dendrites and highlight their central role in modifying and transforming received information into neuronal output.

1.2 Dendrite Morphology/Channels/Passive Properties

Neurons typically have extremely complex dendritic trees. Dendrites can be categorized into distinct domains depending on their cell type and morphological characteristics; pyramidal neurons in the hippocampus and cortex have basal dendrites that project from the base of the soma whereas apical dendrites project from the somatic apex, and oblique and tuft dendrites extend from the proximal and distal region of the apical dendrite, respectively (Fig. 2). Since these dendritic domains are located in spatially distinct regions of the brain structure in which they reside, they often receive different streams of information input. For example, the tuft dendrites of cortical pyramidal neurons are located in the upper layers of the cortex that receive long-range feedback input from other cortical areas and the thalamus including the posterior medial nucleus (POm) of the thalamus [31], the secondary somatosensory cortex [32], and parahippocampal structures [33]. Basal dendrites, however, receive the majority of synaptic inputs [34]

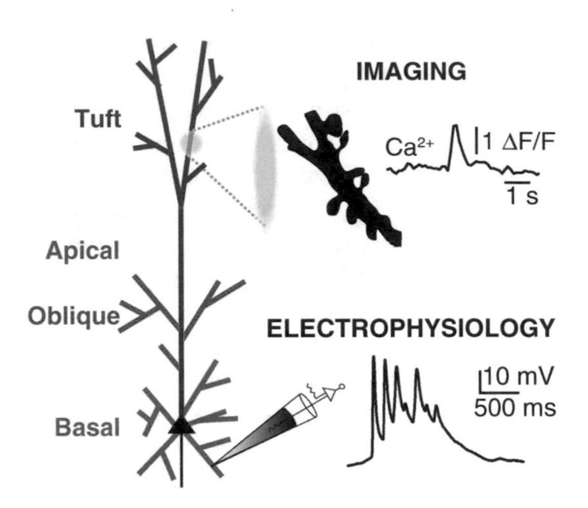

Fig. 2 Dendritic nomenclature and schematic of the recording techniques

which largely carry feed forward information [35]. Hippocampal principal neurons in the CA1 area also receive segregated input onto particular regions of their dendritic tree, with the basal and proximal apical dendrites receiving input primarily from CA3 pyramidal neurons (Schaffer collateral pathway), whereas the apical tuft receives direct input from the entorhinal cortex (perforant pathway) and the thalamic nucleus reuniens [36]. How, and even whether, these different pathways are integrated and ultimately combine information has been a source of debate for decades. This question can now be addressed with the recent advances in molecular biology, in particular the development of optogenetics [37, 38], which enables subsets of neurons to be targeted and controlled by light [39]. Known patterns of axonal presynaptic activity can be reproduced with photoactivation, and the resulting dendritic excitability in the postsynaptic neurons can be recorded, revealing the importance of synaptic input sequences in dendritic integration [40]. This, however, is still an emerging field and there is no doubt that many great discoveries are yet to be revealed.

1.3 Combining Imaging and Patch Clamp for Dendritic Recordings

Despite over 100 years of intensive investigations (Fig. 1), we are just starting to unravel the complexity of dendritic function and the role they play in transforming synaptic input. Our understanding of dendritic functioning has lagged behind other fields of neuroscience research. This is largely due to the difficulty in recording from the very thin dendritic structures, especially in vivo, which has only been achieved in a handful of studies [41–44]. However, recent advances in neuroscience techniques have now opened this field of research. Here we examine how the role of dendrites can be probed using electrophysiological and imaging techniques. Firstly, details of performing dendritic electrophysiological recordings are described; secondly, florescence imaging techniques used for monitoring dendritic activity are explained; and lastly the past, present, and future of dendritic recordings are explored.

This chapter is not designed to be an exhaustive review of all the tools that can be used to investigate dendritic function, but highlights various electrophysiological and imaging techniques which have proved to be fundamental in developing our understanding of dendritic function and its importance in neuronal integration and computation.

2 Dendritic Electrophysiological Recordings

2.1 In Vitro Patch-Clamp Electrophysiology

There are no significant differences in the techniques used to perform patch-clamp recordings from either the soma [45–47] or fine structures such as dendrites [12]. However, due to the decreasing visibility of such small structures with increasing distance from the slice surface, both the quality of the slices and the optics of the microscope need to be optimized [12, 48]. In addition, vibrations and drifts in the experimental setup should be completely eliminated; this aspect is especially true for the micromanipulators.

2.2 Slicing

During the slicing procedure, the orientation of the brain slice should be parallel to the axis of the dendritic tree. Therefore, the most suitable plane must be identified for the various brain regions (see [48]). This is especially true during simultaneous somatic and dendritic recordings, where both the soma and the dendrite need to be visibly identified and accessible by the pipette. In this case, there will probably be only one or two viable slices per hemisphere. Access to neurons in plane is further complicated when dual recordings from different brain areas are needed to activate presynaptic neurons and record postsynaptic responses; however, this limitation can now be overcome with the use of optogenetic techniques (see Introduction and Discussion).

Many fine-tuning techniques have been developed to improve the quality of slices for patching dendrites in vitro; in general, they aim to reduce excitotoxicity during the slicing procedure and therefore improve the health of neurons to be examined. Many of these modifications consist of changes to the regular artificial cerebrospinal fluid (ACSF) used during the slicing procedures, including using ice-cold ACSF to slow down the metabolic processes (but see [49]), modifying the Ca^{2+}:Mg^{2+} ratio from 2:1 in the regular ACSF to 1:6 (up to 0.5:7) to limit glutamate- and depolarization-induced Ca^{2+} influxes [50] and adding glutamatergic antagonists, such as kynurenic acid to the slicing solution [51]. In addition, further improvements can be obtained by substituting NaCl with iso-osmolar sucrose [52] in the cutting solution; this substitution reduces the Na^{+} influx and neuronal swelling from passive Cl^{-} influx [53] and appears to better preserve GABA-mediated synaptic transmission in the hippocampus [54]. Further modifications that have shown to be beneficial in maintaining healthy brain slices include the addition of sodium ascorbate and sodium pyruvate to the slicing and storage solution to protect neurons from oxidation [47]. In addition, it is advisable to perfuse adult animals (especially when older than 6 weeks of age) through the ascending aorta with an oxygenated ice-cold slicing solution, to allow a rapid cooling of the brain and the elimination of harmful substances in the blood [47, 55]. Another trick to obtain healthier neurons, especially in adult animals, is to incubate the slices for 30–60 min at physiological temperatures (35–37 °C) just after the slicing procedure [47, 55]. The higher temperature accelerates the decay of partially damaged neurons, leaving only the healthier neurons to be able to completely recover from the slicing procedure.

Strategies are also employed during the slicing procedure to not only increase the number of healthy neurons but also maintain the neuron morphology. This includes obtaining brain slices with surfaces as smooth as possible to keep even the finest processes close to the surface intact. To this end, it is recommended that the *z*-axis vibration of the vibratome be checked regularly and kept to a minimum. Before commencing the slicing procedure, it is also helpful to remove the pia to avoid the thick membrane from dragging with the blade and damaging the slice surface. In addition, some labs use sapphire or ceramic blades which slice in smoother fashion than the regular stainless steel blades. These techniques are especially advantageous in older animals, which, due to the need for dendritic arbors to be fully mature, are often used in experiments involving dendrites.

2.3 Visualizing, Identifying, and Recording from Dendrites

Visualization of fine structures such as dendrites is complicated. For instance, dendrites from healthy neurons lack contrast and are often very hard to distinguish from the surrounding slice tissue, whereas dead branches are more easily visible as they appear “crunchy” and tridimensional [48]. Various methods are used to enhance contrast in brain slices, namely differential interference contrast (DIC) with infrared illumination [56], and oblique illumination achieved using an oblique condenser or Dodt Gradient Contrast [57]. Oblique illumination can be advantageous when fluorescent microscopy is employed together with electrophysiology, because it does not require optics above the objective [57]; however in most microscopes with DIC optics, the analyzer can be easily removed when fluorescent illumination is employed. To further aid visualization of dendrites, magnifiers (2–4×) are often placed between the objective and the video camera.

Once a viable dendrite has been identified, the pipette should be lowered to the slice surface; at this point the voltage offset is adjusted to zero and a voltage test pulse (5–10 mV) is injected into the electrode to monitor the pipette resistance. Positive pressure is then applied to the electrode, which not only improves the visualization of the dendrite by clearing the surrounding debris and neuroglia but also ensures the pipette tip is free of debris until it comes in contact with the cell of interest. The pipette is then lowered through the tissue; once at the same plane as the dendrite, the best strategy, in our hands, is to position the pipette just above the dendrite before advancing it diagonally forward. It is important at this point to keep focus of the pipette tip. When approached with an electrode with applied positive pressure, flat dendrites will extend and allow the formation of a dimple, whereas dead dendrites will "jump" from the pipette tip. Dimple formation is accompanied by an increase in the resistance of the electrode and, once formed, it is advantageous to keep advancing for a few microns until the dimple grows larger before releasing the positive pressure. This way, when the pressure is released, the dendritic membrane will rebound towards the pipette and create a tight seal without requiring (or requiring very little) negative pressure. Numerous strategies can then be employed to promote the formation of a tight seal (>giga-ohm resistance) between the pipette and dendritic membrane. Formation of this "giga-seal" is facilitated by the slow hyperpolarization of the membrane patch to −70 mV, probably due to electrostatic interactions between the electrode and the membrane. This approach is preferable to applying slow and prolonged negative pressure, which increases the amount of membrane drawn into the tip of the pipette. Another trick to increase the probability of giga-seal formation is to fill the tip of the recording electrode with an intracellular solution that does not contain the ATP-regenerating system and backfill with the regular solution (see Solutions below); the lower osmolarity in the recording electrode facilitates seal formation. Once a giga-seal is obtained, the pipette should be retracted to allow the dendrite to relax to its original position and recordings can be performed in cell-attached configuration (see below) or, most commonly, the whole-cell configuration is achieved by briefly applying strong suction to break into the dendrite. After whole-cell configuration is obtained, the series resistance generally tends to increase in the first few minutes; in this case the electrode tip should be cleared by using either gentle positive pressure or suction into the tubing. It is very important at this point to keep monitoring the tip of the pipette as well as the resistance of the electrode. When using all these tricks, good whole-cell recordings from dendrites can be maintained for over an hour.

During patch-clamp electrophysiology, voltage or current is typically passed through the patch pipette to alter neuronal excitability in a stepwise manner. Although this has revealed a lot about single-cell physiology (for a review, see [58]), there are other methods used to capture the dynamic nature of neurons. One such experimental configuration is the dynamic clamp, where the recorded membrane potential [$V(t)$] is used to control the amount of current [$I(t) = g(t)(V(t) - E_{rev})$] to be injected in a neuron given a certain conductance waveform [$g(t)$] and reversal potential [E_{rev}] [59, 60]. This technique requires accurate recording and fast sampling of the membrane voltage and rapid computation of the current to be injected [61]. To minimize series resistance artifacts in these situations (especially if access resistance is >15 MΩ), it is helpful to use two adjacent patch electrodes (placed at ≤ 20 μm), one to inject the calculated current and the second one to independently record the membrane voltage [62]. In this case, we have found that the most successful approach for dual patching is to place both electrodes (arranged at 180° from one another) just above the surface of the slice before obtaining whole-cell configuration with both electrodes (using a similar technique as the one described above) in rapid succession (not more than 5 min apart). Obviously vibrations will have to be even more controlled in this situation, in order to avoid losing the recordings.

When investigating dendritic function, the investigator typically wishes to also record associated activity at another site, typically at the soma. To perform dual recordings, the best strategy is to obtain whole-cell configuration in the "easiest" location first before advancing to the second "harder" location. This usually involves recording from the soma first, since the larger size means it is not only easier to record from but the recording will also be able to withstand greater movement of the slice while obtaining the dendritic seal. Naturally, this experimental configuration is difficult to achieve because it is typically hard to identify a connected dendrite. Identifying dendrites can be aided by the high K^+ intracellular solution flowing from the approaching pipette, which causes the membrane potential of the entire neuron to depolarize. Therefore, a depolarization recorded with the somatic electrode is a good indication that the dendritic pipette is approaching a dendrite originating from the same neuron. Unfortunately this trick is not foolproof, i.e., the soma might depolarize even if you are not targeting a dendrite emerging from it because its dendrites are located close to the targeted dendrite. However, if the soma does not depolarize while approaching the dendrite with a positive pressure, you can be almost sure that they do not belong to the same neuron.

Many dendrites are too small to be visualized under infrared microscopy and therefore fluorescence-assisted patching is often used in these situations. This involves obtaining a whole-cell recording first from the soma with a pipette containing a fluorescent dye such as Alexa. Once the dye has diffused, the epifluorescence image of the dendrite is overlaid to the IR image, such that the dendrite of interest can be targeted for patching [63]. This technique has allowed recording from very thin dendrites, such as the tuft [26] and the basal arborization [64] of neocortical layer 5 pyramidal neurons.

2.4 Recording Configurations

All the various configurations of the patch-clamp technique can be employed on dendrites, depending on the question asked. In particular, *cell-attached* patches have been used to examine the density and features of voltage-dependent ion channels along larger, mainly apical, dendrites [16]. To correctly determine the kinetics of ion channels as a function of the voltage, the resting dendritic membrane potential also needs to be measured by breaking-in to achieve whole-cell configuration at the end of the experiment [14]. Cell-attached patch-clamp configuration has also been used to analyze changes in the activity of single ion channels following the activation [65] or inhibition [66] of protein kinases. The *outside-out* configuration has been used to map the density and features of ligand-gated receptors (such as glutamatergic receptors [67]). For both these configurations, particular care should be used to keep the size of the patch pipette tip constant, as well as approximately the same amount of suction to form the seal. These tricks reduce the variations in membrane patch size and should definitely be employed when measuring current densities [14, 67]. The vast majority of dendritic studies, however, have been performed using the *whole-cell configuration* (see above), in order to study linear and supralinear dynamics of dendritic integration, and this configuration is also commonly used in concert with fluorescence imaging (discussed below). *Perforated patch*, a configuration that is used to maintain the intracellular milieu unperturbed and decrease the wash-out of intracellular components [68], is rarely used on dendrites as it further increases the series resistance due to the pores created in the membrane by antibiotics. Its use exacerbates the problems due to the small electrodes that are normally used for dendritic recordings and the consequent high series resistance that is encountered already in whole-cell recordings (see above).

2.5 Solutions

The solutions used for dendritic recordings are similar to those used for somatic recordings [48]. The slices are bathed in an extracellular solution (ACSF, see above), containing high concentrations of NaCl, $NaHCO_3$, and dextrose and lower concentrations of NaH_2PO_4, $CaCl_2$, and $MgCl_2$, continuously buffered with an O_2 95 % and CO_2 5 % mixture; its composition might be modified to analyze and isolate specific ion currents.

As for the pipette solution, for cell-attached recordings its composition will mimic the ACSF with a high concentration of NaCl, an HEPES buffer and KCl, $MgCl_2$ and $CaCl_2$ in smaller concentrations and will be buffered with NaOH. Changes can be made to isolate the ion channels of interest by adding blockers of the other conductances [14, 69]. In some instances, enzyme activators or inhibitors might be added to locally affect the activity of the channels in the membrane patch [65, 66]. For whole-cell recordings, the intracellular solution will contain a high concentration of potassium salt

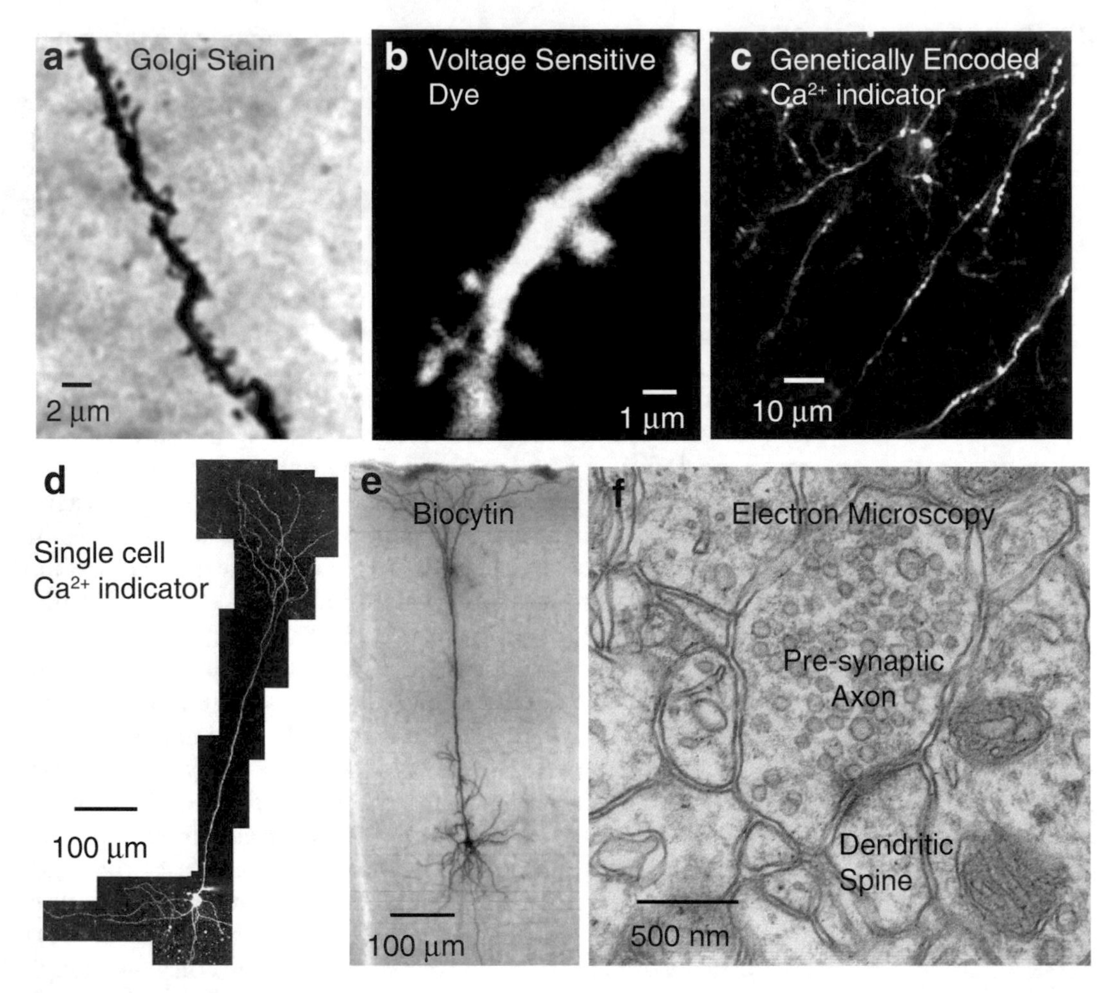

Fig. 3 Examples of different imaging techniques used to record dendritic activity. (**a**) Single dendrite stained with the Golgi staining technique. (**b**) Single layer 5 pyramidal neuron filled with voltage-sensitive dye (JPW3028) via the whole-cell patch pipette in vitro. (**c**) Multiple dendrites from layer 5 pyramidal neurons transfected with genetically encoded Ca^{2+} indicator (GCaMP) in vivo. (**d**) Single entorhinal cortex pyramidal neuron filled with Ca^{2+} indicator (OGB1) via the whole-cell patch pipette in vitro. (**e**) Single cortical pyramidal neuron stained with biocytin. (**f**) Electron microscopy image illustrating a presynaptic axon and postsynaptic dendritic spine

(methanesulfate, methanesulfonate, or gluconate), HEPES, and sometimes a Ca^{2+} buffer (BAPTA or EGTA) and low concentrations of NaCl. A combination of ATP, GTP, and phosphocreatine are also added to maintain the energy balance of the cell and partially counterbalance the wash-out of intracellular factors [55]. Often the intracellular marker biocytin is added to the intracellular solution to allow for post-hoc processing and identification of the neuron from which the recording was made (see Fig. 3).

3 Techniques Used to Image from Dendrites

The art of imaging has opened the field of cellular neuroscience. During the past decade, we have seen an explosion of optical imaging techniques, which has led to many great discoveries. None have been more important than those associated with addressing the activity of small neuronal processes which cannot otherwise be probed by the more traditional electrophysiological techniques (Fig. 3). Optical imaging enables not only fine structures to be assessed, but also activity to be recorded in multiple locations simultaneously both within a single and a network of neurons. This is hugely advantageous when trying to assess the local and spatial spread of a signal within a neuron and/or the relationship with the activity of neighboring neurons.

Early studies used wide-field illumination and charge-coupled detector-based (CCD) cameras to probe the activity of thin dendritic processes, and more recently single- and multi-photon imaging is the technique of choice to image deep into tissue. These techniques and more, along with the use of fluorophores and genetic indicators and how they are incorporated into brain tissue, are discussed in detail below (see Chap. 2 for details on voltage-sensitive dye imaging).

4 Fluorophores

Although not required for all imaging techniques, loading of fluorophores into the neurons of interest is a crucial first step for most studies delving into the role of dendrites. The ideal fluorophore needs to have: (1) a high resting fluorescence to enable visualization of the neuron/dendrite at rest, (2) high signal-to-noise ratio, (3) linear relationship between fluorescence changes and imaged ion concentration/voltage, (4) fast kinetics, and (5) little perturbation to the cellular processing. Depending on the question at hand, there are a variety of different fluorophores and loading techniques. In-depth discussions on every fluorophore/loading technique are out of the scope of this chapter; however the most common techniques and examples of their dendritic application are discussed below.

5 Calcium-Sensitive Indicators

Ca^{2+} is a very important signaling molecule for many cellular processes (for a review, see [70]), and therefore determining intracellular Ca^{2+} dynamics is crucial for understanding brain activity as a whole. Consequently, the past few decades have seen the development of many Ca^{2+} indicators, including synthetic and more

Table 1
Investigations into dendritic activity using various fluorophores and imaging techniques

Indicator	Dendritic application	Relevant references
Cell impermeant		
Oregon Green BAPTA-1	Individual dendrite imaging	[19, 21, 29, 44, 74–77, 79–81, 85, 87, 88, 114]
Oregon Green BAPTA-6		[81]
Calcium Green	Individual dendrite imaging	[24, 42, 44, 79]
Fura-2	Individual dendrite imaging	[156]
Fluo-4/5	Individual dendrite imaging	[157]
Cell permeant		
Oregon Green BAPTA-1AM	Dendritic population imaging	[86, 117, 158]
Magnesium Green	Individual dendrite imaging	[159]
Genetically Encoded Calcium Indicators (GECI)		
GCaMP	Individual dendrite imaging	[29, 30, 92]

recently genetic indicators. When choosing the right indicator, various factors must be considered including the preparation (in vivo/in vitro), the question (single or population imaging), the application (cell permeant/impermeant), as well as the Kd (disassociation constant) of the indicator (Table 1). Details of synthetic and genetic Ca^{2+} indicators are described below.

5.1 Synthetic

Synthetic Ca^{2+} indicators were first developed over half a century ago [71, 72] and have since been continuously developed to improve dye delivery, neuronal labeling, and signal-to-noise ratio. Currently, the most popular synthetic Ca^{2+} indicators were developed in the laboratory of Roger Tsien [73] and involve the hybridization of Ca^{2+}-selective chelators (BAPTA or EGTA). The Oregon Green BAPTA dye family is typically the Ca^{2+} indicator of choice for dendritic imaging (Table 1), due to their high resting fluorescence, large signal-to-noise ratio, and easy delivery. Depending on the cell permeability of the dye, it has to be loaded either extra- or intracellularly (see below) to investigate important dendritic processes such as branch computation [19, 21, 29, 74–80], response to input patterns [81], plasticity [82, 83], and input modulation/association [84–88].

5.2 Genetic

The development of genetically encoded Ca^{2+} indicators (GECIs) by the laboratory of Roger Tsien [89] has greatly facilitated research into dendritic functioning. In particular, GECIs enable particular cell types and even subcellular compartments [90] to be specifically targeted and repetitively imaged over long periods of time [91]. This has led to exciting discoveries in dendritic function, especially in the awake preparation.

Over the past few years, numerous GECIs have been developed based on either changes in fluorescence resonance energy transfer (FRET) efficiency (Cameleon) or changes in the florescence of a single fluorophore (GCaMP). Due to their fast kinetics and high Ca^{2+} sensitivity, GCaMP indicators have been extremely popular in studies measuring dendritic activity [29, 30, 92]. Over the past decade, GCaMP indicators have been systematically improved [93] and advanced from signaling bursts of action potentials [94] to now being able to reliably undergo a change in fluorescence in response to a single action potential, as well as subthreshold voltage changes [95]. These advances in the sensitivity of GCaMP indicators have been important in improving our knowledge of dendritic function, especially in the awake preparation. For example, using GCaMP indicators, Xu and colleagues [30] illustrated that nonlinear dendritic processing is involved in the integration of correlated sensory and motor information during active sensation in the mammalian neocortex.

Due to being long lasting and having high expression levels, the most popular delivery modes of GECIs for dendritic imaging are lenti- [96] and adeno-associated [97] viral vectors. The viral vector chosen depends on the level of safety required (in general, lentiviral vectors require higher laboratory safety regulations) and the size of the virus genome (lentivirus—can contain larger genomes compared to adeno-associated vectors; 9 kb vs. 4.7 kb). One of the greatest advantages of GECIs is their ability to target specific cell types within identified brain regions. Cell type specificity can be achieved using various techniques including the use of cell-type-specific promoters or transgenic Cre recombinase driver lines (combined with compatible recombinase-dependent viral vectors). This is extremely advantageous for dendritic imaging, as it endows the investigator with the knowledge of the exact cell type where the dendritic branch stems from. Despite their obvious advantages including long expression levels (enabling chronic imaging) and cell-type specificity, GECIs have various serious shortcomings including nonspecific labeling, tissue damage during delivery, cytotoxicity, and slow kinetics. Although they are currently outperformed by synthetic Ca^{2+} indicators, GECIs are proving to hold great promise for further research in not only dendritic function but also cellular and network activity.

6 Sodium Indicators

Although the field of dendritic imaging is dominated by Ca^{2+} indicators, imaging Na^{+} fluxes can reveal a lot about dendritic activity since Na^{+} ions are typically the major charge carriers during action potentials and EPSPs. Historically, the fluorescent Na^+ indicator sodium-binding benzofuran isophthalate (SBFI) has been the primary Na^+ imaging dye of choice [98]. In contrast to other Na^+ indicators (for e.g., Sodium Green), SBFI reliably reports changes in intracellular Na^+ concentration in dendrites [99–104]. However, since SBFI can only be excited with wavelengths below 400 nm, specialized equipment is required for excitation and therefore Na^+ imaging is not commonly performed in many laboratories.

7 Green Fluorescent Protein

Since its discovery in 1962 [72], the green fluorescent protein (GFP) from the jellyfish Aequorea victoria has become one of the most widely used proteins in neuroscience. Used as a marker of gene expression and protein targeting, GFP's value lies in its highly visible and efficient emission (for a review, see [105]). This makes it an ideal marker for visualizing dendrites and GFP has therefore been used in a multitude of studies looking into dendritic morphology, both in vivo and in vitro. In particular, morphological changes to the dendritic processes and spines of GFP-infected neurons have been extensively studied in vivo through cranial windows [106] during sensory deprivation [107], memory formation [108, 109], and connectivity [110, 111]. The development of GFP-tagging has dramatically improved our knowledge of dendrites and neurons as a whole, a field which is continuously expanding with the development of specialized GFP-tagging strategies such as tagging pre- and post-synapses (mGRASP [111, 112]) and only dendritic branches [113].

8 Dendrite Fluorophore Loading Techniques

Depending on the particular question at hand, the dendrites of a single neuron or a network of neurons must be filled with a fluorophore. This can be achieved using "single-cell loading" or "bulk loading" as described below.

9 Single-Cell Loading

To image the activity of a single dendritic branch from an identified neuron, a single neuron must be loaded with the fluorophore. There are numerous techniques which have been used to achieve this, mostly involving a glass pipette as the mode of delivery. In vivo single-cell loading was initially achieved with the use of sharp electrodes which punctured the cell of interest and passively diffused the fluorophore intracellularly [42, 44]. Since the advent of the patch-clamp technique (discussed above), a popular mode of fluorophore delivery for individual dendritic imaging is via a whole-cell patch pipette [19, 21, 29, 74–81, 83–85, 87, 88, 114]. Using this method, the fluorophore is able to passively diffuse into a single cell causing little disruption to the cellular membrane. Another mode of loading single cells with fluorophores which causes little to no disruption of the cellular membrane is electroporation. This method involves fluorophore delivery by brief voltage pulses delivered through glass pipettes positioned close to target cells (for more details, see [115]). Since only a loose-seal patch is required, electroporation is typically used when loading multiple single cells, or post cell-attached recordings to label the recorded cell.

10 Bulk Loading

Loading multiple neurons with a fluorescence molecule can be achieved by using cell permeant fluorescent sensors. In brief, these molecules are able to diffuse into a cell where their ester bonds are cleaved, trapping them inside the cell. This enables easy loading of multiple cells by injecting a small bolus of dye into the region where the cells of interest reside. Unfortunately, however, the acetoxymethyl (AM) ester indicator diffuses not only into the cell with ease but also into intracellular compartments such as the endoplasmic reticulum [116] which leads to significant background fluorescence. A common AM dye which has been used to image from populations of dendrites is Oregon Green BAPTA 1AM. Bulk loading this dye into cortical layer 5 causes the dye to permeate all the cells within this layer; however, only the principal cell dendrites extend to the cortical surface; therefore imaging the upper layers (<500 μm depth) captures the Ca^{2+} changes from the dendrites of these neurons only [86, 117].

11 Imaging Techniques

Once loaded with a fluorophore, dendritic fluorescence can be measured via a multitude of different devices depending on the question at hand. The most popular techniques used when assessing dendritic activity are discussed below.

Quantifying the mobility and turnover rate of proteins within dendrites involves bleaching fluorescence-tagged proteins using a technique called fluorescence recovery after photobleaching (FRAP) [118]. In brief, the fluorescence of tagged proteins within a region of interest is bleached using a strong excitation laser and the diffusion of unbleached fluorescence-tagged protein into the region of interest is then measured. This technique has been used extensively in the past 30 years since its development, and was used to address the highly controversial topic of whether dendritic spines act as chemical compartments and isolate proteins from the parent dendrite [119, 120]. In these studies, a fluorescent protein (fluorescein dextran/green fluorescent protein) was bleached in the spine head and the time for diffusion from the dendrite was measured. The measured fluorescence recovery illustrated that spine necks are diffusion barriers.

A technique which is used extensively to visualize or image activity in dendrites is wide-field fluorescence microscopy. During wide-field microscopy, the sample is bathed in light typically from a mercury or xenon source and excitation with specific wavelengths is simply achieved by placement of a filter within the light path. The resulting fluorescence emission can be captured by numerous devices including charge-coupled detector-based (CCD) cameras, complementary metal-oxide-semiconductor-based cameras (CMOS), and photodiode arrays. This is an easily adopted technique which has great temporal resolution, and hence wide-field microscopy has been used in numerous laboratories to record fluorescence in dendrites [121–125]. However, since wide-field microscopy captures all photons from both within the focal plane and also from the surrounding out-of-focus areas, the resolution of the dendrite is drastically limited. A technique which has better spatial resolution (at the expense of temporal resolution) is laser-scanning microscopy (LSM). As the name implies, LSM requires a laser and a scanning system to excite the fluorophore using either one (confocal) or multiple (two-photon) photons. These different LSM techniques utilize different methods to optimize spatial resolution by eliminating out-of-focus light. In brief, confocal microscopy uses a spatial pinhole placed at the confocal plane of the lens [126], whereas two-photon microscopy employs the spatial profile of two-photon absorption [127, 128]. Although they have both been used extensively to image the activity of fluorophore-filled dendrites, the application of two-photon microscopy is widely popular since the longer excitation wavelengths (>800 nm) can probe deeper into brain tissue. This is particularly useful for in vivo preparations where two-photon microscopy has been used to assess the activity of dendrites during sensory perception and behavior [29, 30, 42, 44, 77, 80, 88, 129]. However, recovering the emitted fluorescence represents a limitation even during two-photon microscopy and therefore this technique is still limited to imaging from the upper 500 μm of the brain [130]. Until the attempts being made to develop imaging techniques that can penetrate deeper into brain tissue become mainstream (for a review, see [131]), probing brain function in the intact brain is currently limited to the upper regions of the cortex. Imaging dendritic activity in deeper brain regions is possible; however it requires physical removal of upper brain tissue [75].

12 Beyond the diffraction limit of light microscopy

Since the optical resolution of light microscopy is limited to approximately half of the excitation wavelength, many cellular processes, such as dendritic spines, are beyond the resolution of conventional light microscopy. An early and powerful approach to solving this spatial resolution limitation is electron microscopy (EM) which uses accelerated electrons as the source of illumination to achieve up to 1 nm resolution [132]. Historically, EM has been used to visualize the ultrastructure of neurons [133] revealing many important structural and connectivity functions including the site of synaptic connections [134, 135]. However, conventional EM reveals only a single snapshot or a small region of brain tissue [136]. The development of block-face imaging combined with serial sectioning [137] has enabled entire volumes of brain tissue to be imaged at higher resolution. Although it holds great promise in addressing ultrastructural connectivity of neuronal circuits, currently this technique is limited by the lengthy data analysis requirements. Another technique which is extensively used to image beyond the resolution of light microscopy is stimulated emission depletion (STED) microscopy [138]. This technique uses a doughnut-shaped laser beam for de-exciting the molecules activated by a co-aligned excitation laser (for a review, see [139]). STED microscopy has been used to study dendritic branch and spine morphology and more recently has been performed on live cells [140–143]. Imaging neuronal structures, including dendrites, is crucial to understanding brain function as a whole and therefore numerous approaches are currently being explored to improve and extend the spatial and temporal resolution and sensitivity of current imaging techniques.

13 Discussion/Future Directions

As discussed previously in this chapter, in the last 20 years technological advances in optics and microscopy, electrophysiology, and imaging have opened up the field of research on dendrites, allowing for a greater understanding of their structure and functions [144]. Due to the work of numerous labs, we have learned that the density and features of voltage-dependent channels and receptors can vary widely along dendritic compartments in different neuronal types [16, 145]. This differential distribution endows neurons with a variety of integration modes depending on the location and timing of the input which has a dramatic effect on the neuronal output [20, 23]. More recently, the application of these techniques to in vivo experiments has allowed to link dendritic events to specific behaviors [29, 30]. Other techniques led us to discover that biochemical signals, which are tightly coupled to electrical activity, can be restricted to spines and/or specific compartments created by the complicated morphology of dendrites [146, 147]. This biochemical compartmentalization is at the basis of localized changes in the integration features of dendritic compartments following plasticity [148–150]. In addition, we have learned that, not surprisingly, abnormalities in dendritic morphology and alterations in their excitability are at the basis of many neurological disorders, such as epilepsy, Alzheimer's disease, schizophrenia, and drug addiction [151, 152].

Despite these advances in our knowledge about the function of dendrites, there are still questions that need to be addressed. One such question is why there is a remarkable variety of dendritic morphologies in the brain and how they relate to the function of the various neuronal subtypes. Another similar question involves the reason for different distributions of channels and receptors in the various dendritic compartments and how they are created and maintained from a transcriptional/translational/biochemical point of view. In addition, we would like to completely comprehend the principles that govern the connectivity patterns that govern the interactions between the axons of presynaptic neurons and the dendrites of postsynaptic neurons. Answers to all these questions would ultimately allow us to connect the functions of single neuronal compartments to the activity of neuronal networks, by understanding how morphological and physiological dendritic features relate to higher-level functions.

Some technical developments are needed to unravel these questions. For example, optics and light source enhancements are needed to allow imaging beyond the superficial layers of the neocortex and improved indicators are required for imaging subthreshold membrane potential changes. For example, it would be useful to have more reliable red-shifted Ca^{2+}- and voltage-sensitive indicators; longer wavelength dyes would allow better penetration through brain tissue due to reduced light scattering as well as lower phototoxicity. The development of red-shifted indicators is also vital for experiments with three-color schemes, where it is required to image dendritic processes in combination with uncaging, photoswitching, or optogenetic activation of cells identified with a fluorescent protein [153, 154]. In addition, Ca^{2+}-sensitive indicators could benefit from faster kinetics (especially in the decay phase) and voltage-sensitive dyes could be improved by better photostability and higher sensitivity/signal-to-noise ratio; this last aspect could be also possibly accomplished by the use of optic elements, such as second harmonic generators. A further major area of technological improvements that is evolving daily resides in optogenetics, which allow light-activated proteins such as channelrhodopsins and calcium or voltage sensors to be expressed under specific promoter in individual neurons. We have now a large toolbox that allows us to activate specific brain areas with light-activatable proteins of various colors [39], as well as to record neuronal activity using genetically encoded calcium or voltage indicators [95, 155]. These techniques could be further developed to deliver these proteins to specific subcellular domains, such as identified dendritic compartments. Furthermore, techniques like GRASP (GFP Reconstitution Across Synaptic Partners) can eventually help us unravel the rules of pre- and postsynaptic connectivity patterns and how they ultimately control dendritic integration [111, 112]. Lastly, the ultimate challenge is the development of noninvasive imaging techniques that would allow for studying the functions of dendrites and neurons in humans. The combination of all these techniques will allow us to understand how dendrites, being the site of information transfer between neuronal input and output, contribute to brain function not only in physiological conditions, but also how their disruption can cause or be the result of various neurological diseases.

Acknowledgements

This work was supported by NIH grant NS069714 (to SG) and NHMRC grants 1063533 and 1085708 (to LMP).

References

1. Deiters O (1865) Untersuchungen über Gehirn und Rückenmark des Menschen und der Säugethiere. Vieweg, Braunschweig
2. Cajal SR (1889) The dynamic clamp comes of age. Med Pract 2:341–346
3. Brock LG, Eccles JC, Rall W (1951) Experimental investigations on the afferent fibres in muscle nerves. Proc R Soc Lond Ser B Biol Sci 138:453–475
4. Bullock TH, Hagiwara S (1957) Intracellular recording from the giant synapse of the squid. J Gen Physiol 40:565–577
5. Eccles RM (1955) Intracellular potentials recorded from a mammalian sympathetic ganglion. J Physiol 130:572–584
6. Rall W (1977) Core conductor theory and cable properties of neurons. In: Kandel E, Brookhart J, Mountcastle V (eds) Handbook of physiology, the nervous system, vol 1, 2nd edn, Cellular biology of neurons. Oxford University Press, Oxford, pp 39–97
7. Gray EG (1959) Axo-somatic and axo-dendritic synapses of the cerebral cortex: an electron microscope study. J Anat 93:420–433
8. Gray EG (1959) Electron microscopy of synaptic contacts on dendrite spines of the cerebral cortex. Nature 183:1592–1593
9. Segev I, Rall W (1998) Excitable dendrites and spines: earlier theoretical insights elucidate recent direct observations. Trends Neurosci 21:453–460. doi:10.1016/S0166-2236(98)01327-7
10. Neher E, Sakmann B (1976) Single-channel currents recorded from membrane of denervated frog muscle fibres. Nature 260: 799–802
11. Neher E, Sakmann B, Steinbach JH (1978) The extracellular patch clamp: a method for resolving currents through individual open channels in biological membranes. Pflügers Arch 375:219–228
12. Stuart GJ, Dodt HU, Sakmann B (1993) Patch-clamp recordings from the soma and dendrites of neurons in brain slices using infrared video microscopy. Pflügers Arch 423:511–518
13. Stuart GJ, Sakmann B (1994) Active propagation of somatic action potentials into neocortical pyramidal cell dendrites. Nature 367:69–72. doi:10.1038/367069a0
14. Magee JC, Johnston D (1995) Characterization of single voltage-gated Na + and Ca2+ channels in apical dendrites of rat CA1 pyramidal neurons. J Physiol 487(Pt 1):67–90

15. Stuart G, Häusser M (1994) Initiation and spread of sodium action potentials in cerebellar purkinje cells. Neuron 13:703–712. doi:10.1016/0896-6273(94)90037-X
16. Magee JC (2007) Dendritic voltage-gated ion channels. In: Stuart GJ, Spruston N, Häusser M (eds) Dendrites, 2nd edn. Oxford University Press, Oxford, pp 225–250
17. Dan Y, Poo M-M (2004) Spike timing-dependent plasticity of neural circuits. Neuron 44:23–30. doi:10.1016/j.neuron.2004.09.007
18. Larkum ME, Nevian T (2008) Synaptic clustering by dendritic signalling mechanisms. Curr Opin Neurobiol 18:321–331. doi:10.1016/j.conb.2008.08.013
19. Ariav G, Polsky A, Schiller J (2003) Submillisecond precision of the input-output transformation function mediated by fast sodium dendritic spikes in basal dendrites of CA1 pyramidal neurons. J Neurosci 23:7750–7758
20. Gasparini S, Magee JC (2006) State-dependent dendritic computation in hippocampal CA1 pyramidal neurons. J Neurosci 26:2088–2100. doi:10.1523/JNEUROSCI.4428-05.2006
21. Losonczy A, Magee JC (2006) Integrative properties of radial oblique dendrites in hippocampal CA1 pyramidal neurons. Neuron 50:291–307. doi:10.1016/j.neuron.2006.03.016
22. Milojkovic BA, Wuskell JP, Loew LM, Antic SD (2005) Initiation of sodium spikelets in basal dendrites of neocortical pyramidal neurons. J Membr Biol 208:155–169. doi:10.1007/s00232-005-0827-7
23. Larkum ME, Zhu JJ, Sakmann B (1999) A new cellular mechanism for coupling inputs arriving at different cortical layers. Nature 398:338–341. doi:10.1038/18686
24. Schiller J, Schiller Y, Stuart G, Sakmann B (1997) Calcium action potentials restricted to distal apical dendrites of rat neocortical pyramidal neurons. J Physiol 505(Pt 3):605–616
25. Branco T, Häusser M (2011) Synaptic integration gradients in single cortical pyramidal cell dendrites. Neuron 69:885–892. doi:10.1016/j.neuron.2011.02.006
26. Larkum ME, Nevian T, Sandler M et al (2009) Synaptic integration in tuft dendrites of layer 5 pyramidal neurons: a new unifying principle. Science 325:756–760. doi:10.1126/science.1171958
27. Polsky A, Mel BW, Schiller J (2004) Computational subunits in thin dendrites of pyramidal cells. Nat Neurosci 7:621–627. doi:10.1038/nn1253
28. Lavzin M, Rapoport S, Polsky A et al (2012) Nonlinear dendritic processing determines angular tuning of barrel cortex neurons in vivo. Nature 490:397–401. doi:10.1038/nature11451
29. Palmer LM, Shai AS, Reeve JE et al (2014) NMDA spikes enhance action potential generation during sensory input. Nat Neurosci 17:383–390. doi:10.1038/nn.3646
30. Xu N, Harnett MT, Williams SR et al (2012) Nonlinear dendritic integration of sensory and motor input during an active sensing task. Nature 492:247–251. doi:10.1038/nature11601
31. Rubio-Garrido P, Pérez-de-Manzo F, Porrero C et al (2009) Thalamic input to distal apical dendrites in neocortical layer 1 is massive and highly convergent. Cereb Cortex 19:2380–2395. doi:10.1093/cercor/bhn259
32. Cauller LJ, Clancy B, Connors BW (1998) Backward cortical projections to primary somatosensory cortex in rats extend long horizontal axons in layer I. J Comp Neurol 390:297–310
33. Witter MP, Groenewegen HJ (1986) Connections of the parahippocampal cortex in the cat. III. Cortical and thalamic efferents. J Comp Neurol 252:1–31. doi:10.1002/cne.902520102
34. Larkman AU (1991) Dendritic morphology of pyramidal neurones of the visual cortex of the rat: I. Branching patterns. J Comp Neurol 306:307–319. doi:10.1002/cne.903060207
35. Felleman DJ, Van Essen DC (1991) Distributed hierarchical processing in the primate cerebral cortex. Cereb Cortex 1:1–47
36. Amaral DG, Lavenex P (2006) Hippocampal neuroanatomy. In: Andersen P, Morris R, Bliss T, O'Keefe J (eds) Hippocampus book. Oxford University Press, Oxford, pp 37–114
37. Kim J-M, Hwa J, Garriga P et al (2005) Light-driven activation of beta 2-adrenergic receptor signaling by a chimeric rhodopsin containing the beta 2-adrenergic receptor cytoplasmic loops. Biochemistry 44:2284–2292. doi:10.1021/bi048328i
38. Nagel G, Szellas T, Huhn W et al (2003) Channelrhodopsin-2, a directly light-gated cation-selective membrane channel. Proc Natl Acad Sci U S A 100:13940–13945. doi:10.1073/pnas.1936192100
39. Fenno L, Yizhar O, Deisseroth K (2011) The development and application of optogenetics. Annu Rev Neurosci 34:389–412. doi:10.1146/annurev-neuro-061010-113817

40. Losonczy A, Zemelman BV, Vaziri A, Magee JC (2010) Network mechanisms of theta related neuronal activity in hippocampal CA1 pyramidal neurons. Nat Neurosci 13:967–972. doi:10.1038/nn.2597
41. Buzsáki G, Penttonen M, Nádasdy Z, Bragin A (1996) Pattern and inhibition-dependent invasion of pyramidal cell dendrites by fast spikes in the hippocampus in vivo. Proc Natl Acad Sci U S A 93:9921–9925
42. Helmchen F, Svoboda K, Denk W, Tank DW (1999) In vivo dendritic calcium dynamics in deep-layer cortical pyramidal neurons. Nat Neurosci 2:989–996. doi:10.1038/14788
43. Smith SL, Smith IT, Branco T, Häusser M (2013) Dendritic spikes enhance stimulus selectivity in cortical neurons in vivo. Nature 503:115–120. doi:10.1038/nature12600
44. Svoboda K, Helmchen F, Denk W, Tank DW (1999) Spread of dendritic excitation in layer 2/3 pyramidal neurons in rat barrel cortex in vivo. Nat Neurosci 2:65–73. doi:10.1038/4569
45. Edwards FA, Konnerth A, Sakmann B, Takahashi T (1989) A thin slice preparation for patch clamp recordings from neurones of the mammalian central nervous system. Pflügers Arch 414:600–612
46. Hamill OP, Marty A, Neher E et al (1981) Improved patch-clamp techniques for high-resolution current recording from cells and cell-free membrane patches. Pflügers Arch 391:85–100
47. Moyer JR, Brown TH (2002) Patch-clamp techniques applied to brain slices. Patch-clamp analysis. Humana Press, Totowa, NJ, pp 135–194
48. Davie JT, Kole MHP, Letzkus JJ et al (2006) Dendritic patch-clamp recording. Nat Protoc 1:1235–1247. doi:10.1038/nprot.2006.164
49. Kirov SA, Petrak LJ, Fiala JC, Harris KM (2004) Dendritic spines disappear with chilling but proliferate excessively upon rewarming of mature hippocampus. Neuroscience 127:69–80. doi:10.1016/j.neuroscience.2004.04.053
50. Feig S, Lipton P (1990) N-methyl-D-aspartate receptor activation and Ca2+ account for poor pyramidal cell structure in hippocampal slices. J Neurochem 55:473–483
51. Ganong AH, Lanthorn TH, Cotman CW (1983) Kynurenic acid inhibits synaptic and acidic amino acid-induced responses in the rat hippocampus and spinal cord. Brain Res 273:170–174
52. Aghajanian GK, Rasmussen K (1989) Intracellular studies in the facial nucleus illustrating a simple new method for obtaining viable motoneurons in adult rat brain slices. Synapse 3:331–338. doi:10.1002/syn.890030406
53. Rothman SM (1985) The neurotoxicity of excitatory amino acids is produced by passive chloride influx. J Neurosci 5:1483–1489
54. Kuenzi FM, Fitzjohn SM, Morton RA et al (2000) Reduced long-term potentiation in hippocampal slices prepared using sucrose-based artificial cerebrospinal fluid. J Neurosci Methods 100:117–122
55. Magee JC, Avery RB, Christie BR, Johnston D (1996) Dihydropyridine-sensitive, voltage-gated Ca2+ channels contribute to the resting intracellular Ca2+ concentration of hippocampal CA1 pyramidal neurons. J Neurophysiol 76:3460–3470
56. Dodt H-U, Zieglgänsberger W (1990) Visualizing unstained neurons in living brain slices by infrared DIC-videomicroscopy. Brain Res 537:333–336. doi:10.1016/0006-8993(90)90380-T
57. Dodt H-U, Frick A, Kampe K, Zieglgänsberger W (1998) NMDA and AMPA receptors on neocortical neurons are differentially distributed. Eur J Neurosci 10:3351–3357. doi:10.1046/j.1460-9568.1998.00338.x
58. Spruston N, Stuart G, Hausser M (2007) Dendritic integration. In: Stuart GJ, Spruston N, Häusser M (eds) Dendrites, 2nd edn. Oxford University Press, Oxford, pp 351–399
59. Robinson HPC, Kawai N (1993) Injection of digitally synthesized synaptic conductance transients to measure the integrative properties of neurons. J Neurosci Methods 49:157–165. doi:10.1016/0165-0270(93)90119-C
60. Sharp AA, O'Neil MB, Abbott LF, Marder E (1993) The dynamic clamp: artificial conductances in biological neurons. Trends Neurosci 16:389–394. doi:10.1016/0166-2236(93)90004-6
61. Prinz AA, Abbott LF, Marder E (2004) The dynamic clamp comes of age. Trends Neurosci 27:218–224. doi:10.1016/j.tins.2004.02.004
62. Gasparini S, Migliore M, Magee JC (2004) On the initiation and propagation of dendritic spikes in CA1 pyramidal neurons. J Neurosci 24:11046–11056. doi:10.1523/JNEUROSCI.2520-04.2004
63. Ledergerber D, Larkum ME (2010) Properties of layer 6 pyramidal neuron apical dendrites. J Neurosci 30:13031–13044. doi:10.1523/JNEUROSCI.2254-10.2010
64. Nevian T, Larkum ME, Polsky A, Schiller J (2007) Properties of basal dendrites of layer 5

pyramidal neurons: a direct patch-clamp recording study. Nat Neurosci 10:206–214. doi:10.1038/nn1826

65. Colbert CM, Johnston D (1998) Protein kinase C activation decreases activity-dependent attenuation of dendritic Na + current in hippocampal CA1 pyramidal neurons. J Neurophysiol 79:491–495
66. Gasparini S, Magee JC (2002) Phosphorylation-dependent differences in the activation properties of distal and proximal dendritic Na + channels in rat CA1 hippocampal neurons. J Physiol 541:665–672. doi:10.1113/jphysiol.2002.020503
67. Andrasfalvy BK, Magee JC (2001) Distance-dependent increase in AMPA receptor number in the dendrites of adult hippocampal CA1 pyramidal neurons. J Neurosci 21:9151–9159
68. Horn R, Marty A (1988) Muscarinic activation of ionic currents measured by a new whole-cell recording method. J Gen Physiol 92:145–159
69. Hoffman DA, Magee JC, Colbert CM, Johnston D (1997) K+ channel regulation of signal propagation in dendrites of hippocampal pyramidal neurons. Nature 387:869–875. doi:10.1038/43119
70. Berridge MJ (1998) Neuronal calcium signaling. Neuron 21:13–26
71. Ashley CC, Ridgway EB (1968) Simultaneous recording of membrane potential, calcium transient and tension in single muscle fibers. Nature 219:1168–1169
72. Shimomura O, Johnson FH, Saiga Y (1962) Extraction, purification and properties of aequorin, a bioluminescent protein from the luminous hydromedusan, Aequorea. J Cell Comp Physiol 59:223–239
73. Tsien RY (1980) New calcium indicators and buffers with high selectivity against magnesium and protons: design, synthesis, and properties of prototype structures. Biochemistry 19:2396–2404
74. Gordon U, Polsky A, Schiller J (2006) Plasticity compartments in basal dendrites of neocortical pyramidal neurons. J Neurosci 26:12717–12726. doi:10.1523/JNEURO SCI. 3502-06.2006
75. Grienberger C, Chen X, Konnerth A (2014) NMDA receptor-dependent multidendrite Ca (2+) spikes required for hippocampal burst firing in vivo. Neuron 81:1274–1281. doi:10.1016/j.neuron.2014.01.014
76. Hill DN, Varga Z, Jia H et al (2013) Multibranch activity in basal and tuft dendrites during firing of layer 5 cortical neurons in vivo. Proc Natl Acad Sci U S A 110:13618–13623. doi:10.1073/pnas.1312599110
77. Jia H, Rochefort NL, Chen X, Konnerth A (2010) Dendritic organization of sensory input to cortical neurons in vivo. Nature 464:1307–1312. doi:10.1038/nature08947
78. Larkum ME, Waters J, Sakmann B, Helmchen F (2007) Dendritic spikes in apical dendrites of neocortical layer 2/3 pyramidal neurons. J Neurosci 27:8999–9008. doi:10.1523/JNEUROSCI.1717-07.2007
79. Schiller J, Major G, Koester HJ, Schiller Y (2000) NMDA spikes in basal dendrites of cortical pyramidal neurons. Nature 404:285–289. doi:10.1038/35005094
80. Varga Z, Jia H, Sakmann B, Konnerth A (2011) Dendritic coding of multiple sensory inputs in single cortical neurons in vivo. Proc Natl Acad Sci U S A 108:15420–15425. doi:10.1073/pnas.1112355108
81. Takahashi H, Magee JC (2009) Pathway interactions and synaptic plasticity in the dendritic tuft regions of CA1 pyramidal neurons. Neuron 62:102–111. doi:10.1016/j.neuron.2009.03.007
82. Holthoff K, Kovalchuk Y, Konnerth A (2006) Dendritic spikes and activity-dependent synaptic plasticity. Cell Tissue Res 326:369–377. doi:10.1007/s00441-006-0263-8
83. Losonczy A, Makara JK, Magee JC (2008) Compartmentalized dendritic plasticity and input feature storage in neurons. Nature 452:436–441. doi:10.1038/nature06725
84. Gasparini S, Losonczy A, Chen X et al (2007) Associative pairing enhances action potential back-propagation in radial oblique branches of CA1 pyramidal neurons. J Physiol 580: 787–800. doi:10.1113/jphysiol.2006.121343
85. Harnett MT, Makara JK, Spruston N et al (2012) Synaptic amplification by dendritic spines enhances input cooperativity. Nature 491:599–602. doi:10.1038/nature11554
86. Murayama M, Pérez-Garci E, Nevian T et al (2009) Dendritic encoding of sensory stimuli controlled by deep cortical interneurons. Nature 457:1137–1141. doi:10.1038/nature07663
87. Waters J, Helmchen F (2004) Boosting of action potential backpropagation by neocortical network activity in vivo. J Neurosci 24:11127–11136. doi:10.1523/JNEURO SCI.2933-04.2004
88. Waters J, Larkum M, Sakmann B, Helmchen F (2003) Supralinear Ca2+ influx into dendritic tufts of layer 2/3 neocortical pyramidal neurons in vitro and in vivo. J Neurosci 23:8558–8567

89. Miyawaki A, Llopis J, Heim R et al (1997) Fluorescent indicators for Ca2+ based on green fluorescent proteins and calmodulin. Nature 388:882–887. doi:10.1038/42264
90. Dreosti E, Odermatt B, Dorostkar MM, Lagnado L (2009) A genetically encoded reporter of synaptic activity in vivo. Nat Methods 6:883–889. doi:10.1038/nmeth.1399
91. Margolis DJ, Lütcke H, Schulz K et al (2012) Reorganization of cortical population activity imaged throughout long-term sensory deprivation. Nat Neurosci 15:1539–1546. doi:10.1038/nn.3240
92. Harnett MT, Xu N-L, Magee JC, Williams SR (2013) Potassium channels control the interaction between active dendritic integration compartments in layer 5 cortical pyramidal neurons. Neuron 79:516–529. doi:10.1016/j.neuron.2013.06.005
93. Akerboom J, Chen T-W, Wardill TJ et al (2012) Optimization of a GCaMP calcium indicator for neural activity imaging. J Neurosci 32:13819–13840. doi:10.1523/JNEUROSCI.2601-12.2012
94. Mao T, O'Connor DH, Scheuss V et al (2008) Characterization and subcellular targeting of GCaMP-type genetically-encoded calcium indicators. PLoS One 3:e1796. doi:10.1371/journal.pone.0001796
95. Chen T-W, Wardill TJ, Sun Y et al (2013) Ultrasensitive fluorescent proteins for imaging neuronal activity. Nature 499:295–300. doi:10.1038/nature12354
96. Dittgen T, Nimmerjahn A, Komai S et al (2004) Lentivirus-based genetic manipulations of cortical neurons and their optical and electrophysiological monitoring in vivo. Proc Natl Acad Sci U S A 101:18206–18211. doi:10.1073/pnas.0407976101
97. Monahan PE, Samulski RJ (2000) Adeno-associated virus vectors for gene therapy: more pros than cons? Mol Med Today 6:433–440
98. Minta A, Tsien RY (1989) Fluorescent indicators for cytosolic sodium. J Biol Chem 264:19449–19457
99. Callaway JC, Ross WN (1997) Spatial distribution of synaptically activated sodium concentration changes in cerebellar Purkinje neurons. J Neurophysiol 77:145–152
100. Knöpfel T, Anchisi D, Alojado ME et al (2000) Elevation of intradendritic sodium concentration mediated by synaptic activation of metabotropic glutamate receptors in cerebellar Purkinje cells. Eur J Neurosci 12:2199–2204
101. Lasser-Ross N, Ross WN (1992) Imaging voltage and synaptically activated sodium transients in cerebellar Purkinje cells. Proc Biol Sci 247:35–39. doi:10.1098/rspb.1992.0006
102. Mittmann T, Linton SM, Schwindt P, Crill W (1997) Evidence for persistent Na + current in apical dendrites of rat neocortical neurons from imaging of Na + -sensitive dye. J Neurophysiol 78:1188–1192
103. Myoga MH, Beierlein M, Regehr WG (2009) Somatic spikes regulate dendritic signaling in small neurons in the absence of backpropagating action potentials. J Neurosci 29:7803–7814. doi:10.1523/JNEUROSCI.0030-09.2009
104. Rose CR, Kovalchuk Y, Eilers J, Konnerth A (1999) Two-photon Na + imaging in spines and fine dendrites of central neurons. Pflügers Arch 439:201–207
105. Tsien RY (1998) The green fluorescent protein. Annu Rev Biochem 67:509–544. doi:10.1146/annurev.biochem.67.1.509
106. Holtmaat A, Bonhoeffer T, Chow DK et al (2009) Long-term, high-resolution imaging in the mouse neocortex through a chronic cranial window. Nat Protoc 4:1128–1144. doi:10.1038/nprot.2009.89
107. Zuo Y, Yang G, Kwon E, Gan W-B (2005) Long-term sensory deprivation prevents dendritic spine loss in primary somatosensory cortex. Nature 436:261–265. doi:10.1038/nature03715
108. Liston C, Cichon JM, Jeanneteau F et al (2013) Circadian glucocorticoid oscillations promote learning-dependent synapse formation and maintenance. Nat Neurosci 16:698–705. doi:10.1038/nn.3387
109. Xu T, Yu X, Perlik AJ et al (2009) Rapid formation and selective stabilization of synapses for enduring motor memories. Nature 462:915–919. doi:10.1038/nature08389
110. Chen JL, Villa KL, Cha JW et al (2012) Clustered dynamics of inhibitory synapses and dendritic spines in the adult neocortex. Neuron 74:361–373. doi:10.1016/j.neuron.2012.02.030
111. Druckmann S, Feng L, Lee B et al (2014) Structured synaptic connectivity between hippocampal regions. Neuron 81:629–640. doi:10.1016/j.neuron.2013.11.026
112. Kim J, Zhao T, Petralia RS et al (2012) mGRASP enables mapping mammalian synaptic connectivity with light microscopy. Nat Methods 9:96–102. doi:10.1038/nmeth.1784
113. Kameda H, Furuta T, Matsuda W et al (2008) Targeting green fluorescent protein to dendritic membrane in central neurons. Neurosci Res 61:79–91. doi:10.1016/j.neures.2008.01.014

114. Gasparini S (2011) Distance- and activity-dependent modulation of spike back-propagation in layer V pyramidal neurons of the medial entorhinal cortex. J Neurophysiol 105:1372–1379. doi:10.1152/jn.00014.2010
115. Nevian T, Helmchen F (2007) Calcium indicator loading of neurons using single-cell electroporation. Pflügers Arch 454:675–688. doi:10.1007/s00424-007-0234-2
116. Silver RA, Whitaker M, Bolsover SR (1992) Intracellular ion imaging using fluorescent dyes: artefacts and limits to resolution. Pflügers Arch 420:595–602
117. Palmer LM, Schulz JM, Murphy SC et al (2012) The cellular basis of GABA(B)-mediated interhemispheric inhibition. Science 335:989–993. doi:10.1126/science.1217276
118. Reits EA, Neefjes JJ (2001) From fixed to FRAP: measuring protein mobility and activity in living cells. Nat Cell Biol 3:E145–E147. doi:10.1038/35078615
119. Majewska A, Tashiro A, Yuste R (2000) Regulation of spine calcium dynamics by rapid spine motility. J Neurosci 20:8262–8268
120. Svoboda K, Tank DW, Denk W (1996) Direct measurement of coupling between dendritic spines and shafts. Science 272:716–719
121. Antic SD (2003) Action potentials in basal and oblique dendrites of rat neocortical pyramidal neurons. J Physiol 550:35–50. doi:10.1113/jphysiol.2002.033746
122. Djurisic M, Popovic M, Carnevale N, Zecevic D (2008) Functional structure of the mitral cell dendritic tuft in the rat olfactory bulb. J Neurosci 28:4057–4068. doi:10.1523/JNEUROSCI.5296-07.2008
123. Milojkovic BA, Zhou W-L, Antic SD (2007) Voltage and calcium transients in basal dendrites of the rat prefrontal cortex. J Physiol 585:447–468. doi:10.1113/jphysiol.2007.142315
124. Palmer LM, Stuart GJ (2009) Membrane potential changes in dendritic spines during action potentials and synaptic input. J Neurosci 29:6897–6903. doi:10.1523/JNEUROSCI.5847-08.2009
125. Ross WN, Werman R (1987) Mapping calcium transients in the dendrites of Purkinje cells from the guinea-pig cerebellum in vitro. J Physiol 389:319–336
126. Fine A, Amos WB, Durbin RM, McNaughton PA (1988) Confocal microscopy: applications in neurobiology. Trends Neurosci 11:346–351
127. Denk W, Strickler JH, Webb WW (1990) Two-photon laser scanning fluorescence microscopy. Science 248:73–76
128. Göppert-Mayer M (1931) Über Elementarakte mit zwei Quantensprüngen. Ann Phys 401:273–294. doi:10.1002/andp.19314010303
129. Gentet LJ, Kremer Y, Taniguchi H et al (2012) Unique functional properties of somatostatin-expressing GABAergic neurons in mouse barrel cortex. Nat Neurosci 15:607–612. doi:10.1038/nn.3051
130. Helmchen F, Denk W (2005) Deep tissue two-photon microscopy. Nat Methods 2:932–940. doi:10.1038/nmeth818
131. Wilt BA, Burns LD, Wei Ho ET et al (2009) Advances in light microscopy for neuroscience. Annu Rev Neurosci 32: 435–506. doi:10.1146/annurev.neuro.051508.135540
132. Sosinsky GE, Giepmans BNG, Deerinck TJ et al (2007) Markers for correlated light and electron microscopy. Methods Cell Biol 79:575–591. doi:10.1016/S0091-679X(06)79023-9
133. Palay SL, Palade GE (1955) The fine structure of neurons. J Biophys Biochem Cytol 1:69–88
134. Palay SL (1956) Synapses in the central nervous system. J Biophys Biochem Cytol 2:193–202
135. De Robertis ED, Bennett HS (1955) Some features of the submicroscopic morphology of synapses in frog and earthworm. J Biophys Biochem Cytol 1:47–58
136. Harris KM, Weinberg RJ (2012) Ultrastructure of synapses in the mammalian brain. Cold Spring Harb Perspect Biol. 4(5). pii: a005587. doi: 10.1101/cshperspect.a005587
137. Denk W, Horstmann H (2004) Serial block-face scanning electron microscopy to reconstruct three-dimensional tissue nanostructure. PLoS Biol 2:e329. doi:10.1371/journal.pbio.0020329
138. Hell SW, Wichmann J (1994) Breaking the diffraction resolution limit by stimulated emission: stimulated-emission-depletion fluorescence microscopy. Opt Lett 19:780–782
139. Tønnesen J, Nägerl UV (2013) Superresolution imaging for neuroscience. Exp Neurol 242:33–40. doi:10.1016/j.expneurol.2012.10.004
140. Ding JB, Takasaki KT, Sabatini BL (2009) Supraresolution imaging in brain slices using stimulated-emission depletion two-photon laser scanning microscopy. Neuron 63:429–437. doi:10.1016/j.neuron.2009.07.011
141. Nägerl UV, Willig KI, Hein B et al (2008) Live-cell imaging of dendritic spines by STED

microscopy. Proc Natl Acad Sci U S A 105:18982–18987. doi:10.1073/pnas.0810028105

142. Takasaki KT, Ding JB, Sabatini BL (2013) Live-cell superresolution imaging by pulsed STED two-photon excitation microscopy. Biophys J 104:770–777. doi:10.1016/j.bpj.2012.12.053
143. Tønnesen J, Katona G, Rózsa B, Nägerl UV (2014) Spine neck plasticity regulates compartmentalization of synapses. Nat Neurosci 17:678–685. doi:10.1038/nn.3682
144. Stuart G, Spruston N, Hausser M (2007) Dendrites, 2nd edn. Oxford University Press, Oxford
145. Silver R, Farrant M (2007) Neurotransmitter-gated channels in dendrites. In: Stuart GJ, Spruston N, Hausser M (eds) Dendrites, 2nd edn. Oxford University Press, Oxford, pp 190–223
146. Carter A, Sabatini B (2007) Spine calcium signaling. In: Stuart GJ, Spruston N, Häusser M (eds) Dendrites, 2nd edn. Oxford University Press, Oxford, pp 287–308
147. Helmchen F (2007) Biochemical compartmentalization in dendrites. In: Stuart GJ, Spruston N, Häusser M (eds) Dendrites, 2nd edn. Oxford University Press, Oxford, pp 251–285
148. Frick A, Magee J, Johnston D (2004) LTP is accompanied by an enhanced local excitability of pyramidal neuron dendrites. Nat Neurosci 7:126–135. doi:10.1038/nn1178
149. Mainen ZF, Abbott LF (2007) Functional plasticity at dendritic synapses. In: Stuart GJ, Spruston N, Häusser M (eds) Dendrites, 2nd edn. Oxford University Press, Oxford, pp 465–498
150. Makara JK, Losonczy A, Wen Q, Magee JC (2009) Experience-dependent compartmentalized dendritic plasticity in rat hippocampal CA1 pyramidal neurons. Nat Neurosci 12:1485–1487. doi:10.1038/nn.2428
151. Bernard C, Shah M, Johnston D (2007) Dendrites and disease. In: Stuart GJ, Spruston N, Hausser M (eds) Dendrites, 2nd edn. Oxford University Press, Oxford, pp 531–550
152. Palmer LM (2014) Dendritic integration in pyramidal neurons during network activity and disease. Brain Res Bull 103:2–10. doi:10.1016/j.brainresbull.2013.09.010
153. Deisseroth K, Schnitzer MJ (2013) Engineering approaches to illuminating brain structure and dynamics. Neuron 80:568–577. doi:10.1016/j.neuron.2013.10.032
154. Oheim M, van't Hoff M, Feltz A et al (2014) New red-fluorescent calcium indicators for optogenetics, photoactivation and multicolor imaging. Biochim Biophys Acta 1843:2284–2306. doi:10.1016/j.bbamcr.2014.03.010
155. Cao G, Platisa J, Pieribone VA et al (2013) Genetically targeted optical electrophysiology in intact neural circuits. Cell 154:904–913. doi:10.1016/j.cell.2013.07.027
156. Garaschuk O, Griesbeck O, Konnerth A (2007) Troponin C-based biosensors: a new family of genetically encoded indicators for in vivo calcium imaging in the nervous system. Cell Calcium 42:351–361. doi:10.1016/j.ceca.2007.02.011
157. Kitamura K, Häusser M (2011) Dendritic calcium signaling triggered by spontaneous and sensory-evoked climbing fiber input to cerebellar Purkinje cells in vivo. J Neurosci 31:10847–10858. doi:10.1523/JNEUROSCI.2525-10.2011
158. Murayama M, Larkum ME (2009) Enhanced dendritic activity in awake rats. Proc Natl Acad Sci U S A 106:20482–20486. doi:10.1073/pnas.0910379106
159. Major G, Polsky A, Denk W et al (2008) Spatiotemporally graded NMDA spike/plateau potentials in basal dendrites of neocortical pyramidal neurons. J Neurophysiol 99:2584–2601. doi:10.1152/jn.00011.2008

Chapter 6

Patch-Clamp Recording from Myelinated Central Axons

Maarten H.P. Kole and Marko A. Popovic

Abstract

Axons perform the main fundamental electrical operations of neurons. Emerging near the soma, axons integrate synaptic potentials, convert these into action potentials, and conduct the output signal to the presynaptic terminals. With the establishment of patch-clamp recording techniques in brain slices in combination with high-resolution microscopy, it has now become possible to visually target patch-clamp electrodes to various domains of the axon. This chapter provides an overview of the methodology for obtaining patch-clamp recordings from axons, with a focus on their unmyelinated regions, including the axon initial segment and axonal cut endings. Axonal patch-clamp recordings are a prerequisite for the study of the biophysics and diversity of axonal voltage-gated ion channels; in particular, high-temporal resolution, low-noise voltage recordings offer detailed insights into the fast computational properties of central nervous system axons.

Key words Axon, Myelin, Bleb, Voltage clamp, Action potential

1 Introduction

1.1 Axonal Recordings

Many of our current biophysical insights into ionic membrane conductances have been obtained by electrophysiological recordings from axons. In the mid-twentieth century, recordings from the large-diameter giant squid axon established the mathematical and physiological basis of action potential generation and conduction [1]. With the establishment of patch-clamp recording techniques in brain slices in combination with high-resolution video microscopy, patch-clamp recordings could be targeted to mammalian neurons and their small subcellular compartments, including dendrites [2, 3]. More recently, it has also become possible to reliably record from the small-diameter axons of mammals (~1 μm in diameter). Several developments have been critical to this achievement, such as the continual improvement of optical techniques for live fluorescence imaging, including confocal and two-photon microscopy; improvements in slice preparation methods [4]; and the observation that swollen cut endings of axons, i.e., axon blebs, provide stable sites for recording [5]. Fluorescence-guided, intracellular patch-clamp

Alon Korngreen (ed.), *Advanced Patch-Clamp Analysis for Neuroscientists*, Neuromethods, vol. 113,
DOI 10.1007/978-1-4939-3411-9_6, © Springer Science+Business Media New York 2016

recordings of axons are particularly suitable for the study of the temporally and spatially precise action potential (AP) initiation, modulation, and propagation (including subthreshold signals) through the axon, as well as the ligand- and voltage-gated ion channels which define axonal excitability. Excellent papers summarizing the methods for obtaining patch-clamp recordings from visualized presynaptic boutons or unmyelinated hippocampal axonal branches are available elsewhere [6, 7]. The aim of this chapter is to provide practical and key theoretical insights into the techniques for obtaining current- and voltage-clamp recordings from myelinated axons in the central nervous system (CNS). The ion channels expressed in myelinated axons are highly clustered and localized to specific domains, including the axon initial segment (AIS), presynaptic terminals, the juxtaparanode, and nodes of Ranvier. The latter two sites have remained so far inaccessible to direct patch recordings, due to their small dimensions and/or coverage by glial cells. The practical steps for recently successful patch-clamp recordings from axonal cut ends lacking myelin due to the slice-cutting procedure will be illustrated, providing unique insights into the biophysical properties of axonal voltage-gated ion channels.

1.2 The Axon Bleb

Axon blebs, also called "*axonal retraction bulbs*" or "*axonal cut ends*," form as an immediate consequence of the brain slicing procedure and are characterized by oval-shaped swellings at the cut end of axons. Neurons survive efficiently when axons are cut beyond 20 μm from the soma, and although peripheral nervous system axonal blebs initiate regenerative sprouting, this typically does not occur in the CNS [8]. The molecular mechanisms of bleb formation are not fully understood but the first phase of membrane resealing is a calcium-dependent process [9]. During the first 30 min following transection, there is a dramatic neuronal depolarization associated with the ionic leak through the damaged membrane bilayer [10]. In conjunction with this depolarization, calcium ions flow into the axon, causing disassembly of the neurofilaments and microtubules via calcium-dependent protease cleavage activity [9]. After the axonal membrane successfully reseals, the retraction bulb may further increase in size. In peripheral nerves, end-bulbs form active domains of accumulation for both Nav and Kv7 ion channels, from where spontaneous ectopic AP firing can originate [11]. Axon cutting in vivo and in vitro is often used to study the mechanisms of axonal injury and regeneration. For electrophysiological studies, however, axonal blebs provide a unique intracellular access point to the axonal membrane. To date, there is no evidence that active domain formation occurs in CNS axons, consistent with their lack of anchoring filament structures such as Ankyrin G or β-IV spectrin. Whole-cell, cell-attached, and outside-out recordings from AIS end-bulbs indicate, however, that Nav channel expression is nevertheless site-dependent [12–14]. Furthermore, recent work on Kv7 channels in the AIS shows that the distance-dependent ion channel density gradient is similar between the intact and cut-end initial segment [15]. These observations suggest that in the absence of anchoring protein and neurofilaments, CNS axon retraction bulbs passively accumulate, via diffusion, the voltage-gated ion channels from near intact sites proximal to the swelling. More research, however, is required to understand the precise molecular identity of axon swellings.

2 Materials

2.1 Preparation of Slices

To obtain axonal recordings from layer 5 axons, we cut slices parasagittally. Neocortical columns are perpendicular to the pia and the large primary layer 5 axons, projecting to the corpus callosum and subcortical areas, often have a trajectory in the same plane as the primary dendrite (Fig. 1). Therefore, when slicing with an angle of 15°, oriented to the parasagittal plane of the hemisphere, a range of orientations of axons and apical dendrites is obtained relative to the slice surface, allowing for a selection of

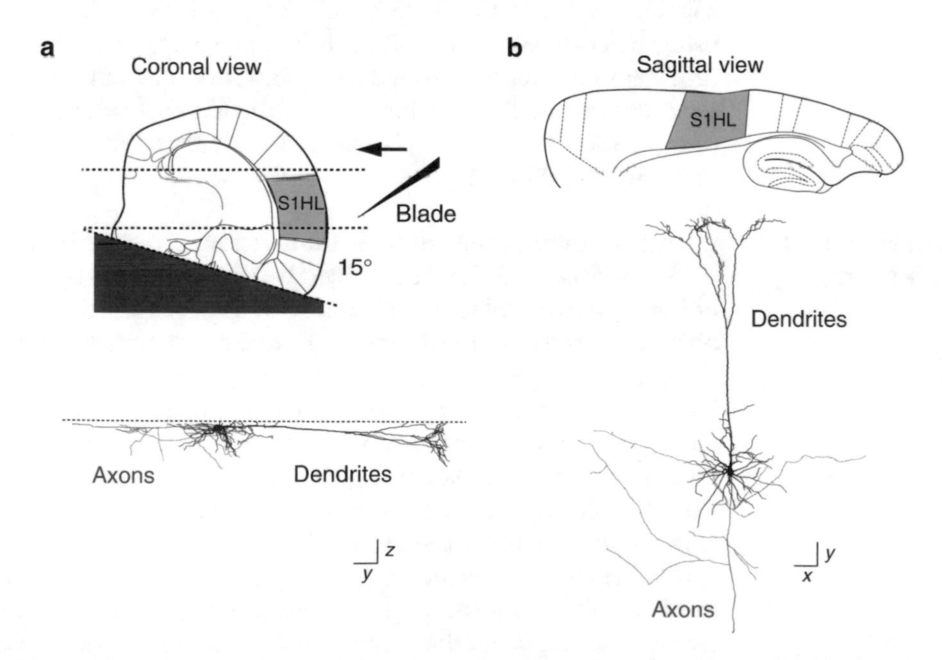

Fig. 1 (**a**) *Top*, schematic coronal view of the brain during slice preparation, illustrating the 15° angle of the cutting block generating a range of slices with distinct angles of the dendro-soma-axonal axis in relation to the slice surface. Hemisphere is glued on the midline. The slice collection range is about 2 mm. *Bottom*, *x*-projections of a layer 5 neuron morphology illustrating the critical role of surface cutting of axons (and dendrites). (**b**) Sagittal schematic view of the area in which layer 5 pyramidal neurons are targeted for axonal recordings (*grey*), approximating the primary somatosensory cortex of the hindlimb (S1HL). *Bottom*, *z*-projection of the same layer 5 neuron as in (**a**)

the required orientation of axons. Typically, 2–3 slices contain cells enabling soma-axon recording distances of up to ~1 mm, while preserving the dendritic tree up to the distal tuft (Fig. 1).

The basic requirements for successful axonal recordings are comparable to those published for dendritic recordings [16], where an undamaged slice surface and superb visibility of fine structural details are major prerequisites. The top surface of the brain slice (5–20 μm) suffers the least from scattering of transmitted light and most stable recordings from the AIS or axonal blebs are often made in the superficial range. Various strategies for obtaining healthy brain slices have been discussed in much detail elsewhere [4, 16] or on online platforms (http://www.brainslicemethods.com/). In general, slice-cutting solutions, optimized for specific brain regions and species, aim to reduce activity levels and prevent neurotoxicity. The cutting solution is usually cooled to between 1 and 4 °C, 50–100 % of NaCl is exchanged with iso-osmolar sucrose, the Ca:Mg ratio (e.g., 1:6) is decreased, and nonspecific glutamate receptor blockers such as kynurenic acid (1 mM) are added. For young adult rats (4–8 weeks of age) an ice-cold specific cutting solution is often sufficient. The standard artificial cerebrospinal fluid (ACSF) contains (in mM): 125 NaCl, 3.0 KCl, 0.5 $CaCl_2$, 7 $MgCl_2$, 25 $NaHCO_3$, 1.25 NaH_2PO_4, and 25 D-glucose. When using older mice or rats (>3 months), pre-perfusion with the cutting solution is recommended to rapidly cool the brain. For cutting slices we use a dedicated Vibratome with adjustable *z*-axis deflection (e.g., Leica VT 1200S or Campden Instruments Vibrating Microtome 5000/7000) [4]).

2.2 Visualization of Myelinated Axons

A large magnification with high-numerical aperture objective (e.g., a 63× or 60×, 1.0–1.1 N.A. objective) with twofold optical magnification is essential to visualize the details of axonal structures. For obtaining contrast, the infrared differential interference contrast (IR-DIC), or gradient-contrast techniques such as oblique or Dodt illumination all provide excellent image detail, but when combining with fluorescence image acquisition (e.g., in fluorescence-assisted patching), the latter two methods are recommended. IR-DIC requires an analyzer (which is a linear polarizer) in the light path above the objective, reducing both the transmitted and fluorescence light levels. It is easy to compensate for this light loss in the DIC by increasing the transmitted light intensity without any adverse effects on the preparation. However, to compensate for the loss in the fluorescence signals, higher excitation light levels are needed, which can lead to bleaching and, even more importantly, phototoxicity. For this reason, when using DIC contrast, care must be taken to separate the DIC image from the fluorescence image onto two separate CCD cameras, so that the fluorescence doesn't go through the analyzer. On the other hand, in oblique and gradient-contrast techniques the aperture plane of the condenser is adjusted for spatial filtering and can be used with or without IR filters. Hence, these techniques have more light availability and also make combined use with fluorescence while using one CCD camera easier while no analyzer is needed.

A high spatial resolution (e.g., 1392 × 1040 pixels) and high sensitivity CCD camera allow sufficiently large areas to be scanned for healthy axons, and brief exposure times (<20 ms) for performing live fluorescence imaging without generating phototoxicity. While many commercial microscope systems are available to overlay the bright-field and fluorescence images, we currently use the open-source platform μ-manager (http://www.micro-manager.org/) in combination with an Arduino board (http://www.arduino.org), enabling computer-controlled integration of microscope ports, camera, and shutters.

For selection and targeting of fine structures like axons, it is very important to work at the highest possible optical resolution. Resolution (r) in microscopy is dependent on the used wavelength (λ) and the combined N.A. throughput of the condenser and objective ($r = 1.22\ \lambda / 2$ N.A.). For transmitted DIC at 750 nm and high-quality N.A. 1.0 objectives, the theoretical optical resolution is about ~0.45 μm. However, when observing slices, the resolution quickly drops with slice depth due to transmitted light interaction with the thick acute slice tissue. Increasing the transmitted wavelength reduces this interaction (such as the infrared 900 nm filter); however, that will subsequently worsen the resolution at the surface. In order to visualize the fine structures located below the slice surface and the small pipette tips used to patch these structures, the use of fluorescence-labeled cells and pipettes in combination with fluorescence-assisted techniques that reject or do not produce out-of-focus fluorescence while maintaining adequate speed is recommended, such as high-speed confocal (spinning-disk confocal) or two-photon imaging. Both techniques provide good contrast when working with fine structures, with two-photon microscopy being superior when working with deep structures and spinning disk for more superficial ones (<30 μm).

2.3 Patch Pipettes

For axonal recordings, the diameter of the axon shaft or cut end (1–5 μm) limits the choice of pipette-tip size. Small pipette tips facilitate Giga-ohm seal formation but also limit the success rate of whole-cell access and increase the access resistance (R_a). Determining a suitable pipette-tip size is thus an important optimization step. While axonal recordings can be obtained from many different mammalian species, the axon diameters in rats are typically ~2-times greater than those in mice—considerably increasing the success rate. For axonal blebs we routinely use pipettes with resistances (R_{Pip}) ranging from 4 to 7 MΩ (standard borosilicate glass with 1.5 mm o.d. and 0.86 mm i.d.), which provide long-term stability and access resistances typically below 20 MΩ. Thick-wall glass capillaries also reduce stray pipette capacitances. For direct recordings from the AIS of large neurons (~1.2–1.7 μm diameter at the distal end), however, smaller pipettes (R_{Pip} >12 MΩ) are required [17]. Small-diameter, unmyelinated axons of interneurons have been recorded with pipette resistances of up to 30 MΩ [13].

3 Methods

3.1 Targeting Axons

The advantage of recording from myelinated primary axons, in contrast to unmyelinated ones, is they can be readily identified, due to the densely aligned myelin sheaths interfering with most of the transmitted light, emerging as a dark band of 1–2 μm in diameter in the bright-field image (Fig. 2a, b).

The first step is establishing which slice contains cortical neurons with healthy, large-diameter axons (>1 μm) and myelin sheaths without swellings (Fig. 2a–c). The second is careful bright-field microscopic examination of cellular structures to select end-bulbs or AISs for recording. For example, only a handful of the hundreds of axon swellings in a slice lack myelin at the distal cut end and are also sufficiently large to allow for long-term patch-clamp recordings. Experience shows that when axon blebs have a healthy appearance the soma is equally healthy, but not vice versa. Typically, about ~4 unmyelinated end-bulbs are present in a single healthy cortical brain slice. Unmyelinated axonal blebs are characterized by a low-contrast spherical appearance connected to a highly contrasted longitudinal structure (the myelinated axon). Once a bleb is identified, the trajectory of the high-contrast myelinated axon can be used as a visual trace retrogradely to its somatic origin (Fig. 2). Confirming the soma-bleb connection a few times in the bright-field image by rescanning and cyclical refocusing is recommended as layer 5 myelinated axons often run in bundles and twist. To unequivocally confirm the link between a soma and its axon, neurons are filled with intracellular fluorescent dyes such as Alexa Fluor® 488 or 594 (50–100 μM), combined with epifluorescence and brightfield imaging (Sect. 2.2). Using these guidelines, axo-somatic recording distances of >800 μm can be obtained (Figs. 2b and 3a).

Alternative methods have been used to strip off myelin from axons. In the peripheral nervous system, lysophosphatidylcholine (also known as lysolecithin), a chemical leading to the phagocytosis of myelin, has been successfully used to make both the nodal and internodal membrane regions accessible to patch-clamp recording [18, 19]. Furthermore, some authors reported that the myelin sheath can be physically removed by sealing a patch pipette onto the myelin and then slowly withdrawing the pipette. The stripped

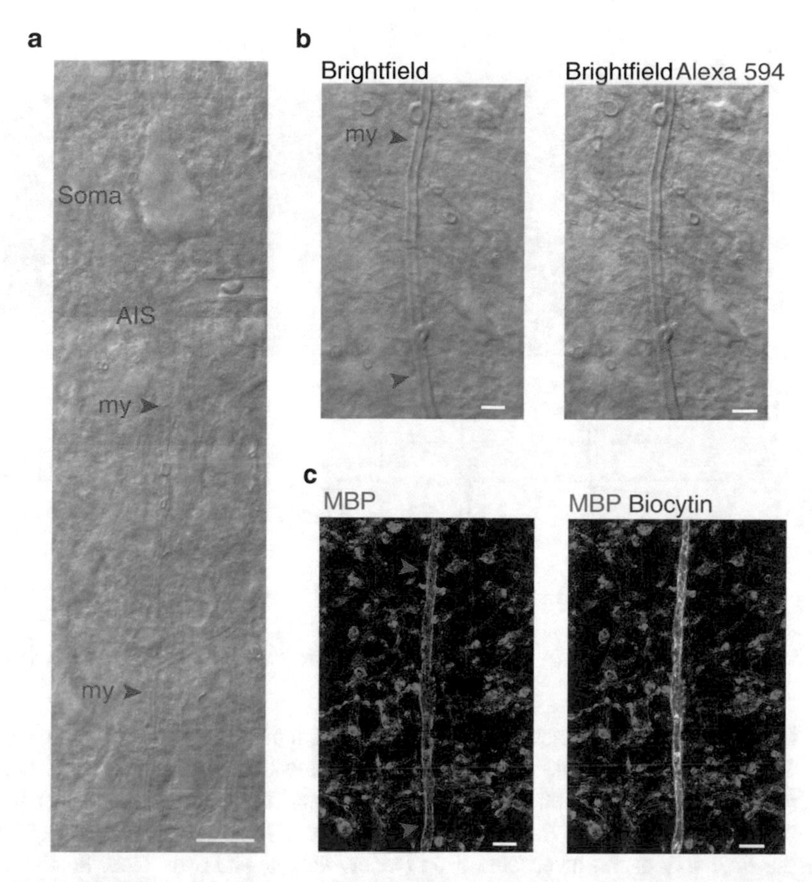

Fig. 2 (**a**) Oblique bright-field contrast image of a layer 5 pyramidal neuron with axon. *Yellow arrows* indicate the trajectory of the AIS and myelinated (my) regions. Scale bar, 20 μm. (**b**) *Left*, high-magnification image (60× and twofold optical magnification) of the primary layer 5 axon. Healthy myelin sheaths appear as a dark band in the bright-field image (*yellow arrows*). Scale bars, 2 μm. *Right*, overlay of bright-field and epifluorescence of Alexa Fluor® 594 fill of the axons. (**c**) *Left*, a primary layer 5 axon stained for myelin basic protein (MBP). *Right*, overlay of streptavidin-biocytin and MBP fluorescence signals. Scale bars, 3 μm

unmyelinated end is subsequently approachable with a second pipette. Axon bleb formation has also been experimentally facilitated by a targeted knife-cut following slice preparation [20].

3.2 Obtaining Whole-Cell Configuration at Axons

Typically, dual whole-cell recording is initiated by approaching the soma in whole-cell voltage-clamp mode and maintaining the cell near its resting membrane potential (~–65 mV, Fig. 3a). The second pipette is subsequently brought to the axonal structure while rectangular test pulses (–10 mV, 10 ms at 50 Hz, Fig. 3b) are continuously applied to monitor R_{Pip}. While cell bodies can be approached with high positive pressures (>100 mbar), axon blebs

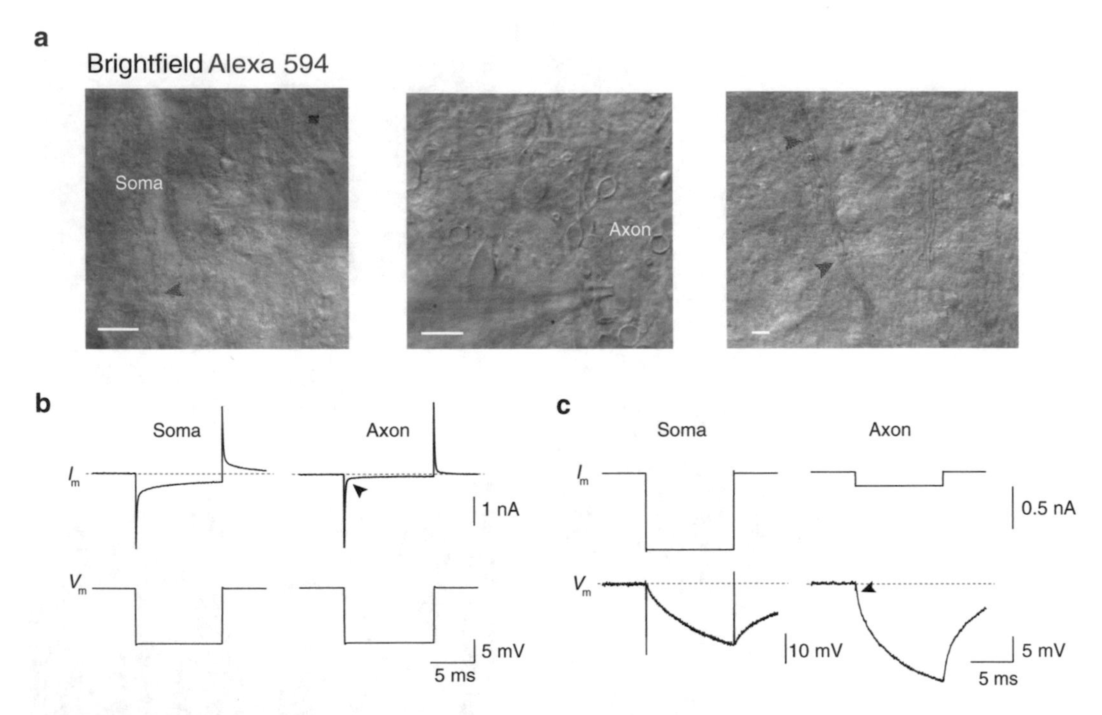

Fig. 3 (**a**) *Left*, a low magnification (40× objective) of the somatic recording electrode with Alexa Fluor 594. *Yellow arrow* shows the onset of the myelinated axons. Scale bar, 20 μm. Middle, high-magnification (60× and two-fold zoom) images showing an unmyelinated axon bleb near the surface of the slice during patch-clamp recording. *Right*, lower magnification of overlaid fluorescence and bright-field images indicating the trajectory of the myelinated axon (*yellow arrows*) and axon bleb. Scale bars, 4 μm. (**b**) Voltage-clamp responses to 10 mV test pulses generating capacitance current transients of the axon bleb in (**a**). *Left*, somatic response of the same cell. *Right*, test pulse example traces of the axon bleb. Note the much smaller capacitive and resistive transients in the axon compared to the soma (*arrow*). Capacitive transients contain both cell and uncompensated pipette responses. (**c**) Test pulse responses of the same cell in current-clamp mode at the soma and the axon. Voltage recordings were fully compensated for the pipette capacitive transients and bridge balance. Note the small capacitive transient of the axonal recording (*arrow*)

are smaller, located near the slice surface, and therefore requiring much lower pressures (typically ~30 mbar). Once a high-resistance seal (>2 GΩ) has been established, the stability of the seal should be judged both visually and electrically, while negative suctions are applied for accessing whole-cell configuration. Note that axonal whole-cell access is associated with only a very small conductivity (arrow, Fig. 3b), due to high local input impedances (~300 MΩ), as compared to the soma (<30 MΩ).

3.3 Neutralization

In current-clamp mode the voltage input signals are filtered by a time constant, calculated as the access resistance multiplied by the uncompensated capacitance values, resulting in undesirable low-pass filtering. Capacitance neutralization in axons, often suffering from high access resistances, therefore requires careful attention. A fast recording bandwidth in the range of 10–30 kHz is desirable if the aim is to record and quantify AP properties. Ideally, the patch-clamp amplifier should have a true voltage-follower headstage circuit in which the voltage output corresponds to the voltage input, allowing accurate recording of APs without current loss [21]. Amplifiers such as the BVC-700A (Dagan Corporation) or Axo-Clamp from Molecular Devices (2A, 2B and 900A) have true voltage-follower circuitry. Other computer-controlled amplifiers such as EPC-10 or EPC-800 (HEKA Elektroniks) or MultiClamp 700B (Molecular Devices) use mostly modified circuitry, also enabling fast current-clamp recording. Computer-controlled amplifiers have the advantage of lower instrumental noise due to the lack of intrinsic analogue circuitry (e.g., for resistive and capacitance compensation) and a large range of Bessel filter settings (up to 30–100 kHz) allowing the recording of the fast rise times of axonal action potentials (dV/dt up to 2,0 kV s^{-1}).

The pipette and distributed system capacitances are often large (~7 pF) compared to the bleb membrane capacitance (~1 pF), providing a capacitive load and *RC* filter of the voltage responses. Capacitance neutralization to compensate for the combined capacitances of the pipette, including the stray capacitance of the pipette wall and amplifier circuitry, is therefore of critical importance. Depending on the amplifier, cancelation can be done in voltage-clamp prior to switching to current-clamp mode. For voltage-follower amplifiers, canceling the capacitance is done in current-clamp mode by carefully increasing the capacitance compensation, typically set at the point just below the threshold for ringing, where voltage transients at the current onset become oscillatory (Fig. 3c). Using thick-wall pipettes (1 mm), Sylgard coating and/or low solution levels in the bath and pipette will reduce the combined capacitances and improve recording speed. Once the capacitance is compensated, bridge-balance neutralization is performed, to compensate for the voltage drop across the access resistance of the injection pipette (normal range: 20–50 MΩ).

3.4 Anatomical Reconstruction

Biocytin, a membrane impermeable molecule, can be added to the intracellular solution (2–4 mg ml^{-1}) to assess the morphological characteristics of the axonal tree. Visualization procedures often use the avidin-biotinylated horseradish peroxidase complex for the immunocytochemical processing of biocytin. Diaminobenzidine is used as the chromogen, causing the black reaction product which enables light microscopy and/or electron microscopy for the ultrastructural level. Detailed procedures can be found in this book (e.g. Chapter 2) and elsewhere [17, 22]. To visualize the internodal membrane under the myelin sheath or the much thinner axon collaterals, high concentrations of Triton X-100 (1–2 %) or repeated cycles of freezing and thawing should be used as permeabilization steps. High-resolution confocal or two-photon microscopy of the thin axonal arborization can be obtained when using the streptavidin-conjugated fluorophores to visualize biocytin (Fig. 2c). This latter approach can in particular be useful when combined with other primary antibodies for axonal markers (e.g., Ankyrin G), allowing assessment of the subcellular distribution of ion channels in the AIS and nodes [15]. Light, confocal or two-photon microscopy analysis of 3D stacks can further be assessed and reconstructed with three-dimensional software packages (e.g., Neurolucida, Microbrightfield Inc.).

4 Notes on Axonal Recordings

4.1 Voltage Clamp of Axonal Voltage-Gated Ion Channels

Recording of voltage-gated currents in non-isopotential and highly branched structures, such as axons and dendrites, using a single whole-cell electrode, is typically associated with loss of voltage- and space-clamp control [23, 24]. In addition, the presence of voltage-gated channels in myelinated axons is highly confined to the axon initial segment (AIS), juxtaparanode, and nodes of Ranvier—often distally from the recording electrode when recording from axonal blebs—limiting quantitative analysis of current kinetics and voltage dependence. For example, recording of pharmacologically isolated fast Nav currents in whole-cell mode from the axon bleb shows that action currents are generated in nearby nodes (Fig. 4a). Since the passive cable properties of central myelinated axons are not known, theoretical correction methods developed for the distortion of voltage control in dendrites [25, 26] are not yet possible to apply to axons. Fast activating Nav channels are, however, expressed at sufficiently high densities in axon blebs for recording in excised cell-attached and outside-out modes, each allowing for quantitative analysis [12, 13] (Fig. 4b, c). Direct cell-attached recordings from distal axonal blebs in comparison with neuronal bodies reveal that Nav channels inactivate at more hyperpolarized potentials ($V_{Half} = -71$ mV), as compared to the somatic Nav channels ($V_{Half} = -56$ mV) [15]. Theoretical and experimental work show that the space- and voltage-clamp efficacies for ionic currents strongly depend on their time course of activation [27]. The main nodal potassium current, the so-called M-current mediated by Kv7.2/7.3 channels, is slowly activating and non-inactivating. Recently, a direct comparison between excised patch recordings of M-currents from the AIS and whole-cell recordings from axon blebs was made to examine their voltage dependence and kinetics [15]. While these currents are exclusively expressed in nodal domains, distal from the recording electrode, their slow time constant of activation (~20 ms) allows for their

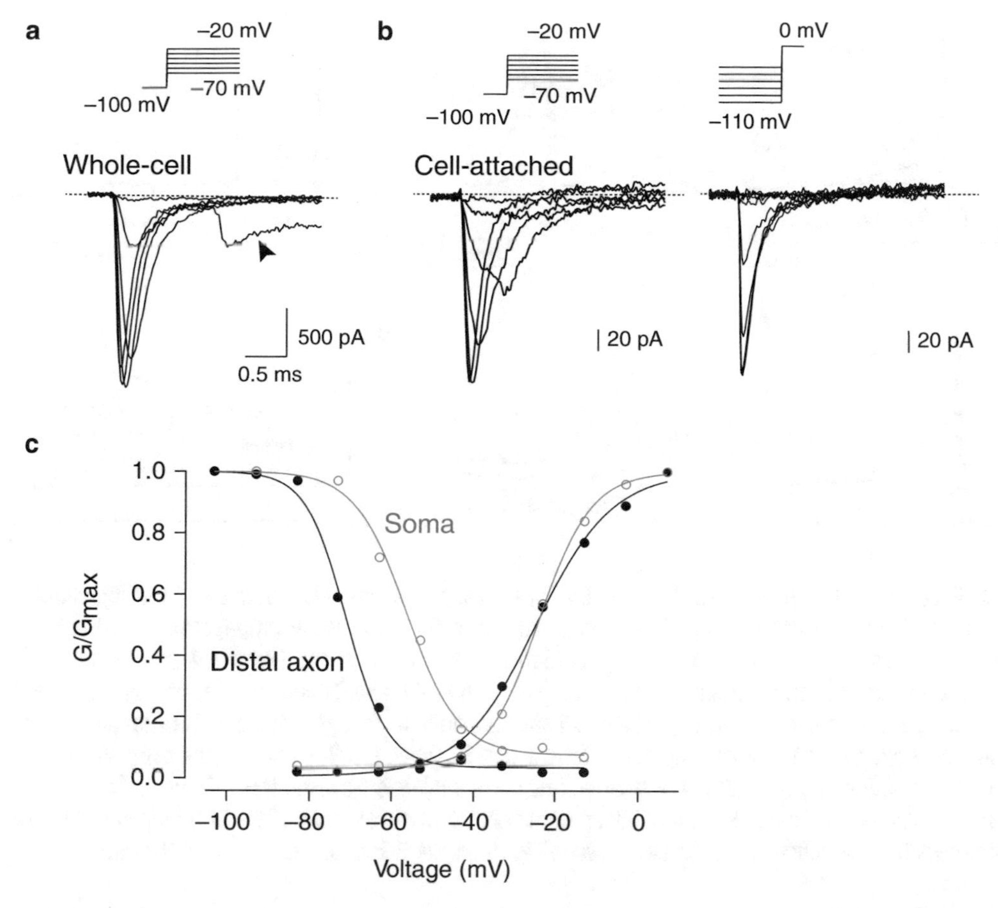

Fig. 4 Voltage-clamp recording of Nav currents in axonal blebs and soma. (**a**) Example of a voltage dependence of activation protocol applied in whole-cell mode from a distal axon bleb. Note that Nav currents cannot be accurately clamped in the whole-cell configuration as indicated by the spontaneous action currents (*black arrow*) probably initiated in upstream nodal regions. (**b**) Cell-attached patch-clamp configuration provides a good voltage control for Nav currents. Currents were recorded with a 10–25 kHz Bessel filter and 50–100 kHz sampling frequency at 35 °C. (**c**) Voltage dependence of activation and inactivation curves from cell-attached patches at the soma (*grey, open circles*) and distal axonal blebs (*black, closed circles*)

activation even in the whole-cell axon bleb recordings, where the error in the voltage dependence of activation is minimal.

4.2 Axonal Conduction

In dual whole-cell current-clamp configuration, action potential propagation can be examined in either the orthodromic direction by a current injection at the soma (Fig. 5a) or the antidromic direction by a current injection at the axon (Fig. 5b). As compared to the soma, axonal APs are brief in half-width (~300 μs) and generate small after-depolarization potentials [15, 17], due to the high Kv channel density and small membrane capacitance of myelinated axons. An AP evoked by axonal current injection is furthermore subjected to strong Nav inactivation, as visible by the small amplitude of the axonal AP compared to the orthodromic one

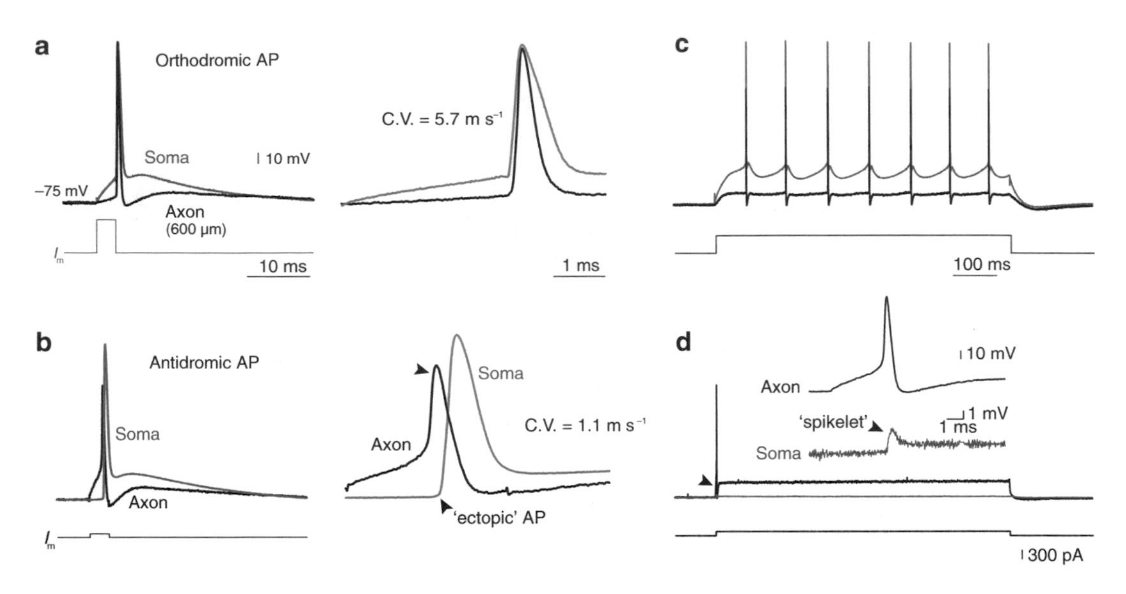

Fig. 5 Action potential initiation and conduction in myelinated axons. (**a**) Somatic current injection-evoked (*grey* trace, 3 ms, 1.8 nA) action potential. Initiated within the AIS, the AP propagates orthodromically and rapidly (~5.7 m s^{-1}) into the axon bleb, recorded at 630 μm from the soma (*black*). (**b**) In a different simultaneous soma-axon recording (380 μm from the soma) the axonal current injection evokes a nodal AP propagating antidromically and slowly into the somato-dendritic region. The shape of the somatic AP reflects a typical ectopic AP. Note the small amplitude of the axonal AP when locally evoked in the axon due to rapid Nav channel inactivation. (**c**) Long (700 ms) steady-state current injection-evoked train of APs, initiated in the axo-somatic region. (**d**) Steady-state current injection-evoked AP in the axon bleb fails to be repetitively activated. Inset shows the magnification of the same recording. Note the failed antidromic AP at the soma (spikelet)

(Fig. 5b). At the soma, the passive current spread from the axonal AP evokes a typical ectopic action potential; the result of an AP propagating antidromic and evoking an AIS AP (Fig. 5b). Large differences in local AP generation and propagation APs also arise when long steady potentials are compared. APs evoked with somatic current injection, initiated at the AIS, are reliably followed in axons (Fig. 5c). Furthermore, while orthodromic propagation is fast in myelinated axons (>3 m s^{-1}), axonally evoked APs propagate in the antidromic direction much slower (<1 m s^{-1}) and may fail to charge the large axo-somatic region to voltage threshold in the AIS, as reflected by a "spikelet" of only a few millivolts in the somatic membrane voltage (Fig. 5c). These data are consistent with the large differences in Nav inactivation in the soma and axon (Fig. 4c) and, more generally, show that while axons have active and passive properties favoring the *propagation* of fast voltage signals, they are poor sites for *initiation* of APs.

4.3 Axonal Action Potential Initiation

The primary site of AP initiation in neurons is the axon initial segment (AIS) [28]. In order to understand the underlying mechanisms of operation of this unique structure, direct electrophysiological recording, enabling high-temporal resolution

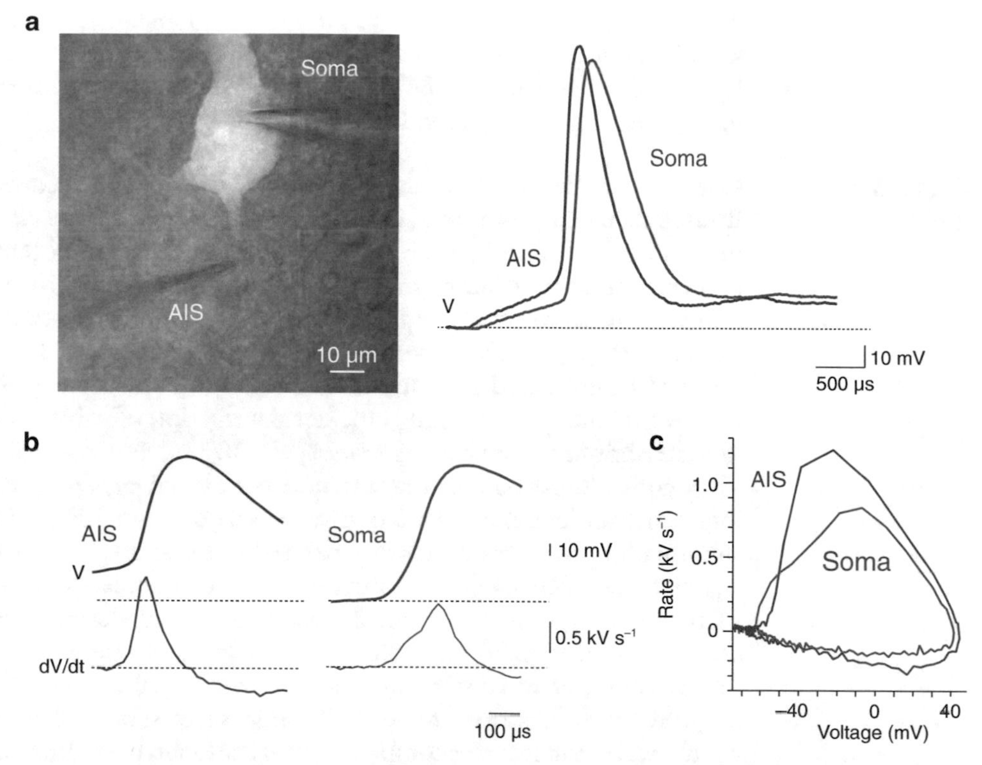

Fig. 6 Action potential initiation in the AIS. (**a**) *Left*, Two-photon microscopy fluorescence-targeted simultaneous voltage recording at the soma (cyan) and AIS (*red*). Fluorescence images are overlaid with the transmitted light (*grey*). Somatic electrode filled with X-Rhod-5F (200 μM) and axonal electrode was filled with Alexa 488 (200 μM). *Right*, simultaneous action potential recording of the AIS (*red*) and soma (*grey*). Note that the AIS rising phase precedes the somatic action potential. (**b**) Voltage and first time derivative traces of the action potential, recorded at the AIS (*red*, 42 μm distance from the soma) and the soma (*grey*). (**c**) Phase plot representation of the action potential shown in (**b**). Note the lower voltage threshold and bi-phasic rising phase of the somatic voltage, due to invasion by the AIS action potential

recording, is a highly valuable tool. The cylindrical and small-diameter anatomical features of the AIS (ranging from 1.2 to 1.7 μm at the distal end) require the use of small-diameter pipettes. While sealing is often not a problem, optimization of the pipette shape (small pipette opening keeping the taper as short as possible) is required to allow for stability during the break-in with negative suctions stage. Figure 6 shows a dual whole-cell recording from the cell body and AIS, at a distance of 22 μm from the soma. Next generation photo-detectors (e.g. Hamamatsu photon counting) and high-speed scanners enable two-photon fluorescence-assisted patching with less light reducing the chances for the photodynamic damage. In the demonstrated example, the cell body is filled with a red dye X-Rhod-5F, which facilitates the second electrode positioning and whole-cell access to the AIS. For easier visualization, the AIS electrode is filled with Alexa 488 green dye, which possesses a non-overlapping emission spectrum with respect to the red dye.

The final image is a composite of the simultaneously recorded signals from green and red fluorescence detectors (placed above the objective and under the condenser) and transmitted infrared light detectors (both placed below the condenser).

4.4 Concluding Remarks

Over the last decade, fluorescence-guided patch-clamp recordings from subcellular locations in axons significantly advanced the understanding of the spatiotemporal initiation of action potentials and current flow within neurons. These recordings have proven to be useful tools for characterizing, with high-temporal resolution, axonal voltage-gated ion channel currents [28], revealing the existence of ligand-gated neurotransmitter receptors in axons [29] as well as the unexpected highly efficient conduction of subthreshold depolarizations within axonal arbors [17, 30].

Caution must be taken by interpreting axonal cut-end recordings. Since the length constants of myelinated axons is >600 μm [17], sodium and potassium currents generated in distal axonal sites may significantly influence the somatically recorded responses during, in particular, long time courses of depolarization. For example, axons cut at a distance <150 μm from somata with otherwise intact dendritic trees generate a distinct intrinsic excitability [10]. Electrophysiological recordings from neurons in brain slices should also more routinely be examined for completeness of the axonal tree. Vice versa, if axonal recordings are made from cut ends of myelinated axons, it is essential to examine where these recordings are made.

The rapidly advancing field of microscopy continues to increase the spatial resolution for live visualization of neuronal structures, using superresolution stimulated emission depletion (STED) or biophysical methods such as scanning-ion conductance microscopy [31]. When combined with conventional electrophysiological patch-clamp techniques, these new optical methods may allow for more powerful, nanometer-targeted recordings from axonal structures. It is therefore likely that in the next coming decade, probing axonal functions will uncover many new features of axons, including their intrinsic electrotonic structure, active properties, and computational features.

Acknowledgments

Maarten H.P. Kole is grantee of an ERC Starting Grant (FP7 framework, Grant #261114) and National MS Society Research Grant (RG 4924A1/1). The authors are thankful to Stefan Hallerman and Charles Cohen for critical reading and valuable comments to the manuscript and Sharon de Vries for support in the preparation of the figures.

References

1. Hodgkin AL, Huxley AF (1952) A quantitative description of membrane current and its application to conduction and excitation in nerve. J Physiol 117:500–544
2. Hamill OP, Marty A, Neher E, Sakmann B, Sigworth FJ (1981) Improved patch-clamp techniques for high-resolution current recording from cells and cell-free membrane patches. Pflugers Arch 391:85–100
3. Stuart GJ, Dodt HU, Sakmann B (2004) Patch-clamp recordings from the soma and dendrites of neurons in brain slices using infrared video microscopy. Pflugers Arch 423:511–518
4. Geiger JRP, Bischofberger J, Vida I, Fröbe U, Pfitzinger S, Weber HJ, Haverkampf K, Jonas P (2002) Patch-clamp recording in brain slices with improved slicer technology. Pflugers Arch 443:491–501. doi:10.1007/s00424-001-0735-3
5. Monsivais P, Clark BA, Roth A, Häusser M (2005) Determinants of action potential propagation in cerebellar Purkinje cell axons. J Neurosci 25:464–472. doi:10.1523/JNEUROSCI.3871-04.2005
6. Bischofberger J, Engel D, Li L, Geiger JRP, Jonas P (2006) Patch-clamp recording from mossy fiber terminals in hippocampal slices. Nat Protoc 1:2075–2081. doi:10.1038/nprot.2006.312
7. Sasaki T, Matsuki N, Ikegaya Y (2012) Targeted axon-attached recording with fluorescent patch-clamp pipettes in brain slices. Nat Protoc 7:1228–1234. doi:10.1038/nprot.2012.061
8. Cajal SRY (1959) Degeneration & regeneration of the nervous system. Hafner Pub. Co., New York
9. Xie XY, Barrett JN (1991) Membrane resealing in cultured rat septal neurons after neurite transection: evidence for enhancement by Ca(2+)-triggered protease activity and cytoskeletal disassembly. J Neurosci 11:3257–3267
10. Kole MHP (2011) First node of Ranvier facilitates high-frequency burst encoding. Neuron 71:671–682. doi:10.1016/j.neuron.2011.06.024
11. Roza C, Castillejo S, Lopez-García JA (2011) Accumulation of Kv7.2 channels in putative ectopic transduction zones of mice nerve-end neuromas. Mol Pain 7:58. doi:10.1186/1744-8069-7-58
12. Kole MHP, Ilschner SU, Kampa BM, Williams SR, Ruben PC, Stuart GJ (2008) Action potential generation requires a high sodium channel density in the axon initial segment. Nat Neurosci 11:178–186. doi:10.1038/nn2040
13. Hu H, Jonas P (2014) A supercritical density of Na(+) channels ensures fast signaling in GABAergic interneuron axons. Nat Neurosci 17(5):686–693. doi:10.1038/nn.3678
14. Hu W, Tian C, Li T, Yang M, Hou H, Shu Y (2009) Distinct contributions of Na(v)1.6 and Na(v)1.2 in action potential initiation and backpropagation. Nat Neurosci 12:996–1002. doi:10.1038/nn.2359
15. Battefeld A, Tran BT, Gavrilis J, Cooper EC, Kole MHP (2014) Heteromeric Kv7.2/7.3 channels differentially regulate action potential initiation and conduction in neocortical myelinated axons. J Neurosci 34:3719–3732. doi:10.1523/JNEUROSCI.4206-13.2014
16. Davie JT, Kole MHP, Letzkus JJ, Rancz EA, Spruston N, Stuart GJ, Häusser M (2006) Dendritic patch-clamp recording. Nat Protoc 1:1235–1247. doi:10.1038/nprot.2006.164
17. Kole MHP, Letzkus JJ, Stuart GJ (2007) Axon initial segment Kv1 channels control axonal action potential waveform and synaptic efficacy. Neuron 55:633–647. doi:10.1016/j.neuron.2007.07.031
18. Jonas P, Bräu ME, Hermsteiner M, Vogel W (1989) Single-channel recording in myelinated nerve fibers reveals one type of Na channel but different K channels. Proc Natl Acad Sci U S A 86:7238–7242
19. Safronov BV, Kampe K, Vogel W (1993) Single voltage-dependent potassium channels in rat peripheral nerve membrane. J Physiol 460:675–691
20. Hu W, Shu Y (2012) Axonal bleb recording. Neurosci Bull 28:342–350. doi:10.1007/s12264-012-1247-1
21. Magistretti J, Mantegazza M, Guatteo E, Wanke E (1996) Action potentials recorded with patch-clamp amplifiers: are they genuine? Trends Neurosci 19:530–534
22. Marx M, Günter RH, Hucko W, Radnikow G, Feldmeyer D (2012) Improved biocytin labeling and neuronal 3D reconstruction. Nat Protoc 7:394–407. doi:10.1038/nprot.2011.449
23. Williams SR, Mitchell S (2008) Direct measurement of somatic voltage clamp errors in central neurons. Nat Neurosci 11(7):$ 790–798. doi:10.1038/nn.2137
24. Taylor RE, Moore JW, Cole KS (1960) Analysis of certain errors in squid axon voltage clamp

measurements. Biophys J 1:161–202. doi:10.1016/S0006-3495(60)86882-8

25. Schaefer AT, Helmstaedter M, Sakmann B, Korngreen A (2003) Correction of conductance measurements in non-space-clamped structures: 1. Voltage-gated K+ channels. Biophys J 84:3508–3528. doi:10.1016/S0006-3495(03)75086-3
26. Major G, Evans JD, Jack JJB (1993) Solutions for transients in arbitrarily branching cables. Biophys J 65:450–468. doi:10.1016/S0006-3495(93)81038-5
27. Williams SR, Wozny C (2011) Errors in the measurement of voltage-activated ion channels in cell-attached patch-clamp recordings. Nat Commun 2:242. doi:10.1038/ncomms1225
28. Kole MHP, Stuart GJ (2012) Signal processing in the axon initial segment. Neuron 73:235–247. doi:10.1016/j.neuron.2012.01.007
29. Sasaki T, Matsuki N, Ikegaya Y (2011) Action-potential modulation during axonal conduction. Science 331:599–601. doi:10.1126/science.1197598
30. Alle H, Geiger JRP (2006) Combined analog and action potential coding in hippocampal mossy fibers. Science 311:1290–1293. doi:10.1126/science.1119055
31. Novak P, Gorelik J, Vivekananda U, Shevchuk AI, Ermolyuk YS, Bailey RJ, Bushby AJ, Moss GWJ, Rusakov DA, Klenerman D, Kullmann DM, Volynski KE, Korchev YE (2013) Nanoscale-targeted patch-clamp recordings of functional presynaptic ion channels. Neuron 79:1067–1077. doi:10.1016/j.neuron.2013.07.012

Chapter 7

Analysis of Transsynaptic Attentional Neuronal Circuits with Octuple Patch-Clamp Recordings

Daniel R. Wyskiel, Trevor C. Larry, Xiaolong Jiang, Guangfu Wang, and J. Julius Zhu

Abstract

Deciphering interneuronal circuitry is essential to understanding brain functions yet remains a daunting task in neurobiology. To facilitate the dissection of complex cortical neuronal circuits, a process requiring analysis of synaptic interconnections and identification of cell types of interconnected neurons, we have developed a simultaneous quadruple-octuple whole-cell recordings technique that allows physiological analysis of synaptic interconnection among up to eight neurons and anatomical identification of the majority of recorded neurons. Using this method, we have recently revealed two transsynaptic disinhibitory and inhibitory circuits connecting layer 1–3 interneurons with pyramidal neurons in both supragranular and infragranular cortical layers of the rat neocortex. Here, we outline the technique that permits decoding the complex cortical interneuronal circuits involved in controlling salience detection.

Key words Multiple whole-cell recordings, Interneurons, Circuits, Coincidence detection, Salience

1 Background

The cerebral cortex is a multilayered structure responsible for many higher-order functions, including those involved in attention [1–4]. Layer 1 of the cerebral cortex (L1) is strategically positioned to regulate the pathways involved in attention as it receives projections from both higher-order thalamic nuclei and higher-order cortical areas, two regions known to be critical for the selection of salient information [5–13]. These inputs generate direct excitatory postsynaptic potentials in L1 [14–16], and the excitation is selectively and dramatically enhanced during attentional tasks [17–19]. Moreover, L1 receives dense innervations from neuromodulatory systems that may robustly modulate neuronal activity in L1 [19–22].

L1 is composed of almost entirely GABAergic interneurons that fall into two general groups [14, 16, 23–29]. One group of L1 interneurons has heterogeneous dendritic morphology and

Alon Korngreen (ed.), *Advanced Patch-Clamp Analysis for Neuroscientists*, Neuromethods, vol. 113,
DOI 10.1007/978-1-4939-3411-9_7,

axons projecting into deeper layers of the cortex, whereas the other group of L1 interneurons is multipolar, aspiny interneurons with highly ramified axons projecting horizontally within L1. These two groups of interneurons generally fire adapting non-late-spiking and non-adapting late-spiking action potentials, respectively. However, exceptions to this cell morphology-firing pattern correlation have been reported [25], and confirmed with a much larger L1 interneuron sample [16]. On the other hand, these two groups of interneurons can be unambiguously defined as single-bouquet cells (SBCs) and elongated neurogliaform cells (ENGCs) based on their visually distinguishable axonal arborization patterns [16], following the recently proposed nomenclature [30].

Pyramidal neurons, the primary excitatory neurons in the cortex, contain long apical dendrites which contribute to the synaptic integration that forms the basis of neuronal computational power [13, 31–33]. L2/3 and L5 pyramidal neurons, the major cortical output neurons, send apical dendrites that terminate in L1, where they are innervated by attention-related thalamic and cortical inputs [34–38]. These modulatory synaptic inputs to distal apical dendrites of L2/3 and L5 pyramidal neurons can induce dendritic action potentials if of sufficient amplitude [39–42]. Moreover, the induction of dendritic action potentials can be greatly facilitated by the back-propagation of action potentials elicited by concurrent L4 sensory inputs [40, 42, 43], resulting in dendritic complex spikes and bursts of somatic/axonal action potentials [42, 44–47]. Because the dendritic complex spikes and bursts of somatic/axonal action potentials secure further processing of the concurrent synaptic signals [32, 48], they act as a coincidence detection mechanism for salient synaptic inputs. Therefore, interneurons from L1 may regulate initiation of dendritic complex spikes in L2/3 and L5 pyramidal neurons and thereby effectively control the coincidence detection or salience selection mechanisms.

The functional significance of L1 neurons and how they are integrated into the cortical circuit remain poorly understood. Specifically, the morphological differences between SBCs and ENGCs suggest different regulatory functions. However, due to the extensive, intricate synaptic organization between the vastly diverse interneurons and pyramidal neurons, deciphering complex neuronal circuits, such as those L1 interneuron-involved salience selection circuits, has been a daunting task. To combat this challenge, we have developed a method that allows for stable whole-cell recordings from up to eight neurons simultaneously and leads to morphological recovery and subsequent cell type identification of >85 % of recorded interneurons and ≥ 99 % of pyramidal neurons [16, 29].

Using the simultaneous quadruple-octuple whole-cell recordings technique we developed (Fig. 1), we have investigated L1 interneurons and their connections between L2/3 and L5 neurons [16, 29].

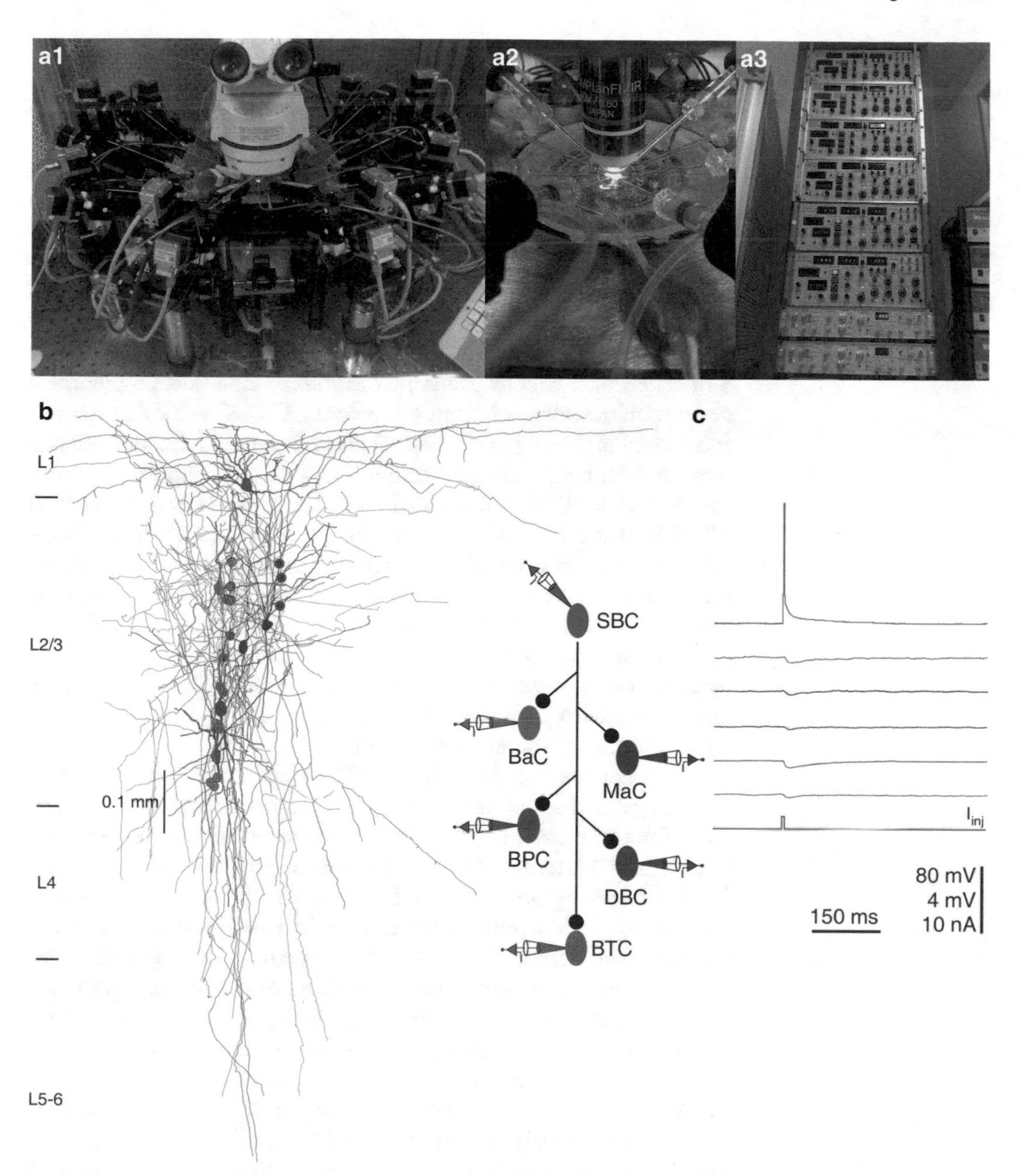

Fig. 1 Cortical L1 $\rightarrow$ L2/3 interneuronal circuits. (**a**$_{1-3}$) Images show a typical octuple recording setting. (**b**) Reconstruction of L1 SBC (*pink*), L2/3 BaC (*cyan*), L2/3 MaC (*red*), L2/3 BPC (*green*), L2/3 DBC (*blue*), and L2/3 BTC (*yellow*) recorded simultaneously from an acute cortical slice. The double colored dots indicate the putative synaptic contacts based on anatomical reconstruction. The schematic drawing shows symbolically the synaptic connections. (**c**) Single action potentials elicited in SBC evoked uIPSPs in postsynaptic BaC, MaC, BPC, DBC, and BTC. Scale bars apply to all recording traces with 80 and 4 mV bars applied to traces with and without action potentials, respectively

Our analysis reveals that SBCs predominantly make unidirectional inhibitory connections (SBC→) with L2/3 interneurons, whereas ENGCs frequently form reciprocal inhibitory and electric connections (ENGC↔) with L2/3 interneurons. Axon arborization analysis identifies seven general interneuron subtypes in L2/3: Martinotti cells (MaCs), neurogliaform cells (NGCs), bitufted cells (BTCs), bipolar cells (BPCs), basket cells (BaCs), double-bouquet cells (DBCs), and chandelier cells (ChCs). SBCs innervate all seven L2/3 interneuron subtypes. By contrast, ENGCs preferentially innervate three of these subtypes, MaCs, NGCs, and BTCs. Simultaneous recordings from L1, L2/3, and L5 neurons show that SBC → L2/3 interneuronal circuits primarily disinhibit L2/3 and L5 pyramidal neurons in the same columns. Conversely, ENGC ↔ L2/3 interneuronal circuits directly inhibit L2/3 and L5 pyramidal neurons in the same and/or neighboring columns. These results support the notion that SBCs and ENGCs form distinct interneuronal circuits with both L2/3 interneurons and L2-5 pyramidal neurons such that SBC → L2/3 interneuronal circuits disinhibit whereas ENGC ↔ L2/3 interneuronal circuits inhibit L2/3 and L5 pyramidal neurons. Furthermore, SBC-led interneuronal circuits disinhibit primarily the dendritic-somato-axonal axis of a small number of L2/3 and L5 pyramidal neurons in the same columns, whereas ENGC-led interneuronal circuits primarily inhibit the distal apical dendrite of much more of these cells in multiple columns.

The contrasting SBC- and ENGC-led interneuronal circuits with L2/3 and L5 pyramidal neurons suggest different functional roles for these circuits. Indeed, action potentials elicited in SBCs inhibit L2/3 interneurons and relieve the suppression of complex dendritic spiking and somatic bursting in L5 pyramidal neurons. Conversely, action potentials initiated in ENGCs recruit and synchronize the activity in L2/3 interneurons, and potentiate the suppression of complex dendritic spiking and somatic bursting in L5 pyramidal neurons. In intact brains, paired recordings from SBCs and L5 pyramidal neurons show that spontaneous or whisker-evoked synaptic events can trigger action potentials in SBCs that enhance dendritic complex spiking in L5 pyramidal neurons. Conversely, paired recordings from ENGCs and L5 pyramidal neurons show that spontaneous or whisker-evoked synaptic events can trigger action potentials in ENGCs that suppress dendritic complex spiking in L5 pyramidal neurons. Altogether, these results consistently testify to the coexistence of two distinct L1-interneuron-led attention-related interneuronal circuits: SBC → L2/3 interneuron → L2-5 pyramidal neuronal circuits disinhibit the coincidence detection mechanism, whereas ENGC ↔ L2/3 interneuron → L2-5 pyramidal neuronal circuits inhibit the coincidence detection mechanism. These two distinct interneuronal circuits can transform L1 inputs into complementary "filters" by differentially regulating the output of L2/3 and L5 pyramidal neurons. Thus, the two interneuronal circuits seem to work synergistically with the dendritic coincidence detection mechanism in pyramidal neurons to filter out "noise" in the incoming information to achieve effective salience selection.

2 Materials

To ensure examination of mature and stabilized cortical inhibitory neurons and circuits [49, 50], rodents postnatal 20 days or older should be used. When preparing acute brain slices, anesthetics such as sodium pentobarbital are required. Immediately upon extraction of the brain, an artificial cerebrospinal fluid (ACSF), saturated with carbogen (95 % O_2/5 % CO_2), will be needed. ACSF is prepared with the following ingredients: NaCl, KCl, NaH_2PO_4, $NaHCO_3$, $MgCl_2$, dextrose, and $CaCl_2$. This solution should be cold (0–4 °C) for the extraction and slicing procedure. A microslicer with minimal vibrations in the vertical axis is preferable to obtain healthy acute cortical slices [51]. After slices are obtained, a beaker containing ACSF saturated with carbogen and heated to 37.0 $\pm$ 0.5 °C will be needed to incubate the slices.

Electrophysiological recordings require two internal solutions (one for current clamp and one for voltage clamp) containing different combinations of the following components: cesium methanesulfonate, HEPES, $MgCl_2$, Na_2ATP, Na_3GTP, sodium phosphocreatine, EGTA, spermine, biocytin, potassium gluconate, KCl, and MgATP.

Eight amplifiers are needed to record from up to eight neurons simultaneously. We used Axopatch 200B and/or Axoclamp 2A/B amplifiers (Molecular Devices, Sunnyvale, CA), which are ideal for current and voltage recordings, respectively. Because no single commercially available interface board is able to supply enough A/D and D/A channels for eight amplifiers, at least two interface boards are needed. In our setup, two ITC-18 interface boards (HEKA Instruments Inc, Bellmore, NY) are used to achieve simultaneous A/D and D/A conversions of current, voltage, command, and triggering signal for up to eight amplifiers. In addition, custom-written Igor-based software programs are used to synchronize two ITC-18 boards, operate eight Axon amplifiers, and perform online and offline data analysis. Stable long-lasting recordings are crucial to fill sufficient biocytin to recover the complex axonal arborization of recorded interneurons. In our setup, L&N Mini motorized manipulators (Luigs & Neumann Feinmechanik and Elektrotechnik, Ratingen, Germany) are used because of their excellent stability and compactness. Furthermore, their headstage rails are specially designed to minimize the loss of other recordings when the exchange of patch pipettes becomes necessary.

Immunohistochemistry procedures are carried out with several toxic reagents. The slices are first fixed in a saline solution containing acrolein, paraformaldehyde, and phosphate (used to buffer the solution). After 24 h, the slices are processed with avidin-biotin-peroxidase and some require additional fixation with OsO_4 and uranyl acetate. A computerized reconstruction system (Neurolucida, Williston, VT) is needed to reconstruct recorded neurons and analyze their morphological properties. For electron microscopy, an ultramicrotome is required for preparing ultrathin tissue sections, and an electron microscope, such as a JEOL-1230 transmission electron microscope (Japan Electron Optic, Tokyo, Japan), is required for visualization of synapses.

3 Methods

All procedures for animal surgery and maintenance are performed following experimental protocols in accordance with US National and Institutional Guidelines for the Care and Use of Laboratory Animals.

3.1 In Vitro Multiple Whole-Cell Patch-Clamp Recordings

3.1.1 Preparing the Artificial Cerebrospinal Fluid (ACSF)

Proper preparation of ACSF is essential to the success of whole-cell patch-clamp recordings. To produce 10 l of ACSF solution, add 69.54 g NaCl (119 mM), 1.86 g KCl (2.5 mM), 21.84 g $NaHCO_3$ (26 mM), and 1.2 g Na_2PO_4 (1 mM) in 9.5 l of double-distilled water. The solution is stirred during the addition of these reagents, and then filled to the 10 l mark with additional double distilled water. This solution is stable and can be stored for months until it is needed, at which time 4.5 g of dextrose, 1 ml of 1 M $MgCl_2$, and 2 ml of 1 M $CaCl_2$ are added to 1 l of the solution.

3.1.2 Acute Cortical Brain Slice Preparation

As mentioned above, male and female rodents at least 20 days old are to be used for the preparation of acute cortical slices. The animals are anesthetized by an intraperitoneal injection of sodium pentobarbital at a ratio of 90 mg/kg of body weight. Once deep anesthesia has been established, the animals are decapitated, the skin and skull cut open, and then pulled back with curved, blunt forceps. Icy ACSF (0–4 °C) is immediately poured onto the brain to limit cell death. The brain region of interest is extracted and put directly into a beaker of oxygenated and iced ACSF to cool. Apply a thin layer of glue to the specimen plate. After a brief placement of the brain block on a piece of filter paper to remove excess ACSF, it can be glued to the specimen plate of the microslicer. Icy ACSF is immediately poured to cover the tissue. Parasagittal slices 350 μm thick are cut from the tissue block at an angle (<~4°) closely parallel to apical dendrites of L5 pyramidal neurons, which should retain the majority of distal ascending and descending axonal trees of L1-3 interneurons that project into L1 and L5-6. The brain slices are gently transferred to a beaker of oxygenated ACSF, kept in a water bath at 37.0 ± 0.5 °C, for ~30 min before recordings.

3.1.3 Whole-Cell Patch-Clamp Recording

For whole-cell patch-clamp recordings, the determination of patch-pipette solutions depends upon the desired measurement: current or voltage. For current recordings, fill the pipette with 135 mM cesium methanesulfonate, 10 mM HEPES, 2.5 mM $MgCl_2$, 4 mM Na_2ATP, 0.4 mM Na_3GTP, 10 mM sodium phosphocreatine, 0.6 mM EGTA, 0.1 mM spermine, and 0.5 % biocytin. For voltage recordings, 120 mM potassium gluconate, 10 mM HEPES, 4 mM KCl, 4 mM MgATP, 0.3 mM Na_3GTP, 10 mM sodium phosphocreatine, and 0.5 % biocytin are used in the recording pipette. The addition of biocytin to these solutions allows for diffusion of the compound from the pipette into neurons during the recording so that histological staining and proceeding morphological analysis can be achieved thereafter. The resistance of the pipette is between 3–7 MΩ.

To perform electrophysiological recordings, transfer a cortical slice to the recording chamber on the stage of an upright microscope. The recording bath should be continuously perfused with ACSF, saturated with carbogen, and held at 34.0 ± 0.5 °C. The cortical slice is held in place on the chamber using a platinum ring covered with a grid of nylon strings, similar to one described by Edwards et al. in 1989 [52]. Ideally, the entire apical dendrite of a pyramidal neuron can be visually tracked with little adjustment of the focus. It is a good practice to identify and select healthy neurons to be recorded before placing the electrodes slightly above the tissue. Paramount to formation of tight GΩ seals is the maintenance of clean pipette tips. As the pipettes are lowered through the tissue, positive pressure should be applied to the pipettes so as to push the tissue away from the tip of the pipette, which also helps to clean off the membrane debris around the neurons as the electrodes approach them. Tight GΩ seal recordings can be achieved following the previously established procedure [52]. In order to establish multiple recordings from eight neurons simultaneously, many different exercises should be adhered to (for the detailed methodology, see [53, 54]). First and foremost, it is imperative that the rig be properly assembled and maintained to keep it fully functional (a detailed description on the setup needed for in vitro recordings can be found in Molecular Devices Axon Guide). Furthermore, continuous practice is needed to achieve the skills necessary to complete it in an effective and timely manner.

3.1.4 Identification of Inhibitory Synaptic Events

Although it is often most helpful to determine all the cells you wish to patch prior to starting, in some cases this may not be preferred. If a particular cell type is desired, such as a fast-spiking interneuron, it may be beneficial to examine the intrinsic electrophysiological properties before selecting the next cells. It should be noted that although somatic morphology and membrane and firing characteristics help in identifying inhibitory interneurons, there is no certainty that the cell is inhibitory unless it inhibits a postsynaptic neuron. Another method one might also employ is different "searching" techniques to find a connected neuron before patching as described by Feldmeyer and colleagues [55], and see [53]. Regardless of the above strategies, once all the cells have been selected it is best to immerse each electrode into the bath, zero any offsets, establish the desired positive pressure, and lower each one to just above the slice. Properly aligned and stable well-functioning micromanipulators along with plenty of working area underneath the objective will greatly facilitate sequentially positioning each electrode in a timely manner. After each successful patch, make sure the electrodes from previously established patches have not drifted away from their neuron.

Once the cells have been patched and their membrane and firing properties characterized, inject a depolarizing current step to elicit an action potential in one of the neurons and monitor the postsynaptic cells for a response. Importantly, one must know the reversal potential for the response desired and adjust the holding potential accordingly. When examining GABAergic responses, if the calculated reversal potential is around −80 mV, holding the neuron closer to −55 mV will maximize the inhibitory response. In some cases, multiple action potentials or a burst of spikes might be needed in the presynaptic neuron to obtain a reliable response in the postsynaptic neuron. In this case, inject a series or train of depolarizing pulses into the presynaptic neuron. Ideally, to clearly determine the synaptic response, at least 50 sweeps (each sweep representing a stimulated presynaptic neuron) will be needed to verify the connection.

To study specific currents, select channel antagonists may be added to isolate the currents of interest. For example, AMPA-sensitive glutamate receptor (-R) antagonist DNQX and NMDA-R antagonist DL-AP5 can be used to eliminate excitatory synaptic transmission, and thus limit cell–cell communication to primarily inhibitory synapses.

3.2 Histology and Electron Microscopy

After the electrophysiological results are obtained, the cortical slices may be immersed in 3 % acrolein/4 % paraformaldehyde in a 0.1 M phosphate-buffered saline at 4 °C to preserve tissue morphology. After the 24 h incubation, the slices can be further processed using the avidin-biotin-peroxidase method to elucidate the morphologies of the recorded cells, and then reconstructed using a computerized reconstruction system like Neurolucida. For electron microscopic examination, the slices should be further sectioned and postfixed. To do so, the slice products from the first staining are re-sectioned into 60 μm sections, immersed in 1 % OsO_4, counterstained with 1 % uranyl acetate, and embedded into resin. From here, ultrathin slices are taken from small areas of interest, specifically to examine putative synaptic bouton sites. From this analysis, identification of inhibitory synapses can be determined by examining details of the synaptic cleft, including: (1) presence of membranes with parallel alignment forming synaptic clefts that are wider in the middle and narrower at one or both edges; (2) absence of a prominent postsynaptic density; and (3) presence of multiple flattened synaptic vesicles with at least one docked at the presynaptic membrane.

4 Exemplar Results and Analysis

Figure 1 illustrates representative results using the techniques outlined above. A preliminary step in our study of inhibitory interneuronal circuit analysis is the identification of inhibitory synaptic connections. By including DNQX and DL-AP5 in the bath solution, the unitary inhibitory postsynaptic currents (uIPSCs) or potentials (uIPSPs) were isolated and measured. After recordings, the axonal arborization of each recorded neuron was reconstructed. The reconstruction provided anatomical confirmation of physiologically identified inhibitory synaptic connections, as well as identification of the cell type of recorded cells. Based on previously established axonal arborization-based interneuronal classification schemes [16, 30, 56, 57], we could unambiguously classify L1-3 interneurons into 9 general groups: SBCs and ENGCs in L1, and MaCs, NGCs, BTCs, BPCs, BaC, DBCs, and ChCs in L2/3. Specifically, L1 SBCs have heterogeneous dendritic morphology and a characteristic vertically descending horsetail-like axonal bundle with short side branches. In contrast, L1 ENGCs have their dense axonal arborization elongated horizontally and restricted largely within L1. L2/3 MaCs have their axonal arborization specialized to project mainly towards L1. L2/3 NGCs have their arborization formed into a highly symmetrical and spherical dendritic field and axonal arbor. L2/3 BTCs have given rise to two primary dendrites from opposite poles and an axon forming wide vertical and horizontal projections. L2/3 BPCs have an axon commonly emerging from one of the primary dendrites and forming a narrow descending band that crosses multiple layers. L2/3 BaCs have a basket-like axonal arborization. L2/3 ChCs have an axon forming characteristic chandelier-like terminals with short vertical rows of boutons. These interneurons can often be identified with direct visual assessment if the majority of their axonal arborization is recovered. The axonal length density analysis can verify the cell type identification since these seven types of L2/3 interneurons differ significantly in their axonal arborization patterns, reflecting presumably the participation of distinct circuit connections.

5 Conclusion

In order to analyze complex behavior at the circuit level, the identification and role of each participating neuron must be known. Because of the vast diversity of interneurons, in their morphology, chemical and electrophysiological characteristics, as well as their distinct synaptic properties and organization, deciphering neuronal circuitry and its effect on behavior has been challenging. Therefore, to analyze the circuits involved in attention, which are known to contain many diverse interneurons, we have employed a simultaneous quadruple-octuple whole-cell recordings method we developed. Furthermore, because the testable connectivity pattern, C, increases exponentially with the increase of the number of simultaneously recorded neurons, n, or $C = 4^{n(n-1)/2}$, simultaneous quadruple-octuple whole-cell recordings dramatically increase the yield and chance to detect the complex synaptic interconnections. Therefore, analyzing transsynaptic connections of 4–8 cell-type identified neurons promises to decode the organization of complex cortical inhibitory circuits used in different types of behavior.

References

1. Gilbert CD, Wiesel TN (1983) Functional organization of the visual cortex. Prog Brain Res 58:209
2. Mountcastle VB (1997) The columnar organization of the neocortex. Brain 120(Pt 4):701
3. Kastner S, Ungerleider LG (2000) Mechanisms of visual attention in the human cortex. Annu Rev Neurosci 23:315
4. Douglas RJ, Martin KA (2007) Mapping the matrix: the ways of neocortex. Neuron 56:226
5. Robinson DL, Petersen SE (1992) The pulvinar and visual salience. Trends Neurosci 15:127
6. Tomita H, Ohbayashi M, Nakahara K, Hasegawa I, Miyashita Y (1999) Top-down signal from prefrontal cortex in executive control of memory retrieval. Nature 401:699
7. Pascual-Leone A, Walsh V (2001) Fast backprojections from the motion to the primary visual area necessary for visual awareness. Science 292:510
8. Gilbert CD, Sigman M (2007) Brain states: top-down influences in sensory processing. Neuron 54:677
9. van Boxtel JJ, Tsuchiya N, Koch C (2010) Consciousness and attention: on sufficiency and necessity. Front Psychol 1:217
10. Baluch F, Itti L (2011) Mechanisms of top-down attention. Trends Neurosci 34:210
11. Purushothaman G, Marion R, Li K, Casagrande VA (2012) Gating and control of primary visual cortex by pulvinar. Nat Neurosci 15:905
12. van Gaal S, Lamme VA (2012) Unconscious high-level information processing: implication for neurobiological theories of consciousness. Neuroscientist 18(287)
13. Larkum M (2013) A cellular mechanism for cortical associations: an organizing principle for the cerebral cortex. Trends Neurosci 36:141
14. Zhu Y, Zhu JJ (2004) Rapid arrival and integration of ascending sensory information in layer 1 nonpyramidal neurons and tuft dendrites of layer 5 pyramidal neurons of the neocortex. J Neurosci 24:1272
15. Zhu JJ (2009) Activity level-dependent synapse-specific AMPA receptor trafficking regulates transmission kinetics. J Neurosci 29:6320
16. Jiang X, Wang G, Lee AJ, Stornetta RL, Zhu JJ (2013) The organization of two new cortical interneuronal circuits. Nat Neurosci 16:210
17. Kuhn B, Denk W, Bruno RM (2008) In vivo two-photon voltage-sensitive dye imaging

reveals top-down control of cortical layers 1 and 2 during wakefulness. Proc Natl Acad Sci U S A 105:7588

18. Cauller LJ, Kulics AT (1991) The neural basis of the behaviorally relevant N1 component of the somatosensory-evoked potential in SI cortex of awake monkeys: evidence that backward cortical projections signal conscious touch sensation. Exp Brain Res 84:607
19. Letzkus JJ et al (2011) A disinhibitory microcircuit for associative fear learning in the auditory cortex. Nature 480:331
20. Christophe E et al (2002) Two types of nicotinic receptors mediate an excitation of neocortical layer I interneurons. J Neurophysiol 88:1318
21. Yuen EY, Yan Z (2009) Dopamine D4 receptors regulate AMPA receptor trafficking and glutamatergic transmission in GABAergic interneurons of prefrontal cortex. J Neurosci 29:550
22. Brombas A, Fletcher LN, Williams SR (2014) Activity-dependent modulation of layer 1 inhibitory neocortical circuits by acetylcholine. J Neurosci 34:1932
23. Chu Z, Galarreta M, Hestrin S (2003) Synaptic interactions of late-spiking neocortical neurons in layer 1. J Neurosci 23:96
24. Wozny C, Williams SR (2011) Specificity of synaptic connectivity between layer 1 inhibitory interneurons and layer 2/3 pyramidal neurons in the rat neocortex. Cereb Cortex 21:1818
25. Kubota Y et al (2011) Selective coexpression of multiple chemical markers defines discrete populations of neocortical GABAergic neurons. Cereb Cortex 21:1803
26. Cruikshank SJ et al (2012) Thalamic control of layer 1 circuits in prefrontal cortex. J Neurosci 32:17813
27. Ma J, Yao XH, Fu Y, Yu YC (2014) Development of layer 1 neurons in the mouse neocortex. Cereb Cortex 24:2604
28. Muralidhar S, Wang Y, Markram H (2013) Synaptic and cellular organization of layer 1 of the developing rat somatosensory cortex. Front Neuroanat 7:52
29. Lee AJ et al (2015) Canonical organization of layer 1 neuron-led cortical inhibitory and disinhibitory interneuronal circuits. Cereb Cortex 25:2114
30. Ascoli GA et al (2008) Petilla terminology: nomenclature of features of GABAergic interneurons of the cerebral cortex. Nat Rev Neurosci 9:557
31. Reyes A (2001) Influence of dendritic conductances on the input-output properties of neurons. Annu Rev Neurosci 24:653
32. Sjostrom PJ, Rancz EA, Roth A, Hausser M (2008) Dendritic excitability and synaptic plasticity. Physiol Rev 88:769
33. Spruston N (2008) Pyramidal neurons: dendritic structure and synaptic integration. Nat Rev Neurosci 9:206
34. Rockland KS, Pandya DN (1979) Laminar origins and terminations of cortical connections of the occipital lobe in the rhesus monkey. Brain Res 179:3
35. DeFelipe J, Farinas I (1992) The pyramidal neuron of the cerebral cortex: morphological and chemical characteristics of the synaptic inputs. Prog Neurobiol 39:563
36. Johnson RR, Burkhalter A (1997) A polysynaptic feedback circuit in rat visual cortex. J Neurosci 17:7129
37. Cauller LJ, Clancy B, Connors BW (1998) Backward cortical projections to primary somatosensory cortex in rats extend long horizontal axons in layer I. J Comp Neurol 390:297
38. Petreanu L, Mao T, Sternson SM, Svoboda K (2009) The subcellular organization of neocortical excitatory connections. Nature 457:1142
39. Schiller J, Schiller Y, Stuart G, Sakmann B (1997) Calcium action potentials restricted to distal apical dendrites of rat neocortical pyramidal neurons. J Physiol (Lond) 505(605)
40. Zhu JJ, Connors BW (1999) Intrinsic firing patterns and whisker-evoked synaptic responses of neurons in the rat barrel cortex. J Neurophysiol 81:1171
41. Zhu JJ (2000) Maturation of layer 5 neocortical pyramidal neurons: amplifying salient layer 1 and layer 4 inputs by Ca^{2+} action potentials in adult rat tuft dendrites. J Physiol (Lond) 526 (571)
42. Larkum ME, Zhu JJ (2002) Signaling of layer 1 and whisker-evoked Ca^{2+} and Na^{+} action potentials in distal and terminal dendrites of rat neocortical pyramidal neurons in vitro and in vivo. J Neurosci 22:6991
43. Antic SD (2003) Action potentials in basal and oblique dendrites of rat neocortical pyramidal neurons. J Physiol 550:35
44. Larkum ME, Zhu JJ, Sakmann B (1999) A new cellular mechanism for coupling inputs arriving at different cortical layers. Nature 398:338
45. Waters J, Larkum M, Sakmann B, Helmchen F (2003) Supralinear Ca^{2+} influx into dendritic tufts of layer 2/3 neocortical pyramidal neurons in vitro and in vivo. J Neurosci 23:8558
46. Larkum ME, Nevian T, Sandler M, Polsky A, Schiller J (2009) Synaptic integration in tuft dendrites of layer 5 pyramidal neurons: a new unifying principle. Science 325:756

47. Xu NL et al (2012) Nonlinear dendritic integration of sensory and motor input during an active sensing task. Nature 492:247
48. Lisman JE (1997) Bursts as a unit of neural information: making unreliable synapses reliable. Trends Neurosci 20:38
49. Huang ZJ, Di Cristo G, Ango F (2007) Development of GABA innervation in the cerebral and cerebellar cortices. Nat Rev Neurosci 8:673
50. Batista-Brito R, Fishell G (2009) The developmental integration of cortical interneurons into a functional network. Curr Top Dev Biol 87:81
51. Geiger JR et al (2002) Patch-clamp recording in brain slices with improved slicer technology. Pflugers Arch 443:491
52. Edwards FA, Konnerth A, Sakmann B, Takahashi T (1989) A thin slice preparation for patch clamp recordings from neurones of the mammalian central nervous system. Pflugers Arch 414:600
53. Wang G et al (2014) An optogenetics- and imaging-assisted simultaneous multiple patch-clamp recordings system for decoding complex neural circuits. Nat Protoc 10:397
54. Davie JT et al (2006) Dendritic patch-clamp recording. Nat Protoc 1:1235
55. Feldmeyer D, Egger V, Lubke J, Sakmann B (1999) Reliable synaptic connections between pairs of excitatory layer 4 neurones within a single "barrel" of developing rat somatosensory cortex. J Physiol 521(Pt 1):169
56. Markram H et al (2004) Interneurons of the neocortical inhibitory system. Nat Rev Neurosci 5:793
57. Kubota Y (2014) Untangling GABAergic wiring in the cortical microcircuit. Curr Opin Neurobiol 26:7

Chapter 8

Optogenetic Dissection of the Striatal Microcircuitry

Gilad Silberberg and Henrike Planert

Abstract

The striatum is the principal input structure of the basal ganglia, comprised almost entirely of inhibitory neurons, which include projection neurons and a small yet diverse population of interneurons. Striatal afferents include glutamatergic inputs from the neocortex and thalamus, and massive dopaminergic input from the substantia nigra *pars compacta*. In order to better understand the operational roles of striatum, it is essential to have a good grasp of its microcircuitry, namely a detailed description of its neuron types and their synaptic connectivity. Traditionally, studying synaptic connectivity between identified neurons was performed using paired and multineuron intracellular recordings in brain slices. The recent introduction of optogenetic methods offers new experimental approaches for microcircuit analysis, one of which is the combination of whole-cell patch-clamp recordings and optogenetic activation of presynaptic neurons. In this chapter we present recent advances in our understanding of the striatal microcircuitry when studied with electrophysiological and optogenetic methods. We first introduce the different neuron types comprising the striatal microcircuitry and describe their basic interconnectivity as inferred from electrophysiological measurements. We then present a few recent studies performed primarily in striatal and corticostriatal slices, where the powerful combination of electrophysiology and optogenetics revised our understanding of striatal functional organization.

Key words Optogenetics, Channelrhodopsin, Patch clamp, Striatum, Slice, In vivo, Striatal microcircuit, Feedforward inhibition

1 Striatal Projection Neurons

The striatum forms the main input nucleus of the Basal Ganglia (BG). It receives excitatory glutamatergic input from cortex and dopaminergic (DAergic) modulation from Substantia Nigra *pars compacta* (SNc) [1–4], as well as ACh neuromodulation from within. Figure 1a depicts afferent input and efferent projections of the striatum, as well as its main neuron and connection types. These were inferred from studies using electrophysiological, anatomical, histochemical, and relatively recently also transgenic techniques.

The medium spiny projection neurons (MSNs) comprise the vast majority of striatal neurons, taking their name from their morphological appearance [5]. MSNs project to external globus

Alon Korngreen (ed.), *Advanced Patch-Clamp Analysis for Neuroscientists*, Neuromethods, vol. 113,
DOI 10.1007/978-1-4939-3411-9_8, © Springer Science+Business Media New York 2016

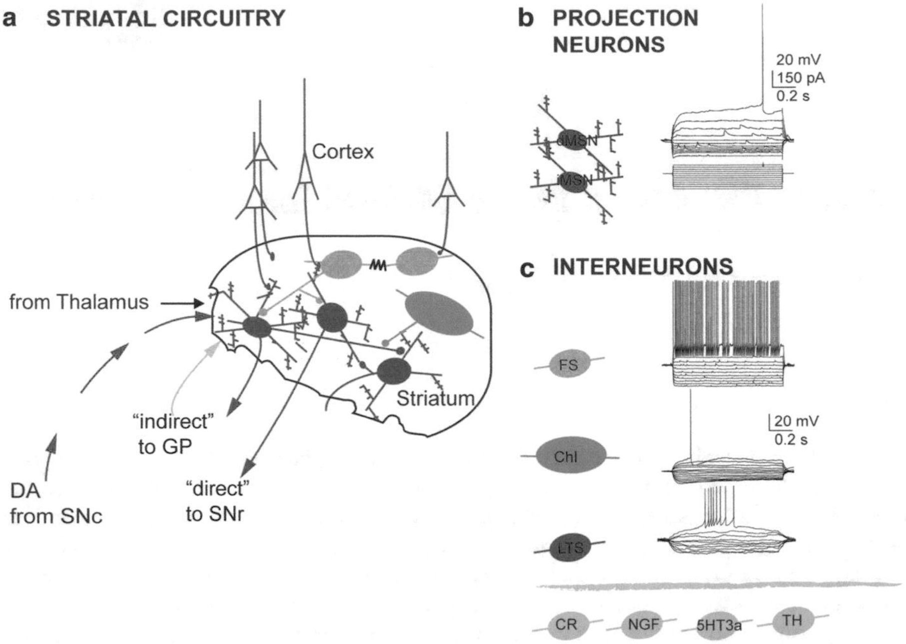

Fig. 1 Main striatal circuitry and neuron types. (a) Striatal afferent and efferent projections and main connectivity. The striatum receives glutamatergic input from cortex and thalamus (*black*), dopaminergic modulation (*yellow*) from Substantia Nigra pars compacta (SNc), and feedback from globus pallidus (GP, *light green*). Its main neuron type, the medium spiny neuron (MSN, *dark green*), sends projections to GP and Substantia Nigra pars reticulata (SNr). Within the striatum, Fast-spiking interneurons (FSIs, *light green*) feed forward to MSNs, and MSNs display collateral inhibition onto other MSNs. In addition, cholinergic neurons (*pink*) feed forward to MSNs via an axo-axonic pathway that has been unraveled with optogenetic methods. (b) Characteristic nonlinear voltage response of an MSN to injection of equal amplitude current steps. MSNs of the different "direct" and "indirect" pathways (dMSNs/iMSNs) have similar morphological and electrophysiological properties. (c) Typical electrophysiological responses of the "classical" striatal interneuron types to current steps: FSIs, cholinergic interneurons, and "low threshold spiking" (LTS) interneurons. Additional subtypes (*grey*) have been characterized based on immunohistochemistry only (Calretinin expressing interneurons, CR), or relatively recently with the help of transgenic mice through their expression of specific proteins and consecutive electrophysiology. Parts of the figure are reproduced, with permission, from Planert et al. [56]

pallidus (GPe, GP in rats), Substantia nigra *pars reticulata* (SN*r*), and internal globus pallidus (GPi, entopeduncular nucleus in rats). They discharge at low rates in vivo [6, 7]. Because MSNs appear similar based on anatomical and electrophysiological grounds, they were long investigated as one electrophysiological subtype [8–11]. Only with the introduction of transgenic mice expressing EGFP under control of the promoter of D1 and D2 receptors, separate descriptions of the MSN types projecting via different pathways have become possible (see below).

In vivo, under anesthesia as well as during slow wave sleep, the membrane of MSNs alternates between a hyperpolarized "down state" and a depolarized "up state" [12–14]. The up state is driven by cortical input and its membrane potential is determined by specific potassium channels [13]. When the animal is awake, however, the membrane potential distribution of MSNs does not appear bimodal but centered around one peak only [15].

In vitro, medium spiny neurons rest at hyperpolarized membrane potential similar to the down state. They show characteristic ramp responses and delayed firing to near-threshold current steps, which are due to slowly inactivating potassium conductances [9]. MSNs furthermore display nonlinear membrane properties with inward and outward rectification, making the input resistance highly dependent on the membrane potential of the neuron, and leading to nonlinear IV relationships (Fig. 1b). Specifically, potassium inward rectifying conductances clamp the membrane potential to hyperpolarized values, whereas depolarization-activated potassium channels lead to outward currents at near-threshold membrane potentials [10].

1.1 The "Direct" and "Indirect" Striatal Pathways

A subset of MSNs projects *directly* to BG output structures substantia nigra *pars reticulata* (SN*r*) and GPi. Others send axons to GP, which in turn projects to the BG output structures. The idea has been put forward that these two *direct* and *indirect* striatofugal pathways via different effect on BG output functionally oppose each other in motor control, such that the direct pathway facilitates and the indirect pathway inhibits movements. This "parallel" [16] direct and indirect pathway organization of striatofugal projections has been highly influential in models of function and dysfunction of the BG in motor control [17, 18]. The opposite effects of dopamine (DA) depletion on the two pathways lead to the assumption that DA affects direct and indirect pathways differentially [19]. From gene expression experiments it was furthermore inferred that the effects are mediated by selective expression of D1 Rs in striatonigral, and D2 Rs in striatopallidal MSNs, respectively [20]. Also the neuropeptides substance P and Dynorphin versus Enkephalin are expressed in direct and indirect pathways, respectively [5].

As described above, MSNs were initially treated as one morphological and electrophysiological subtype. Only relatively recently, transgenic BAC mice have become available [21], allowing the separate investigation of the striatofugal pathways based on their receptor expression, for example through expression of EGFP under the promoters of D1 or D2 receptors. A few studies have addressed electrophysiological properties of the neuron types in the transgenic mouse model or in retrogradely labeled rat MSNs, consistently finding differences in intrinsic excitability [22–25].

2 Striatal Interneurons

Embedded within the vast majority of projection neurons is a small yet diverse population of interneurons, making roughly 5 % of all striatal neurons (Fig. 1a, c). Most of these interneurons are inhibitory (GABAergic) neurons of different types, classified according to their expression of molecular markers, morphology, and electrical properties. One unique type of striatal neuron is the cholinergic interneuron, which is the only type of striatal neuron that is not GABAergic.

2.1 Cholinergic Interneurons (ChI)

Cholinergic interneurons (ChI) comprise ~1 % of striatal neurons and are characterized by a large soma, "sag" response to step current injection, and spontaneous "tonic" discharge in vivo and in slice recordings [26, 27]. ChIs express Choline acetyltransferase (ChAT) and release acetylcholine (ACh) within the striatum, affecting different types of neurons and synapses via both muscarinic and nicotinic receptors [28]. ChIs were presumed to be the sole source of acetylcholine to the striatum; however, an afferent cholinergic pathway has recently been described, arising from the PPN [29]. ChIs are relatively well-studied neuron type due to their tonic activity in vivo, which makes them easily identified in extracellular recordings, and their large somata, enabling their identification in slice recordings even without specific fluorescent markers. ChIs, or TANs, as referred to in in vivo extracellular studies, were shown to decrease their discharge rate in response to reward-related stimuli, simultaneous with increase of midbrain DA neuron discharge [30], suggesting that striatal DA levels are increased concomitantly with the cholinergic pause.

2.2 Parvalbumin (PV) Expressing Interneurons

Parvalbumin (PV) expressing interneurons are the most studied GABAergic striatal interneuron. PV expressing interneurons are also referred to as Fast-Spiking (FS) interneurons due to their electrophysiological properties, including narrow action potentials, deep afterhyperpolarization, non-accommodating or stuttering discharge response to step current injection, and the ability to discharge at high frequencies of several hundred Hz [31]. Morphologically, PV interneurons have aspiny dendrites and dense axonal arbor, which is typically larger than the dendritic tree. Striatal PV interneurons share most of their neuronal properties with neocortical and hippocampal basket cells, which are also PV interneurons. However, PV interneurons in these brain regions are more diverse and include several subtypes, such as the chandelier (axo-axonal) interneurons that have no striatal counterpart [32]. Striatal PV interneurons, in contrast, appear to constitute a relatively homogeneous population although they do not all co-express certain markers such as the serotonin 5HT3a receptor [33].

2.3 Somatostatin (SOM) Expressing Interneurons

Somatostatin (SOM) expressing interneurons are GABAergic interneurons which are also referred to as low threshold spiking (LTS) interneurons [31]. These are interneurons with high input resistance, narrow action potentials, accommodating discharge pattern in response to step current injection, often triggering a calcium-mediated plateau potential, hence referred to in some studies as pLTS interneuron, and rebound action potential discharge following hyperpolarizing current injections [34].

SOM interneurons have smaller and sparser axonal arbors than PV interneurons. Apart from somatostatin they typically express additional molecular markers such as nitric oxide synthase (NOS) and neuropeptide-Y (NPY); however, it was recently shown that NPY is not an exclusive marker for SOM interneurons but also for a different class of interneurons, the neurogliaform cells [35].

2.4 Neurogliaform (NGF) Interneurons

Neurogliaform (NGF) interneurons are the second type of interneurons expressing NPY. They are characterized by their dense and spherical arborization, which was described earlier in the neocortex [36]. Their electrical properties are very different from the other class of NPY expressing neurons (SOM interneurons). Upon injection of step current they display a ramp depolarization preceding action potential discharge, hence referred to as "late spiking." Their discharge pattern is non-accommodating and the input resistance is not as high as of SOM interneurons. Interestingly, cortical NGFs are the main source of GABAb input to pyramidal cells and make a very slow type of GABAa synapse, and the latter has also been observed in the striatum [35, 37, 38].

2.5 Calretinin Interneurons

One of the molecular markers used to initially classify striatal interneurons was the calcium binding protein Calretinin [34, 39]. Calretinin expressing interneurons are GABAergic and form a separate, non-overlapping class of interneuron, accounting for less than 1 % of striatal neurons in rodents [40] and higher proportion in the primate striatum [41]. Whereas other interneuron types such as those expressing ChAT, PV, and SOM/NO have been extensively characterized [31], very little is known about the intrinsic and synaptic properties of CR interneurons.

2.6 Other Interneuron Types

In addition to the interneuron types mentioned above, there have been new subtypes described recently based on BAC transgenic animals expressing EGFP under the promoters of specific genes such as those coding for tyrosine hydroxylase (TH) and the 5HT-3a type receptor (5HT3aR-EGFP) [33, 42, 43]. In these studies labeled neurons were all interneurons, yet they displayed diverse electrical, morphological, and molecular properties. This suggests that such interneuron classes contain several subtypes, some of which overlap with previously defined interneuron classes. Interestingly, in both cases of such interneuron types, the TH and 5HT3a interneurons, there is no evidence for dopamine signaling or the expression of 5ht3a receptors in all labeled interneurons, respectively [33, 43].

3 Intrastriatal Synaptic Connectivity

MSNs of both projection types receive strong cortical as well as thalamic innervation [44], and also FSIs receive glutamatergic synapses from both cortex and thalamus [45]. Within the striatal microcircuitry, two main modes of operation have been described: "Feedforward" inhibition, mainly as the one between FSIs and MSNs [46–48], as well as GABAergic "feedback" inhibition between MSN collaterals [49–51]. Figure 1a depicts these and some of the other main intrastriatal connectivity types described below.

In Fig. 2, different experimental configurations of studying intrastriatal connectivity with electrophysiological methods in the slice preparation are shown (Fig. 2a–d), as well as some example recordings of feedforward and feedback connectivity (Fig. 2e, f). Relatively unspecific information as regarding the presynaptic neuron type is obtained employing presynaptic macro- or microstimulation in combination with patch recording of the postsynaptic neuron and pharmacological blockade of specific afferences (Fig. 2a). Multineuron patch recordings allow the detailed electrophysiological characterization and classification of pre- and postsynaptic neuron, and, when combined with retrograde labeling, also information about the projection target (Fig. 2b). However, as interneuronal populations are small and most of them cannot easily be told apart from the very abundant MSNs in the infrared picture, such a "blind" approach of patching interneurons is very time-consuming. The introduction of transgenic mice expressing fluorescent protein under the promoters of different genes allowed directed patching of interneuronal subpopulations (Fig. 2d–f). However, as mentioned above, it also opened new questions about interneuronal classifications based on electrophysiological, anatomical, and neurochemical grounds. The studies described below allowed the description of basic striatal microcircuitry mostly through multineuron patch-clamp recordings, and sometimes in combination with transgenic techniques.

3.1 Feedforward and Feedback Inhibition in the Striatum

In vitro, FSIs can delay or inhibit MSN discharge [46], and also in vivo the activity of projection neurons inversely covaries with FSI activity [48].

In comparison to FSI-MSN synaptic inhibition, the effect of collateral inhibition between MSNs is rather small as measured at the soma [52, 53]. This small somatic response size argues against a competitive winner-take all mechanism between MSNs [45, 52].

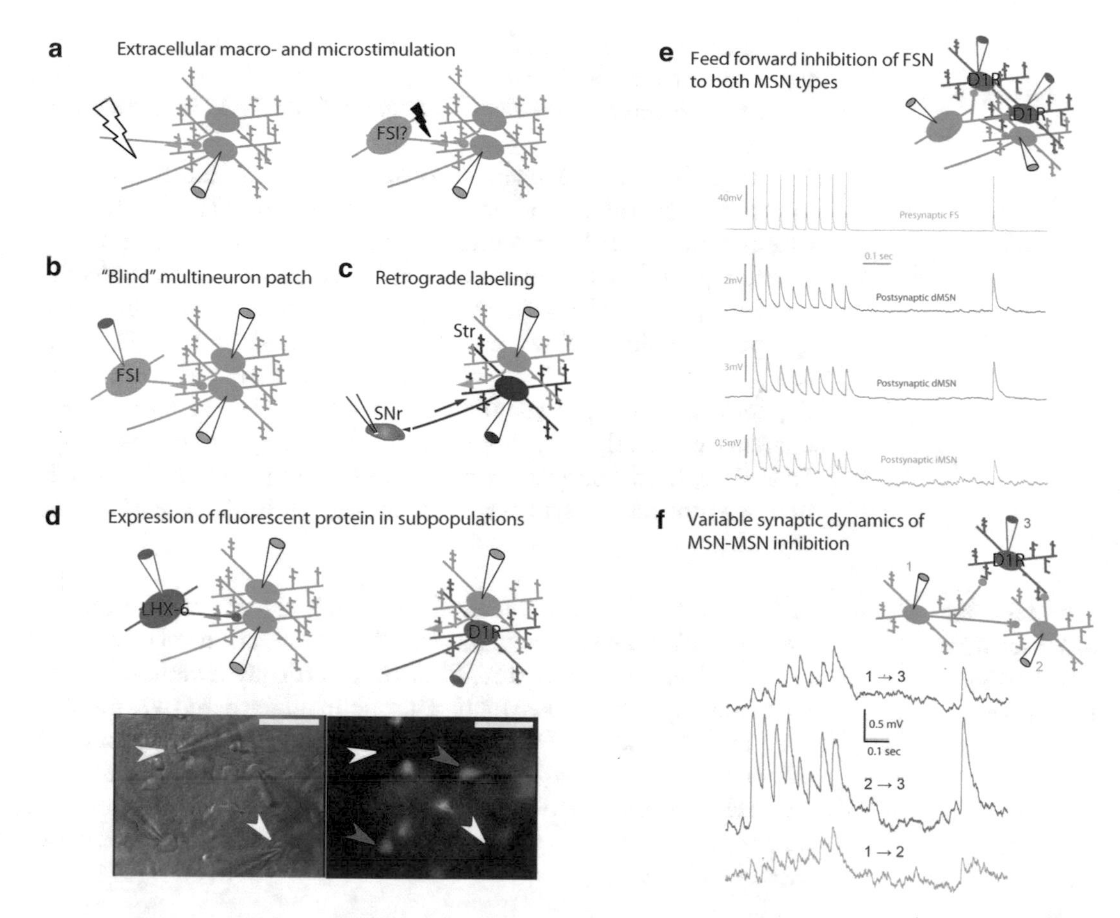

Fig. 2 Intrastriatal connectivity as investigated with electrophysiological methods in the slice preparation. (**a**) Presynaptic macro- or microstimulation of afferences (*green*) and patch recording of the postsynaptic neuron (*grey*). Intrastriatal microstimulation (right scheme) can give information about the amplitude of a unitary response in the postsynaptic neuron, but the electrophysiological signature of the presynaptic neuron remains unknown. (**b**) Multineuron patch recordings allow electrophysiological description of both pre- and postsynaptic neuron. When combined with retrograde labeling of SNr, information about the projection target of MSNs can be inferred (**c**). (**d**) Directed patching of neuronal subpopulations marked with fluorescent protein in slices from transgenic mice, e.g., interneuronal populations in LHX-6-EGFP mice (*left*), or patching of "direct pathway" MSNs in Drd1 EGFP mice (*right*). (**e**, **f**) Example recordings of feedforward and feedback connectivity in the striatum, respectively, measured in Drd1-EGFP transgenic mice, and allowing inferences about the connectivity and dynamics in relation to direct and indirect striatofugal pathways (see also text). Figure 2d, e, f is partially reproduced, with permission, from Planert et al. [56]

Also the convergence of FSIs onto MSNs is large [45], and connectivity as measured in the slice is consequently robust and widespread [46, 54–56], whereas interconnectivity for MSNs is sparse [49, 50, 56–58]. This pattern somewhat resembles the organization of cortical neuronal microcircuits. Here, similar fast-spiking interneurons feed forward onto pyramidal cells seemingly unspecifically and with high connection probabilities [59], while projection neurons themselves are only sparsely connected [60, 61].

Apart from FSIs also LTS neurons have been described to feed forward onto MSNs [46]. High resistance "LTS" neurons in turn appear to receive feedforward synapses from FSIs with relatively low probability [62].

Regarding feedback and feedforward intrastriatal connectivity in relation to direct and indirect pathways or D1/D2 receptor expression, an early immunohistochemical and electron microscopic report of interconnectivity between the different projection neuron types [63] was later complemented with the finding of synaptic coupling of D1 and D2 MSNs in vitro [56, 57]. The study by Taverna et al. concluded rather differential connectivity depending on the presynaptic MSN type, and later results were in agreement with these findings. Both MSN subtypes are targeted by FSIs with high connection probabilities, and even individual FS neurons appear to synapse onto MSNs of both types [54, 56, Fig. 2e].

3.2 Electrical Synapses Between Fast-Spiking Interneurons

A fast mode of communication by which signals can be directly transmitted between neurons is the connection via electrical synapses [64]. In the cortex, FSIs are electrically coupled by these so-called gap junctions, which can synchronize them [47, 65, 66]. Also in the striatum, FSIs interconnect via electrical synapses [46, 47; see Fig. 1a], and they are also connected through chemical synapses, just as neocortical FSIs [54, 65, 66]. They, however, do not appear to be broadly synchronized in vivo and in a network model [67, 68].

3.3 Synaptic Dynamics and Plasticity in Striatal Microcircuits

Synaptic transmission in the nervous system is dynamic, such that synaptic response amplitudes change when receiving multiple presynaptic inputs within tens or hundreds of milliseconds. At synapses between principal cortical neurons for example, response amplitude decreases for consecutive APs, and other cortical synapses show facilitating dynamics [69, 70]. This may allow for dynamic change of the weight of specific synapses within small networks.

One common way of quantifying short-term plasticity is by calculating the ratio of two consecutive synaptic responses (the paired-pulse ratio, PPR). A PPR that is larger than one is an indication for a presynaptic mechanism of vesicle depletion, leading to diminished presynaptic release probability. Tsodyks and Markram developed a model in order to describe the dynamical properties of cortical synapses quantitatively, extracting parameters for both recovery from depression and facilitation [69, 71]. The phenomenological model estimates several parameters to fit the average postsynaptic responses to a train of action potentials (APs). These include the absolute synaptic efficacy of the synapse (i.e., the response if release probability is maximal, this could depend on the postsynaptic receptor density and the number of release sites). The initial PSP amplitude as well as the amplitudes in response to consecutive APs will depend on the release probability of the synapse. The available fraction of resources is reduced with each AP according to the release probability. A time constant describes the time course of recovery of these resources. In a later version of this model, a facilitatory mechanism is added, where the release probability is dynamic as well.

Short-term plasticity within the striatal microcircuitry has been described at the different synaptic contacts [50–53, 56, 72, 73]. When investigating striatal feedforward and feedback connections in identified neurons with the methods of analysis described above, FSIs provide a strong and homogeneously depressing "feedforward" inhibition of both striatonigral/D1 and striatopallidal/D2 MSNs [56; Fig. 2e]. When using experimental data in order to model the postsynaptic effects of the depressing feedforward inhibition that result from naturally occurring FSI discharge patterns, in vivo variability in FSI discharge translates to high variability in postsynaptic amplitudes due to the strong depression of the FSI-MSN synapse [74].

In contrast to depressing FSI-MSN connections, both MSN types receive sparse and variable, depressing and facilitating synaptic "feedback" transmission from other MSNs [56]. In the cortex, the same presynaptic cell can lead to different dynamics of postsynaptic response, depending on the target neuron [71]. However, in the striatum, neither the individual postsynaptic MSN nor its type seems to determine the dynamics of the connection, at least in connections with presynaptic indirect pathway/D2 MSNs (Fig. 2f). Instead, the individual presynaptic MSN (not its projection type) may dictate the dynamics of the response. This may be important in shifting the weights between different interconnected subnetworks depending on the input pattern the postsynaptic neuron receives.

Plastic changes can also occur as long-term effects, for example in response to sustained stimulation as long-term potentiation (LTP) of the synaptic response [75]. Such long-term effects of synaptic activity are often used to model learning in synaptic networks. Depending on the stimulation protocol, or as effects of the timing of pre- and postsynaptic activity, long-term depression (LTD) or spike-timing dependent plasticity (STDP) can occur, respectively. In a multitude of studies, these plastic phenomena have been described in the striatum, often at glutamatergic (corticostriatal) synapses [76–79].

Interestingly, intrastriatal synapses exhibit inhibitory plasticity (iLTD) through depression at GABAergic synapses onto MSNs [80–82]. There is some evidence that this depression can shape striatal output in response to cortical activity [58, 80].

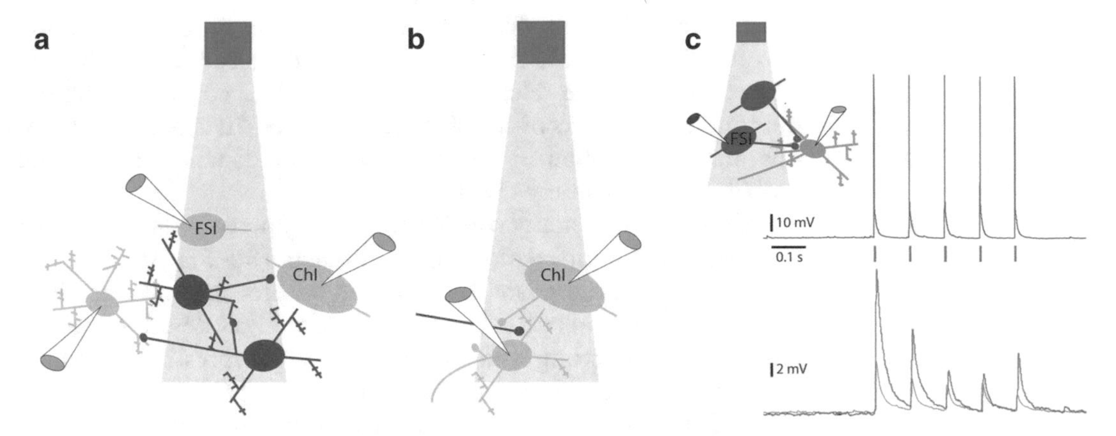

Fig. 3 Neuronal interconnectivity as studied using optogenetic tools. (**a**, **b**) Schemes of activation of neurons and activation of fibers using LED stimulation of the slice preparation. (**c**) *Inset*: example recording of MSN (*grey*) and fluorescent FSI (*red*) in striatum of PV-Cre mouse injected with AAV and expressing mCherry and Channelrhodopsin in FSIs. A train of APs in the presynaptic FSI (*red trace*) elicits a depressing synaptic response (average, *bottom grey trace*). LED stimulation with the same frequency (*light blue*) while ensuring AP response in the FSI leads to large amplitude depressing response in the MSN (*bottom*, *light blue trace*), suggesting activation of multiple FSIs, as described in [62]

3.4 Optogenetics in Striatal Microcircuit Research

The development of optogenetic tools in recent years has significantly expanded the available toolbox for the research of neural circuits, linking specific neuron types and pathways to behavior, brain states, and diseases [83, 84]. In particular, it enables the investigation of local as well as afferent synaptic pathways mediated by specific presynaptic neuronal populations. The specificity is obtained by means of Cre dependent expression of light sensitive channels or pumps (such as Channelrhodopsin and Halorhodopsin) and by targeted topographic injection of the viral constructs into specific brain regions. Investigation of the synaptic pathways can then be performed either under in vivo conditions or in the slice, depending on the question at hand (Fig. 3). In the following section we will present some of the recent advances in our understanding of the striatal microcircuitry that utilized the combination of optogenetics and electrophysiology in slices and in vivo.

3.4.1 In Vitro Striatal Optogenetics

Studying the synaptic connectivity between different types of neurons requires the unambiguous identification of both pre- and postsynaptic neurons. This level of certainty is typically obtained by paired intracellular recordings in which neurons are electrically, morphologically, and chemically characterized during and after recording synaptic connections between them (Fig. 2b–f, see Sect. 5). With the aid of fluorescent microscopy, selection of identified neurons is facilitated if they are labeled with a fluorescent marker such as EGFP or retrograde fluorescent beads [56, 60]. The advantage of optogenetic stimulation in such experiments lies in the possibility to reliably activate specific presynaptic populations and not only single presynaptic neurons (Fig. 3a, see Sect. 5). This advantage is also used in activating specific afferents while recording only from postsynaptic neurons, which is especially helpful when the originating neuron is far away from its target. Traditionally, stimulation of such afferents was achieved by extracellular stimulation (Fig. 2a); however, this method has important limitations due to the nonspecific activation of different axons, both local and afferent. Using optogenetic stimulation of terminals allows the activation of afferents of more specific type and topographic origin (Fig. 3b). One of the first such studies that utilized optogenetic stimulation of a defined neuronal population in the basal ganglia was that of Chuhma and colleagues [85]. In that study the authors created a mouse expressing ChR2 in striatal MSNs to elucidate their intra- and extrastriatal connectivity. By photostimulating in slices the authors could activate local intrastriatal connections as well as the long-range connections between MSNs and their targets at the Substantia Nigra and Globus Pallidus. Using this method allowed for a functional characterization of MSN synaptic connections between MSNs and certain neuron types, and more importantly, showed a high degree of selectivity in postsynaptic targets within the receiving network. In the striatum, MSNs connect to neighboring MSNs and to a lesser degree to neighboring ChIs; however, they do not connect to FS interneurons at all (Fig. 3a). While such data could be accumulated using extensive paired recordings, light activation of many presynaptic terminals of the same type provided a faster and more reliable method to quantify the connectivity. In the SN, photostimulation of MSN synaptic terminals showed that striatal projections specifically targeted GABAergic neurons and avoided DA neurons. Such findings cannot be obtained by paired recordings, and neither earlier anatomical [86] nor electrophysiological [87, 88] studies could provide a conclusive connectivity scheme due to the unspecific electrical stimulation.

By using Cre-dependent expression of channelrhodopsin, several studies could address questions pertaining to the interconnectivity between different types of striatal neurons. In most such studies, optogenetic expression is induced by viral injections, enabling not only the molecular specificity but also region specificity.

Cholinergic interneurons have been of special interest in the investigation of the intrastriatal microcircuitry due to their unique properties, including their non-GABAergic synapses and tonic activity. Using optogenetic stimulation, English and colleagues [89] showed that simultaneous activation of ChIs induced strong and complex disynaptic inhibition onto MSNs. Such inhibition was not observed following stimulation of single ChIs, but required simultaneous activation of several ChIs that was possible only via selective optogenetic activation of ChAT-expressing interneurons. The source of ChI-MSN disynaptic inhibition was fully unraveled a few years later, following the publication of three additional studies, which also used the combination of optogenetics and electrophysiological recordings in striatal slices [90–92]. In the first of these studies, Cragg and colleagues showed that simultaneous activation of ChIs induced DA release in the striatum by activation of DAergic terminals by nicotinic receptors [90]. In a second paper, Sabatini and colleagues showed that photostimulation of nigrostriatal DAergic terminals in striatum co-released GABA, glutamate, and DA [91, 93]. Based on these studies, Kreitzer and colleagues then showed that optogenetic activation of ChIs mediated GABAergic responses in MSNs by activating nigrostriatal DA terminals via an axo-axonic disynaptic pathway [92]. Local inhibition of MSNs is therefore mediated from neighboring MSNs [56, 85], ChIs [89], and FS interneurons [46, 56, 62], (see Fig. 1a for schematic overview). Feedforward inhibition by FS interneurons onto MSNs is particularly robust, with large synaptic amplitudes and high connection probability, similar to the connectivity of cortical FS interneurons onto pyramidal cells [94]. In contrast, the connectivity between FS interneurons and ChIs was tested using optogenetics and multineuron patch-clamp recordings, showing a complete avoidance of ChIs even when located amongst strongly inhibited MSNs [62].

The strongest drive for MSNs arises from glutamatergic inputs by infragranular cortical pyramidal cells. It has been suggested based on anatomical studies that cortical neurons belonging to the pyramidal tract (PT) favor contacting indirect pathway MSNs in the ipsilateral striatum, while intratelencephalic (IT) pyramidal cells target direct pathway MSNs of both hemispheres [95]. This bias in connectivity was recently tested using optogenetic activation of PT and IT pyramidal cells and patch-clamp recordings from direct and indirect pathway MSNs [96]. The authors expressed ChR2 in the different cortical populations using retrograde viral transport from axon terminals, and combined photostimulation with fluorescence identification of the target MSNs. In contrast to the above prediction, the study showed that both direct and indirect pathway MSNs, as classified by their D1 and D2 expression, received excitatory synaptic inputs from PT and IT pyramidal cells.

Long-term synaptic plasticity of intrastriatal connections has not been extensively studied, partly due to the difficulty in accessing specific inhibitory pathways. With the combination of optogenetics, slice electrophysiology, and pharmacology, Lovinger and colleagues could dissect out intrastriatal pathway specific rules of iLTD in MSNs [97]. Synaptic activation was done using optogenetic stimulation of specific presynaptic terminals before and after induction of iLTD by electrical stimulation. Optogenetic activation was thereby used as a more accurate replacement for the extracellular stimulation, enabling the assessment of the contribution of specific presynaptic neuron types, in particular GABAergic synapses from MSNs or FS interneurons.

3.4.2 In Vivo Striatal Optogenetics

Optogenetics is also extensively used in vivo, where light induced activation and silencing alters behavior or activity patterns of recorded neurons. A complete overview of these numerous studies is beyond the scope of this chapter, and we will mention just a few below. One of the first such experiments used Cre transgenic mice for the direct and indirect striatal pathways, showing that optogenetic activation of direct pathway neurons increased locomotion while activation of indirect pathway MSNs caused freezing and an overall reduction in movement [98], thus supporting a long-standing classical model of the direct and indirect BG pathways [16, 19]. The BG are also involved in reward-related functions and addiction, and several studies have used optogenetic manipulation to show the impact of various pathways and neuronal populations related to reward and addiction. In this type of studies, light is delivered via fiber optic in the freely moving animal and the effect of activating or inhibiting certain cell populations is measured for different behavioral tasks [99, 100, see Sect. 5]. For example, the role of ChIs in cocaine-induced behavior was elegantly shown by optogenetically silencing ChI discharge during place preference experiments in rats [100]. A method combining patch-clamp and optogenetic stimulation was recently published, which enables whole-cell recordings from deep structures by using the recording patch pipette also as the light source for optical stimulation [101].

4 Summary and Outlook

In recent years there has been an "explosion" in the number of studies combining slice electrophysiology and optogenetics to study neural circuits. The main advantage of this combination lies in the ability to activate (or silence) specific neuronal populations and afferent pathways, based on their molecular identity and topographic location. There are, however, limitations in the optogenetic approach that should be taken into account in the study design and interpretation of the results. In several studies, including some mentioned above, optogenetic activation evoked synchronous discharge of several neurons in millisecond precision. This type of activation rarely occurs under physiological conditions, as is the case with optogenetic activation of ChIs [89, 90], which might lead to abnormal activation of downstream synaptic pathways. Another consideration is the selectivity in the optogenetic expression, which might cause ambiguity in the results. For example, an afferent cholinergic pathway to striatum was recently discovered

[29], therefore implying that optogenetic activation in ChAT-ChR2 transgenic mice [89] could have stimulated not only striatal ChIs but also cholinergic afferents. Other points to consider in the design of experiments are the intrinsic properties of the optogenetic constructs, such as their inactivation values and frequency limits. The variety of opsins, viral constructs, and transgenic animals is, however, continuously increasing, thus enabling precise choice of tools tailored for the specific questions at hand.

The combination of optogenetics and electrophysiology is a powerful match that is very likely here to stay. In recent years it has dramatically increased our understanding of the functional organization of the striatal microcircuitry. With the development of new transgenic lines and optogenetic tools, we can expect comprehensive descriptions of many neural micro- and macrocircuits, including those of the basal ganglia.

5 Methods

5.1 Slice Preparation and Whole-Cell Recordings, Standard Protocols as Modified from [62] and [56]

Following anesthesia, mouse brains are removed in ice-cold ACSF containing (in mM): 125 NaCl, 25 glucose, 25 $NaHCO_3$, 2.5 KCl, 2 $CaCl_2$, 1.25 NaH_2PO_4, 1 $MgCl_2$. 250 μm thick slices are cut (Leica VT 1000S, Leica Microsystems, Nussloch GmbH, Germany), then transferred to 35 °C for 30–60 min and after that kept at room temperature (22–24 °C). Whole-cell patch recordings are obtained at 35 ± 0.5 °C. Glass electrodes are pulled with a Flaming/Brown micropipette puller P-97 (Sutter Instruments Co, Novato, Ca) and a resistance of 5–10 MΩ. The intracellular solution contains (in mM) 105 K-gluconate, 30 KCl, 10 HEPES, 10 Na-Phosphocreatine, 4 Mg-ATP, 0.3 Na-GTP, and in some experiments 0.3 % neurobiotin/biocytin. Neurons are visualized with IR-DIC microscopy (Zeiss FS Axioskop, Oberkochen, Germany) and fluorescent microscopy using a mercury lamp (HBO 100, Zeiss) and a fluorescent filter cube mounted on the same microscope. Pairs, triplets, and quadruplets of neurons are recorded within a range of 200 μm from mCherry fluorescently labeled somata. Recordings with access resistance above 35 MΩ are discarded. Recordings are amplified using multiclamp 700B (Molecular Devices, CA, USA) and digitized by an ITC-18 (HEKA Elektronik, Lambrecht, Germany) acquisition board. Data is acquired and analyzed using IGOR Pro (Wavemetrics, OR, USA). Neurons are fluorescently labeled by expression of fluorescent proteins (GFP, mCherry, tdTomato, RFP) or by retrograde labeling with fluorescent latex beads. Such bead injections are described in detail in previous publications [56]. In both cases, visualization is done by switching to fluorescent microscopy while the patch-clamp approach and recording is done in IR-DIC mode.

5.2 Photostimulation

Photostimulation in our experiments was generated through a 1 Watt blue LED (wavelength 465 nm) mounted on the microscope oculars and delivered through the objective lens. The LED is fixed at the end of a tube that was inserted instead of the microscope eyepiece. Such tubes are now commercially available from companies such as Mightex, Thorlabs, and others. In a few experiments the same LED was mounted under the slice as described previously [89], providing illumination to the entire field from below. Photostimulation is controlled by a LED driver (Mightex systems, CA, USA) connected to the ITC-18 acquisition board, enabling control over the duration and intensity. The photostimulation diameter through the objective lens is approximately 400 μm with an illumination intensity of 9 mW/mm^2. Other modes of photostimulation include coupling of fiber optic to the microscope and stimulation using laser, as described in [90], or a TTL controlled shutter of the fluorescent light source, as described in [92].

5.3 In Vivo Photostimulation

In vivo *photostimulation* has been described extensively in numerous publications and is beyond the scope of this chapter. In the study described here by the Kreitzer laboratory [98], photostimulation was done via optic fibers that were bilaterally inserted into both striatal hemispheres. Optic fibers were either coupled with silicone probes (Neuronexus Technologies) thus forming an "optrode," or used independently, for optogenetic manipulation of behavior without electrophysiological recordings.

In order to avoid the separate insertion of an optic fiber for optogenetic stimulation in deep patch-clamp recordings, a new method was recently published that enables the glass patch pipette to be used also as the light guide for optogenetic stimulation [101]. The light source for the "optopatcher" (A-M systems, WA, USA) could be a laser or LED coupled to the optic fiber inserted in the pipette holder. The advantage of such a system would be the precise delivery of light to the vicinity of the recorded neuron.

References

1. Kincaid AE, Zheng T, Wilson CJ (1998) Connectivity and convergence of single corticostriatal axons. J Neurosci 18(12):4722–4731
2. Joel D, Weiner I (2000) The connections of the dopaminergic system with the striatum in rats and primates: an analysis with respect to the functional and compartmental organization of the striatum. Neuroscience 96(3): 451–474, doi: S0306-4522(99)00575-8 [pii]
3. Calabresi P, Pisani A, Mercuri NB, Bernardi G (1996) The corticostriatal projection: from synaptic plasticity to dysfunctions of the basal ganglia. Trends Neurosci 19(1):19–24, doi: 0166223696818625 [pii]
4. Moss J, Bolam JP (2008) A dopaminergic axon lattice in the striatum and its relationship with cortical and thalamic terminals. J Neurosci 28(44):11221–11230. doi:10.1523/JNEUROSCI.2780-08.2008, 28/44/11221 [pii]
5. Gerfen CR (2004) Basal ganglia. In: Paxinos G (ed) The rat nervous system, 3rd edn. Elsevier Academic Press, San Diego, London, pp 455–508
6. Wilson CJ, Groves PM (1981) Spontaneous firing patterns of identified spiny neurons in the rat neostriatum. Brain Res 220(1):67–80, doi: 0006-8993(81)90211-0 [pii]

7. Berke JD, Okatan M, Skurski J, Eichenbaum HB (2004) Oscillatory entrainment of striatal neurons in freely moving rats. Neuron 43 (6):883–896. doi:10.1016/j.neuron.2004.08.035, S0896627304005628 [pii]
8. Kawaguchi Y, Wilson C, Emson P (1990) Projection subtypes of rat neostriatal matrix cells revealed by intracellular injection of biocytin. J Neurosci 10(10):3421–3438
9. Nisenbaum ES, Xu ZC, Wilson CJ (1994) Contribution of a slowly inactivating potassium current to the transition to firing of neostriatal spiny projection neurons. J Neurophysiol 71(3):1174–1189
10. Nisenbaum ES, Wilson CJ (1995) Potassium currents responsible for inward and outward rectification in rat neostriatal spiny projection neurons. J Neurosci 15(6):4449–4463
11. Nisenbaum ES, Wilson CJ, Foehring RC, Surmeier DJ (1996) Isolation and characterization of a persistent potassium current in neostriatal neurons. J Neurophysiol 76 (2):1180–1194
12. Wilson C (1993) The generation of natural firing patterns in neostriatal neurons. Prog Brain Res 99:277–297
13. Wilson C, Kawaguchi Y (1996) The origins of two-state spontaneous membrane potential fluctuations of neostriatal spiny neurons. J Neurosci 16(7):2397–2410
14. Mahon S, Deniau JM, Charpier S (2001) Relationship between EEG potentials and intracellular activity of striatal and cortico-striatal neurons: an in vivo study under different anesthetics. Cereb Cortex 11(4):360–373
15. Mahon S, Vautrelle N, Pezard L, Slaght S, Deniau J, Chouvet G, Charpier S (2006) Distinct patterns of striatal medium spiny neuron activity during the natural sleep-wake cycle. J Neurosci 26(48):12587–12595
16. Alexander GE, Crutcher MD (1990) Functional architecture of basal ganglia circuits: neural substrates of parallel processing. Trends Neurosci 13(7):266–271
17. DeLong MR (1990) Primate models of movement disorders of basal ganglia origin. Trends Neurosci 13(7):281–285
18. Mink JW (1996) The basal ganglia: focused selection and inhibition of competing motor programs. Prog Neurobiol 50(4):381–425, doi: S0301-0082(96)00042-1 [pii]
19. Albin R, Young A, Penney J (1989) The functional anatomy of basal ganglia disorders. Trends Neurosci 12(10):366–375
20. Gerfen CR, Engber TM, Mahan LC, Susel Z, Chase TN, Monsma FJ, Sibley DR (1990) D1 and D2 dopamine receptor-regulated gene expression of striatonigral and striatopallidal neurons. Science 250(4986):1429–1432
21. Gong S, Zheng C, Doughty M, Losos K, Didkovsky N, Schambra U, Nowak N, Joyner A, Leblanc G, Hatten M, Heintz N (2003) A gene expression atlas of the central nervous system based on bacterial artificial chromosomes. Nature 425(6961):917–925
22. Cepeda C, André V, Yamazaki I, Wu N, Kleiman-Weiner M, Levine M (2008) Differential electrophysiological properties of dopamine D1 and D2 receptor-containing striatal medium-sized spiny neurons. Eur J Neurosci 27(3):671–682
23. Gertler T, Chan C, Surmeier D (2008) Dichotomous anatomical properties of adult striatal medium spiny neurons. J Neurosci 28 (43):10814–10824
24. Planert H, Berger TK, Silberberg G (2013) Membrane properties of striatal direct and indirect pathway neurons in mouse and rat slices and their modulation by dopamine. PLoS One 8(3):e57054. doi:10.1371/journal.pone.0057054
25. Kreitzer A, Malenka R (2007) Endocannabinoid-mediated rescue of striatal LTD and motor deficits in Parkinson's disease models. Nature 445(7128):643–647
26. Aosaki T, Tsubokawa H, Ishida A, Watanabe K, Graybiel AM, Kimura M (1994) Responses of tonically active neurons in the primate's striatum undergo systematic changes during behavioral sensorimotor conditioning. J Neurosci 14(6):3969–3984
27. Bennett BD, Callaway JC, Wilson CJ (2000) Intrinsic membrane properties underlying spontaneous tonic firing in neostriatal cholinergic interneurons. J Neurosci 20 (22):8493–8503, doi: 20/22/8493 [pii]
28. Oldenburg IA, Ding JB (2011) Cholinergic modulation of synaptic integration and dendritic excitability in the striatum. Curr Opin Neurobiol 21(3):425–432. doi:10.1016/j.conb.2011.04.004
29. Dautan D, Huerta-Ocampo I, Witten IB, Deisseroth K, Bolam JP, Gerdjikov T, Mena-Segovia J (2014) A major external source of cholinergic innervation of the striatum and nucleus accumbens originates in the brainstem. J Neurosci 34(13):4509–4518. doi:10.1523/JNEUROSCI.5071-13.2014
30. Morris G, Arkadir D, Nevet A, Vaadia E, Bergman H (2004) Coincident but distinct messages of midbrain dopamine and striatal tonically active neurons. Neuron 43 (1):133–143. doi:10.1016/j.neuron.2004.06.012

31. Kawaguchi Y (1993) Physiological, morphological, and histochemical characterization of three classes of interneurons in rat neostriatum. J Neurosci 13(11):4908–4923
32. Freund TF (2003) Interneuron diversity series: rhythm and mood in perisomatic inhibition. Trends Neurosci 26(9):489–495. doi:10.1016/S0166-2236(03)00227-3
33. Munoz-Manchado AB, Foldi C, Szydlowski S, Sjulson L, Farries M, Wilson C, Silberberg G, Hjerling-Leffler J (2014) Novel striatal GABAergic interneuron populations labeled in the 5HT3aEGFP mouse. Cereb Cortex. doi:10.1093/cercor/bhu179
34. Kawaguchi Y, Wilson CJ, Augood SJ, Emson PC (1995) Striatal interneurones: chemical, physiological and morphological characterization. Trends Neurosci 18(12):527–535, doi: 0166-2236(95)98374-8 [pii]
35. Ibáñez-Sandoval O, Tecuapetla F, Unal B, Shah F, KoÛs T, Tepper JM (2011) A novel functionally distinct subtype of striatal neuropeptide Y interneuron. J Neurosci 31 (46):16757–16769. doi:10.1523/JNEUROSCI.2628-11.2011, 31/46/16757 [pii]
36. Kawaguchi Y, Kubota Y (1997) GABAergic cell subtypes and their synaptic connections in rat frontal cortex. Cereb Cortex 7 (6):476–486
37. Szabadics J, Tamas G, Soltesz I (2007) Different transmitter transients underlie presynaptic cell type specificity of GABAA, slow and GABAA, fast. Proc Natl Acad Sci U S A 104 (37):14831–14836. doi:10.1073/pnas.0707204104
38. Tamas G, Lorincz A, Simon A, Szabadics J (2003) Identified sources and targets of slow inhibition in the neocortex. Science 299 (5614):1902–1905. doi:10.1126/science.1082053
39. Bennett BD, Bolam JP (1993) Characterization of calretinin-immunoreactive structures in the striatum of the rat. Brain Res 609 (1–2):137–148
40. Tepper JM, Bolam JP (2004) Functional diversity and specificity of neostriatal interneurons. Curr Opin Neurobiol 14 (6):685–692. doi:10.1016/j.conb.2004.10.003, S0959-4388(04)00155-2 [pii]
41. Wu Y, Parent A (2000) Striatal interneurons expressing calretinin, parvalbumin or NADPH-diaphorase: a comparative study in the rat, monkey and human. Brain Res 863 (1–2):182–191
42. Ibáñez-Sandoval O, Tecuapetla F, Unal B, Shah F, Koós T, Tepper JM (2010) Electrophysiological and morphological characteristics and synaptic connectivity of tyrosine hydroxylase-expressing neurons in adult mouse striatum. J Neurosci 30 (20):6999–7016. doi:10.1523/JNEUROSCI.5996-09.2010, 30/20/6999 [pii]
43. Lee S, Hjerling-Leffler J, Zagha E, Fishell G, Rudy B (2010) The largest group of superficial neocortical GABAergic interneurons expresses ionotropic serotonin receptors. J Neurosci 30(50):16796–16808. doi:10.1523/JNEUROSCI.1869-10.2010
44. Doig NM, Moss J, Bolam JP (2010) Cortical and thalamic innervation of direct and indirect pathway medium-sized spiny neurons in mouse striatum. J Neurosci 30 (44):14610–14618. doi:10.1523/JNEUROSCI.1623-10.2010, 30/44/14610 [pii]
45. Wilson CJ (2007) GABAergic inhibition in the neostriatum. Prog Brain Res 160:91–110. doi:10.1016/S0079-6123(06)60006-X, S0079-6123(06)60006-X [pii]
46. Koós T, Tepper JM (1999) Inhibitory control of neostriatal projection neurons by GABAergic interneurons. Nat Neurosci 2 (5):467–472. doi:10.1038/8138
47. Kita H, Kosaka T, Heizmann CW (1990) Parvalbumin-immunoreactive neurons in the rat neostriatum: a light and electron microscopic study. Brain Res 536(1–2):1–15, doi: 0006-8993(90)90002-S [pii]
48. Mallet N, Le Moine C, Charpier S, Gonon F (2005) Feedforward inhibition of projection neurons by fast-spiking GABA interneurons in the rat striatum in vivo. J Neurosci 25 (15):3857–3869
49. Tunstall MJ, Oorschot DE, Kean A, Wickens JR (2002) Inhibitory interactions between spiny projection neurons in the rat striatum. J Neurophysiol 88(3):1263–1269
50. Czubayko U, Plenz D (2002) Fast synaptic transmission between striatal spiny projection neurons. Proc Natl Acad Sci U S A 99 (24):15764–15769. doi:10.1073/pnas.242428599, 242428599 [pii]
51. Guzman JN, Hernandez A, Galarraga E, Tapia D, Laville A, Vergara R, Aceves J, Bargas J (2003) Dopaminergic modulation of axon collaterals interconnecting spiny neurons of the rat striatum. J Neurosci 23 (26):8931–8940
52. Koos T, Tepper JM, Wilson CJ (2004) Comparison of IPSCs evoked by spiny and fast-spiking neurons in the neostriatum. J Neurosci 24(36):7916–7922. doi:10.

1523/JNEUROSCI.2163-04.2004, 24/36/7916 [pii]

53. Gustafson N, Gireesh-Dharmaraj E, Czubayko U, Blackwell KT, Plenz D (2006) A comparative voltage and current-clamp analysis of feedback and feedforward synaptic transmission in the striatal microcircuit in vitro. J Neurophysiol 95(2):737–752. doi:10.1152/jn.00802.2005, 00802.2005 [pii]
54. Gittis AH, Nelson AB, Thwin MT, Palop JJ, Kreitzer AC (2010) Distinct roles of GABAergic interneurons in the regulation of striatal output pathways. J Neurosci 30 (6):2223–2234. doi:10.1523/JNEUROSCI.4870-09.2010, 30/6/2223 [pii]
55. Gittis AH, Leventhal DK, Fensterheim BA, Pettibone JR, Berke JD, Kreitzer AC (2011) Selective inhibition of striatal fast-spiking interneurons causes dyskinesias. J Neurosci 31(44):15727–15731. doi:10.1523/JNEUROSCI.3875-11.2011, 31/44/15727 [pii]
56. Planert H, Szydlowski S, Hjorth J, Grillner S, Silberberg G (2010) Dynamics of synaptic transmission between fast-spiking interneurons and striatal projection neurons of the direct and indirect pathways. J Neurosci 30 (9):3499–3507. doi:10.1523/JNEUROSCI.5139-09.2010, 30/9/3499 [pii]
57. Taverna S, Ilijic E, Surmeier D (2008) Recurrent collateral connections of striatal medium spiny neurons are disrupted in models of Parkinson's disease. J Neurosci 28 (21):5504–5512
58. Adermark L (2011) Modulation of endocannabinoid-mediated long-lasting disinhibition of striatal output by cholinergic interneurons. Neuropharmacology 61 (8):1314–1320. doi:10.1016/j.neuropharm.2011.07.039
59. Packer AM, Yuste R (2011) Dense, unspecific connectivity of neocortical parvalbumin-positive interneurons: a canonical microcircuit for inhibition? J Neurosci 31(37):13260–13271. doi:10.1523/JNEUROSCI.3131-11.2011, 31/37/13260 [pii]
60. Brown SP, Hestrin S (2009) Intracortical circuits of pyramidal neurons reflect their long-range axonal targets. Nature 457 (7233):1133–1136. doi:10.1038/nature07658, nature07658 [pii]
61. Markram H, Lübke J, Frotscher M, Roth A, Sakmann B (1997) Physiology and anatomy of synaptic connections between thick tufted pyramidal neurones in the developing rat neocortex. J Physiol 500(Pt 2):409–440
62. Szydlowski SN, Pollak Dorocic I, Planert H, Carlen M, Meletis K, Silberberg G (2013) Target selectivity of feedforward inhibition by striatal fast-spiking interneurons. J Neurosci 33(4):1678–1683. doi:10.1523/JNEUROSCI.3572-12.2013
63. Yung KK, Smith AD, Levey AI, Bolam JP (1996) Synaptic connections between spiny neurons of the direct and indirect pathways in the neostriatum of the rat: evidence from dopamine receptor and neuropeptide immunostaining. Eur J Neurosci 8(5):861–869
64. Pereda AE, Curti S, Hoge G, Cachope R, Flores CE, Rash JE (2012) Gap junction-mediated electrical transmission: REGULATORY mechanisms and plasticity. Biochim Biophys Acta. doi: S0005-2736(12)00184-8 [pii] 10.1016/j.bbamem.2012.05.026
65. Galarreta M, Hestrin S (1999) A network of fast-spiking cells in the neocortex connected by electrical synapses. Nature 402 (6757):72–75
66. Gibson JR, Beierlein M, Connors BW (1999) Two networks of electrically coupled inhibitory neurons in neocortex. Nature 402 (6757):75–79. doi:10.1038/47035
67. Berke JD (2008) Uncoordinated firing rate changes of striatal fast-spiking interneurons during behavioral task performance. J Neurosci 28(40):10075–10080. doi:10.1523/JNEUROSCI.2192-08.2008, 28/40/10075 [pii]
68. Hjorth J, Blackwell KT, Kotaleski JH (2009) Gap junctions between striatal fast-spiking interneurons regulate spiking activity and synchronization as a function of cortical activity. J Neurosci 29(16):5276–5286. doi:10.1523/JNEUROSCI.6031-08.2009, 29/16/5276 [pii]
69. Tsodyks MV, Markram H (1997) The neural code between neocortical pyramidal neurons depends on neurotransmitter release probability. Proc Natl Acad Sci U S A 94 (2):719–723
70. Thomson AM, Deuchars J (1997) Synaptic interactions in neocortical local circuits: dual intracellular recordings in vitro. Cereb Cortex 7(6):510–522
71. Markram H, Wang Y, Tsodyks M (1998) Differential signaling via the same axon of neocortical pyramidal neurons. Proc Natl Acad Sci U S A 95(9):5323–5328

72. Taverna S, van Dongen YC, Groenewegen HJ, Pennartz CM (2004) Direct physiological evidence for synaptic connectivity between medium-sized spiny neurons in rat nucleus accumbens in situ. J Neurophysiol 91 (3):1111–1121. doi:10.1152/jn.00892.2003, 00892.2003 [pii]
73. Venance L, Glowinski J, Giaume C (2004) Electrical and chemical transmission between striatal GABAergic output neurones in rat brain slices. J Physiol 559(Pt 1):215–230. doi:10.1113/jphysiol.2004.065672, jphysiol.2004.065672 [pii]
74. Klaus A, Planert H, Hjorth JJ, Berke JD, Silberberg G, Kotaleski JH (2011) Striatal fast-spiking interneurons: from firing patterns to postsynaptic impact. Front Syst Neurosci 5:57. doi:10.3389/fnsys.2011.00057
75. Bliss TV, Lomo T (1973) Long-lasting potentiation of synaptic transmission in the dentate area of the anaesthetized rabbit following stimulation of the perforant path. J Physiol 232(2):331–356
76. Calabresi P, Picconi B, Tozzi A, Di Filippo M (2007) Dopamine-mediated regulation of corticostriatal synaptic plasticity. Trends Neurosci 30(5):211–219
77. Kreitzer A, Malenka R (2008) Striatal plasticity and basal ganglia circuit function. Neuron 60(4):543–554
78. Fino E, Venance L (2011) Spike-timing dependent plasticity in striatal interneurons. Neuropharmacology 60(5):780–788. doi:10.1016/j.neuropharm.2011.01.023, S0028-3908(11)00026-8 [pii]
79. Lovinger DM (2010) Neurotransmitter roles in synaptic modulation, plasticity and learning in the dorsal striatum. Neuropharmacology 58(7):951–961. doi:10.1016/j.neuropharm.2010.01.008, S0028-3908(10)00022-5 [pii]
80. Adermark L, Lovinger DM (2009) Frequency-dependent inversion of net striatal output by endocannabinoid-dependent plasticity at different synaptic inputs. J Neurosci 29(5):1375–1380. doi:10.1523/JNEUROSCI.3842-08.2009, 29/5/1375 [pii]
81. Adermark L, Lovinger DM (2007) Retrograde endocannabinoid signaling at striatal synapses requires a regulated postsynaptic release step. Proc Natl Acad Sci U S A 104 (51):20564–20569. doi:10.1073/pnas.0706873104
82. Adermark L, Talani G, Lovinger DM (2009) Endocannabinoid-dependent plasticity at GABAergic and glutamatergic synapses in the striatum is regulated by synaptic activity. Eur J Neurosci 29(1):32–41. doi:10.1111/j.1460-9568.2008.06551.x, EJN6551 [pii]
83. Boyden ES, Zhang F, Bamberg E, Nagel G, Deisseroth K (2005) Millisecond-timescale, genetically targeted optical control of neural activity. Nat Neurosci 8(9):1263–1268. doi:10.1038/nn1525, nn1525 [pii]
84. Yizhar O, Fenno LE, Davidson TJ, Mogri M, Deisseroth K (2011) Optogenetics in neural systems. Neuron 71(1):9–34. doi:10.1016/j.neuron.2011.06.004
85. Chuhma N, Tanaka KF, Hen R, Rayport S (2011) Functional connectome of the striatal medium spiny neuron. J Neurosci 31 (4):1183–1192. doi:10.1523/JNEUROSCI.3833-10.2011, 31/4/1183 [pii]
86. Gerfen CR, Baimbridge KG, Miller JJ (1985) The neostriatal mosaic: compartmental distribution of calcium-binding protein and parvalbumin in the basal ganglia of the rat and monkey. Proc Natl Acad Sci U S A 82 (24):8780–8784
87. Paladini CA, Celada P, Tepper JM (1999) Striatal, pallidal, and pars reticulata evoked inhibition of nigrostriatal dopaminergic neurons is mediated by GABA(A) receptors in vivo. Neuroscience 89(3):799–812
88. Rav-Acha M, Sagiv N, Segev I, Bergman H, Yarom Y (2005) Dynamic and spatial features of the inhibitory pallidal GABAergic synapses. Neuroscience 135(3):791–802. doi:10.1016/j.neuroscience.2005.05.069
89. English DF, Ibanez-Sandoval O, Stark E, Tecuapetla F, Buzski G, Deisseroth K, Tepper JM, Koos T (2012) GABAergic circuits mediate the reinforcement-related signals of striatal cholinergic interneurons. Nat Neurosci 15 (1):123–130. doi:10.1038/nn.2984, nn.2984 [pii]
90. Threlfell S, Lalic T, Platt NJ, Jennings KA, Deisseroth K, Cragg SJ (2012) Striatal dopamine release is triggered by synchronized activity in cholinergic interneurons. Neuron 75(1):58–64. doi:10.1016/j.neuron.2012.04.038, S0896-6273(12)00443-6 [pii]
91. Tritsch NX, Ding JB, Sabatini BL (2012) Dopaminergic neurons inhibit striatal output through non-canonical release of GABA. Nature 490(7419):262–266. doi:10.1038/nature11466, nature11466 [pii]
92. Nelson AB, Hammack N, Yang CF, Shah NM, Seal RP, Kreitzer AC (2014) Striatal cholinergic interneurons Drive GABA release from dopamine terminals. Neuron 82(1):63–70. doi:10.1016/j.neuron.2014.01.023
93. Chuhma N, Mingote S, Moore H, Rayport S (2014) Dopamine neurons control striatal cholinergic neurons via regionally heterogeneous dopamine and glutamate signaling. Neuron 81(4):901–912. doi:10.1016/j.neuron.2013.12.027

94. Karnani MM, Agetsuma M, Yuste R (2014) A blanket of inhibition: functional inferences from dense inhibitory connectivity. Curr Opin Neurobiol 26:96–102. doi:10.1016/j.conb.2013.12.015
95. Lei W, Jiao Y, Del Mar N, Reiner A (2004) Evidence for differential cortical input to direct pathway versus indirect pathway striatal projection neurons in rats. J Neurosci 24 (38):8289–8299
96. Kress GJ, Yamawaki N, Wokosin DL, Wickersham IR, Shepherd GM, Surmeier DJ (2013) Convergent cortical innervation of striatal projection neurons. Nat Neurosci 16 (6):665–667. doi:10.1038/nn.3397
97. Mathur BN, Tanahira C, Tamamaki N, Lovinger DM (2013) Voltage drives diverse endocannabinoid signals to mediate striatal microcircuit-specific plasticity. Nat Neurosci 16(9):1275–1283. doi:10.1038/nn.3478
98. Kravitz A, Freeze B, Parker P, Kay K, Thwin M, Deisseroth K, Kreitzer A (2010) Regulation of parkinsonian motor behaviours by optogenetic control of basal ganglia circuitry. Nature 466(7306):622–626. doi:10.1038/nature09159, nature09159 [pii]
99. Smith KS, Graybiel AM (2013) Using optogenetics to study habits. Brain Res 1511:102–114. doi:10.1016/j.brainres.2013.01.008
100. Witten IB, Lin SC, Brodsky M, Prakash R, Diester I, Anikeeva P, Gradinaru V, Ramakrishnan C, Deisseroth K (2010) Cholinergic interneurons control local circuit activity and cocaine conditioning. Science 330 (6011):1677–1681. doi:10.1126/science.1193771, 330/6011/1677 [pii]
101. Katz Y, Yizhar O, Staiger J, Lampl I (2013) Optopatcher – an electrode holder for simultaneous intracellular patch-clamp recording and optical manipulation. J Neurosci Methods 214(1):113–117. doi:10.1016/j.jneumeth.2013.01.017

Chapter 9

Paired Recordings from Synaptically Coupled Neurones in Acute Neocortical Slices

Dirk Feldmeyer and Gabriele Radnikow

Abstract

The minimal element of a neuronal network is the microcircuit between a single pre- and postsynaptic neurone. To date, a detailed analysis of individual synaptic connections is only possible using paired or multiple recordings from synaptically coupled neurones. No other electrophysiological or optophysiological technique allows a correlated functional and structural characterization of both pre- and postsynaptic neurone, even down to the electron microscopic level. Paired recording studies have shown that a full identification of the neuronal elements in a synaptic microcircuit is necessary to fully understand its connectivity and synaptic dynamics. Furthermore, a description of the dendritic and axonal projection pattern makes it possible to elucidate synaptic connectivity rules, e.g., whether it is random or highly specific. Paired recordings can also be used in combination with pharmacological interventions, e.g., to characterize the role of different ion channel subtypes in synaptic transmission and their modulation. Finally, paired recordings also allow more challenging studies such as a quantal analysis of an identified synaptic connection or the regulation of synaptic transmission by neurotransmitters acting on G protein-coupled receptors. Taken together, paired recordings from synaptically coupled neurones are a powerful technique that helps to describe the properties of synaptic microcircuits, which are the basic building blocks of large-scale neuronal networks in the brain.

Key words Acute brain slices, Paired recordings, Biocytin staining, Postsynaptic potentials, Synaptic contacts, Connectivity, Neuronal morphology

1 Introduction

Paired recordings from synaptically coupled neurones combined with simultaneous dye fillings have been used to elucidate and characterize the properties of small and generally local neuronal microcircuits. The technique allows a correlated investigation of the structural and functional properties of neuronal connections, their development, and modulation.

Initially, paired recordings have been performed using sharp microelectrodes in thick brain slices without optical control. Later, the technique to cut brain slices was significantly improved, infrared illumination of slices was introduced to enhance the visibility of

Alon Korngreen (ed.), *Advanced Patch-Clamp Analysis for Neuroscientists*, Neuromethods, vol. 113,
DOI 10.1007/978-1-4939-3411-9_9, © Springer Science+Business Media New York 2016

neurones, and patch electrodes were used to obtain a better signal-to-noise ratio of synaptic responses.

The major advantage of the technique is that patch-clamp recordings from two (or more) neurones allow a *functional*, electrophysiological characterization of synaptic signals while a morphological identification and classification is possible by the injection of the tracer dye biocytin during the physiological recording. Subsequent histochemical procedures will permit the visualization and subsequent three-dimensional (3D) reconstruction; this is required for a quantitative morphological analysis and a classification of cell types. Synaptic contact sites between the pre- and postsynaptic neurone can be identified under the light microscope and verified on the electron microscopic level. Finally, the specific modulation of synaptic transmission in a defined neuronal microcircuit can be investigated. Up to now, no other experimental technique for studying all of these aspects in an identified synaptic connection is available.

This chapter is aimed at describing the different steps necessary for successful paired recordings from synaptically coupled neurones. We will highlight the use of paired recordings for the correlated analysis of the structural and functional properties of such neuronal microcircuits and will describe different analytic approaches for the classification of pre- and postsynaptic neuronal cell types.

Finally, we will also mention potential limitations and pitfalls of the paired recording technique and the obtained data. In particular, we will address the different approaches in obtaining connectivity estimates, and how these will deviate from true connectivity ratios between two identified neurone types. Furthermore, we will discuss the problem of neurone specificity of synaptic microcircuits, i.e., whether a given presynaptic neurone type targets preferentially or exclusively a specific postsynaptic neurone type.

2 Methods

2.1 Brain Slice Preparation

2.1.1 Optimizing Slicing Conditions for Identified Synaptic Connections

Most studies of neuronal microcircuits and synaptic connectivity have been performed in brain slices from rat and mouse although a few studies using cat and ferret brain slices are available. For each synaptic connection one intends to study, it is important to determine the optimal conditions for preparing brain slices so that the axo-dendritic domain of the pre- and postsynaptic neurone is preserved.

In general, recordings from synaptically coupled neurones are prone to underestimate the true neuronal connectivity. This is even more pronounced in multiple than in dual recordings because the orientation of the pre- and postsynaptic axons and dendrites, respectively, will not be optimal for all connection types so that either deep or superficial axonal projections will be truncated and thus the number of synaptic connections will decrease.

It should be noted that slicing conditions also need to be adjusted for the age of the animal used for brain slicing; conditions optimal for juvenile slices are in most cases not suitable for slices from more mature and adult animals.

In order to determine an optimal slicing protocol, brain slices are made using different cutting angles. Individual pre- and post-synaptic neurones are stained via patch pipettes containing biocytin (3–5 mg/ml) in the pipette (internal) solution. After histochemical processing, stained neurones are inspected for the degree of dendritic and axonal truncations. In addition, their position in the slice (i.e., do dendrites and axons run parallel to the slice surface or does the apical dendrite point into or out of the slice) should be noted. To maximize the synaptic connectivity, both the pre- and postsynaptic neurones should be located deep in the slice (>60 μm and preferably more) to limit the extent of dendritic and axonal truncations. For local synaptic connections (<100 μm soma distance), slices in which neurones are oriented parallel to the surface should be used (Fig. 1a1). For translaminar synaptic connections for which the intersoma distance is well over 100 μm it may be advantageous to use brain slices in which the presynaptic axon and/or the post-synaptic apical dendrite projects slightly into the slice (Fig. 1a2). This procedure is likely to increase the probability of finding synaptically coupled neurones. We found this to be the case, e.g., for the translaminar synaptic connections between L4 spiny and L2/3 pyramidal neurones in the somatosensory cortex. Nevertheless, only few slices will be optimal for finding synaptic connections: The surface of the brain is curved and most apical dendrites are oriented perpendicular to it. When cutting consecutive slices apical dendrites of pyramidal neurones will initially point out of the slices and are likely to be cut. In subsequent slices (marked light red in Fig. 1b) the pyramidal neurones are largely preserved and their apical dendrites run approximately parallel to the slice surface; these are the brain slices used in paired recording experiments. When cutting even deeper, apical dendrites will point into the slice and will eventually be severed at its bottom surface (Fig. 1b). Therefore, in most cases only one to three slices can be used for studying synaptic connections.

Furthermore, slice preparations for studying long-range synaptic interactions between, e.g., a subcortical and a cortical area such as the thalamocortical slices involve cutting the brain not only in the coronal or (para)sagittal plane but also at tilted and/or oblique angles to keep the connectivity to some degree intact (*see* Fig. 1c and [1–4] for somatosensory, auditory, and visual thalamocortical slices, respectively and [5] for corticostriatal slices). A further level of complication arises when the neuronal connectivity between more than two brain areas should be maintained. One such example is the recently developed corticotectal slices in which the pathway between the inferior colliculi (tectum), the thalamus, and the

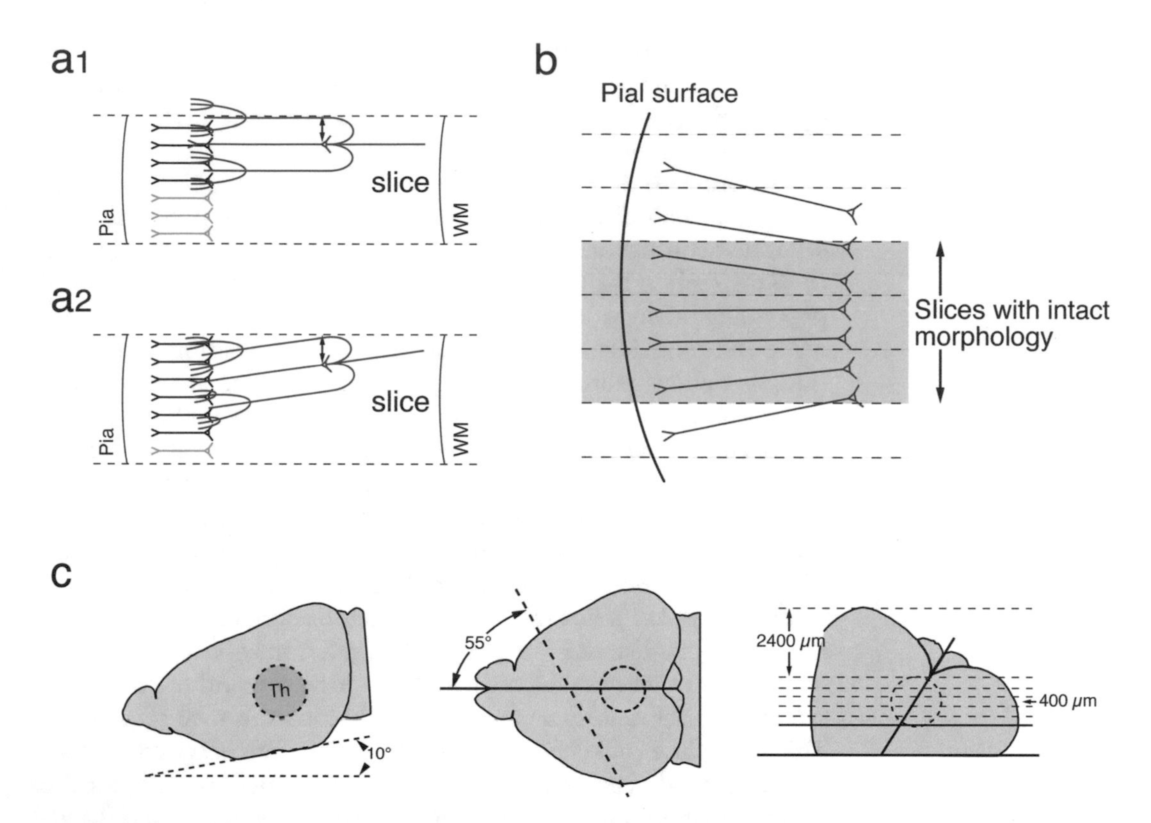

Fig. 1 Slicing procedure and possible artifacts. (**a**) The orientation of neurones in the slice affects the observed synaptic connectivity estimate, in particular for translaminar projections: If a presynaptic neurone is oriented parallel to the slice surface (**a1**), a part of the axon is likely to be cut so that the observed connectivity (in this case with postsynaptic neurones located in more superficial layers) is lower than in (**a2**) where the neurone is at an oblique orientation so that the axon points into the slice. Note that the somata of the presynaptic neurones are at the same depth in the slice but that their synaptic connectivity is likely to differ markedly. The presynaptic pyramidal neurone dendrite is in *red*, its axon in *blue*; synaptically connected neurones are in *black*, unconnected neurones are in *grey*; WM stands for *white* matter. (**b**) Depending on the curvature of the brain surface, the orientation of neurones in a slice changes with the depth of a slice as shown here. In the first two sections, the apical dendrites of pyramidal neurones are oriented in a way that they will be cut. In the following three sections (marked light *red*), the pyramidal neurones are oriented so that their morphology remains largely intact while in all subsequent slices they are again cut. This effect is even more prominent if the axonal domain is more extensive. (**c**) Procedure for cutting thalamocortical, somatosensory slices (see [1] for details). Here, the brain has to be placed on a ramp and cut "semi-coronally" at an angle of 55° to obtain approximately three slices in which the thalamocortical projection is preserved. For thalamocortical slices to different sensory cortices such as the primary auditory and the visual cortex different slicing procedures have to be used (modified from reference [1] with permission from Elsevier)

neocortex is preserved ([6]). However, these preparations are rarely used for paired recordings from synaptically coupled neurones because the probability of finding individual long-range synaptic connections is extremely low.

2.1.2 Slicing Procedure

A highly critical determinant for successful studies of synaptic microcircuits in brain slices is the age of the animal used for the experiment. Generally, neurones appear to be healthier in juvenile brain tissue and dendrite and axon truncations are also less frequent. For this reason, synaptic microcircuits are generally studied in brain slices of rodents aged between 12 and 30 days. However, neuronal connections undergo marked changes in their synaptic properties between the first and fourth postnatal week for example with respect to, e.g., the receptor complement at the synapse, the neurotransmitter release probability, and the time course of the postsynaptic potential (PSP; see [7]). Unfortunately, in more mature or adult brain tissue (from 30-day-old rat or mice), the visibility of cells in the slice is significantly reduced and neurones show a lower survival time making the characterization of synaptic microcircuits more difficult.

For slices from juvenile, up to 4-week-old animals, the following solution (artificial cerebrospinal fluid, aCSF) has been used successfully for cutting brain tissue with a vibrating microtome. Its composition is: 125 mM NaCl, 2.5 mM KCl, 25 mM glucose, 25 mM $NaHCO_3$, 1.25 mM NaH_2PO_4, 1 mM $CaCl_2$, and 5 mM $MgCl_2$ (aerated with 95 % O_2 and 5 % CO_2). Low [Ca^{2+}] together with a high [Mg^{2+}] is used to minimize synaptic activity during the slicing procedure. This reduces excitotoxicity and is essential for healthy and viable slices (see [8–10]). Using these solutions we were able to record from single and synaptically coupled neurones for over 3 h. Prolonged viability of neurones is essential for, e.g., quantal analysis of identified synaptic connections ([11–13]) and long sequences of pharmacological interventions (e.g., [11, 14, 15]). For brain slices from newborn or immature animals, the same slicing solution is used; however, thicker (>400 μm) slices need to be used because the tissue is softer (e.g., axons are largely unmyelinated) and therefore more prone to disintegration.

Brain slices from animals older than 30 postnatal days (P30) can also be cut in a modified, sucrose-containing solution (*see* Sect. 2.4) which has suggested to increase the survival rate of neurones. The composition of this solution is: 206 mM sucrose, 2.5 mM KCl, 25 mM $NaHCO_3$, 1.25 mM NaH_2PO_4, 1 mM $CaCl_2$, 3 mM $MgCl_2$, 3 mM myo-inositol, 2 mM Na-pyruvate, and 0.4 mM ascorbic acid (aerated with 95 % O_2 and 5 % CO_2). For mature animals, increasing oxygen levels (>30 %) during sedation/anesthesia may also improve the viability of neurones. The duration of experiments using slices from adult animals is significantly shorter than for immature animals; therefore the incubation times should not exceed ~30 min and the time between slicing and the start of the experiment kept to a minimum.

Several other slicing protocols and solutions for older animals are available under http://www.brainslicemethods.com/. However, in our hands pure sucrose-based slicing solutions were sufficient for obtaining healthy and viable neurones in brain slices of animals more than ≥2 months old. When using this solution, transcardial perfusion prior to sacrificing the animal appeared not to improve the slice quality significantly but this was not studied systematically.

After an incubation period of 0.5–1 h slices are transferred to the recording chamber. The slice is first visualized under low-power optics (4× objective) to find the brain region of interest. Patching of pre- and postsynaptic neurones is performed under visual control using infrared differential interference contrast (IR-DIC) optics for reduced diffraction and better visibility; for this a high-power (40×) water-immersion objective is used (*see* also Fig. 2a for images of the slice and the pre- and postsynaptic neurones).

2.2 Finding and Identifying Synaptic Connections

For all recordings from synaptically coupled neurones the most crucial step is finding a sufficiently stable synaptic connection between two (or more) identified neurones so that a detailed characterization is possible. This step is dependent on the axonal and dendritic geometry of pre- and postsynaptic neurone. In particular, the axonal domain of different presynaptic neurone types is very variable, ranging from local to translaminar and/or transcolumnar as well as from sparse to dense. This will result in large differences in the connectivity ratio. For example, interneurones with a dense but spatially confined axonal domain such as neocortical basket cells are likely to exhibit a high connection (~70 % and higher) probability with neurones in the reach of their axon ([16–18]). On the other hand, neurones with a sparse axonal domain (i.e., with a low axonal length density per unit volume) generally have connectivity ratios ~5–10 % and rarely exceed this value (e.g., [19–21]). Thus, the connection probability of different neurone types depends on the presynaptic axonal but also on the postsynaptic dendritic projection. To account for these different connectivity ratios, we have developed a "searching" protocol in order to find synaptic connections between two neurone types in brain slices. This procedure differs with respect to the connection type (chemical synapse or gap junction) and—more importantly—the connectivity ratio.

2.2.1 Synaptic Connections with a High Connection Probability

To find synaptic connections with a *high* connectivity ratio (such as most synaptic connections involving GABAergic interneurones), a "*presynaptic*" neurone is patched first in the voltage-recording mode of the whole-cell configuration. Subsequently, a second, potentially postsynaptic neurone is patched in whole-cell mode. A train of action potentials (3–5 at 10 Hz) is elicited by short (5 ms) current injections in the presynaptic cell to test whether postsynaptic potentials can be measured in the second neurone. If no response is found, another potentially postsynaptic neurone is patched with a new patch electrode until a postsynaptic response

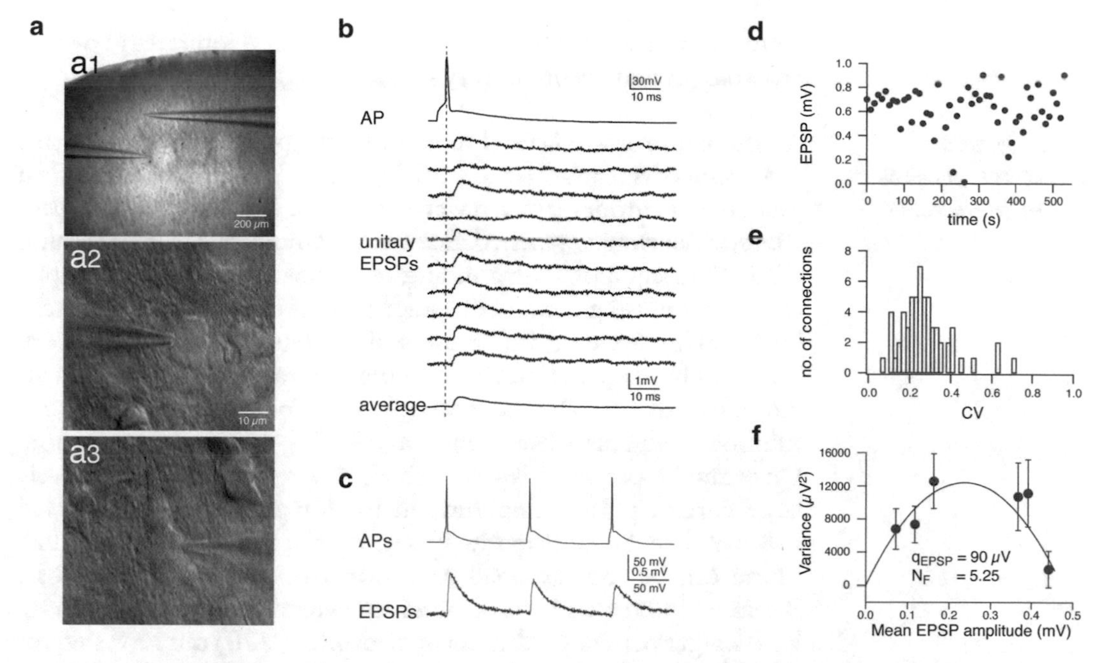

Fig. 2 Functional analysis of neuronal microcircuits with paired recordings. Example of a paired recording from a synaptic connection with a presynaptic spiny stellate neurone in layer 4 (L4) and a postsynaptic pyramidal neurone in layer 2/3 (L2/3). The figure indicates the correlated morphological and functional analysis of synaptic connections that is possible with the paired recording technique. (**a**) IR-DIC light microscopic images of electrode arrangement during paired recording from a presynaptic spiny stellate cell and a postsynaptic pyramidal neurone at a low magnification (**a1**). (**a2**), presynaptic spiny stellate neurone in neocortical layer 4; (**a3**) postsynaptic pyramidal neurone in layer 2/3. (**b**) Presynaptic action potential (*top* trace) and ten successive unitary EPSPs (*middle* traces). The bottom trace shows the average EPSP at this excitatory L4-to-L2/3 connection. The *dashed line* marks the peak of the AP. (**c**) Three presynaptic action potentials (interspike interval, 100 ms) elicit unitary EPSPs that show a marked depression. (**d**) Peak unitary EPSP amplitudes plotted vs. time of the recording. The plot shows the variability of the unitary EPSP during successive AP stimulations in the presynaptic cell. (**e**) CV analysis for the L4-to-L2/3 synaptic connection shown in panel (**a**). The CV analysis shows the normalized standard deviation of the unitary EPSP amplitudes recorded. A low CV indicates a high reliability of the synaptic connection (i.e., a high neurotransmitter release probability), a high CV a low reliability (i.e., a low release probability). (**f**) Quantal analysis of the L4-to-L2/3 synaptic connection shown in panel (**a**). Relationship between the variance of the EPSP and the average unitary EPSP amplitude for an individual L4-to-L2/3 neurone pair. Each data point shows a different probability condition. Error bars indicate the theoretical standard error in the estimate of the variance. Solid line shows the fit to a multinomial model giving quantal EPSP amplitude (q_{EPSP}) of 90 μV and number of functional neurotransmitter release sites (N_F) of 5.25 (see also [11])

can be recorded. For "high probability" synaptic connections, no more than three potentially postsynaptic neurones are tested. If no connection is found, another "presynaptic neurone" is patched. Once a synaptic connection has been established, action potentials are also elicited in the postsynaptic neurone to identify possible reciprocal synaptic connection because "high probability" synaptic connections are frequently (50%) reciprocal. Reciprocal synaptic connections can be homotypic (e.g., inhibitory–inhibitory) or heterotypic (excitatory–inhibitory or vice versa).

2.2.2 Synaptic Connections with a Low Connection Probability

A different protocol is used to identify synaptic connections with a *low* connectivity ratio (i.e., <25 %), most of which are between excitatory neurones (for a review see [22]). Here, a potential *postsynaptic* neurone is patched first in the whole-cell, current-clamp mode. Subsequently, surrounding neurones are tested for synaptic connectivity using a high-resistance (≥10 MΩ), so-called "searching" patch pipette in the "loose seal" cell-attached configuration. To avoid hyperexcitation through high extracellular K^+ and possible excitotoxicity, the "searching" patch pipette is filled with a solution containing Na^+ (instead of K^+) as the major cation. Once the "loose seal" is established, short (~5 ms) but relatively large current pulses (amplitude in the low nA range) are applied that need to be sufficiently strong to elicit an action potential. These can be seen as small excursions on the voltage response. Because "loose seals" can be obtained even when the patch pipette is no longer perfectly clean, many neurones (>20) can be tested for the existence of a synaptic connection without the need to replace the stimulation electrode. It is therefore well suited for low-probability synaptic connections. However, when a seal resistance of ~100 MΩ can no longer be achieved, the "searching" patch pipette needs to be replaced because action potentials can no longer be elicited under this condition.

A synaptic connection is found when the postsynaptic neurone responds with short latency (0.5–4 ms) EPSPs or IPSPs to presynaptic action potentials elicited in the "loose seal" mode. Then, the "searching" patch pipette is removed and the presynaptic cell is repatched in whole-cell mode using patch pipettes with a resistance between 4 and 8 MΩ filled with regular, K^+-based intracellular solution; the same solution is used from the patched postsynaptic neurone. The internal solution is supplemented with 2–5 mg/ml biocytin for subsequent morphological analysis. Then, recordings of action potential and PSPs were made from pre- and postsynaptic neurones, respectively, in the whole-cell, voltage-recording mode (see below; cf. [9, 10]).

2.2.3 Electrically Coupled Neuronal Connections

Paired recordings can also be obtained from electrically coupled neurones. To find these "electrical" connections an approach similar to that for low-probability synaptic connections is used. First, a neurone is patched in the whole-cell mode. Subsequently, a second, potentially gap-junction-coupled neurone is patched in the "loose seal" cell-attached mode. To test for electrical coupling, hyperpolarizing current pulses of 100–200 ms in duration were applied via the loose seal and the response in the first neurone is monitored. If the first neurone showed a small hyperpolarizing response with a duration similar to that of the current pulse, a gap-junction connection is identified and the second neurone is then patched in the whole-cell mode. The subsequent recording procedures are similar to those for neurone pairs with a low connection probability. In electrically coupled neurones, the stimulation-induced response can be elicited also in the opposite direction, i.e., by stimulating the first neurone and recording in the second.

2.3 Functional Characterization of a Synaptic Connection

Paired recordings from synaptically coupled neurones allow a wide variety of functional (Fig. 2) and structural analyses (Fig. 3). The most relevant procedures are listed below. Ca^{2+} imaging of synaptically connected neurones is also feasible using the paired recording technique. It has been shown to be a powerful method when investigating e.g., mechanisms of activity-dependent changes in synaptic transmission (see, e.g., [23, 24]). However, the focus of this chapter is on the combined electrophysiological-morphological characterization of neuronal microcircuits.

2.3.1 Time Course of the Postsynaptic Response

The time course of the postsynaptic current and voltage response are both important determinants of the computational power of a neuronal connection. For example, neurones with a long-lasting PSP show stronger summation and integration of synaptic inputs while short postsynaptic responses are necessary to obtain a high temporal fidelity of repetitive synaptic input. Quantitatively, the unitary EPSP/IPSP time course is described by its rise and decay time. However, it should be noted that the time course of the EPSP/IPSP is affected by dendritic filtering which is a function of the distance between the recording site (generally the neurone cell body) and the synapse location. Thus, the observed PSP summation at the soma and the synaptic integration are dependent on whether synaptic contacts are more proximally or distally located.

When measured in the voltage-clamp mode, the time course of synaptic currents shows also a connection specificity: for example, EPSCs recorded in excitatory neurones decay much more slowly compared to those recorded in inhibitory interneurones. This has been linked to different AMPA-type glutamate receptor subunits in these different neurones (e.g., [25]).

Latencies between the presynaptic action potential and the postsynaptic response (Fig. 2b) may differ even when the axonal and dendritic pathways are of roughly the same length. In layer 4 of the barrel cortex, for example, excitatory synaptic connections between layer 4 (L4) spiny neurones have a latency of about 1 ms while L4 spiny neurone-L4 interneurone and L4 interneurone-L4 spiny neurone connections have a latency of 0.5 ms [18, 26]. This difference in latencies is an important determinant for the temporal window during which excitatory synaptic transmission can take place. Other factors such as the structure of the pre- and postsynaptic site, the neurotransmitter release probability, and/or passive or

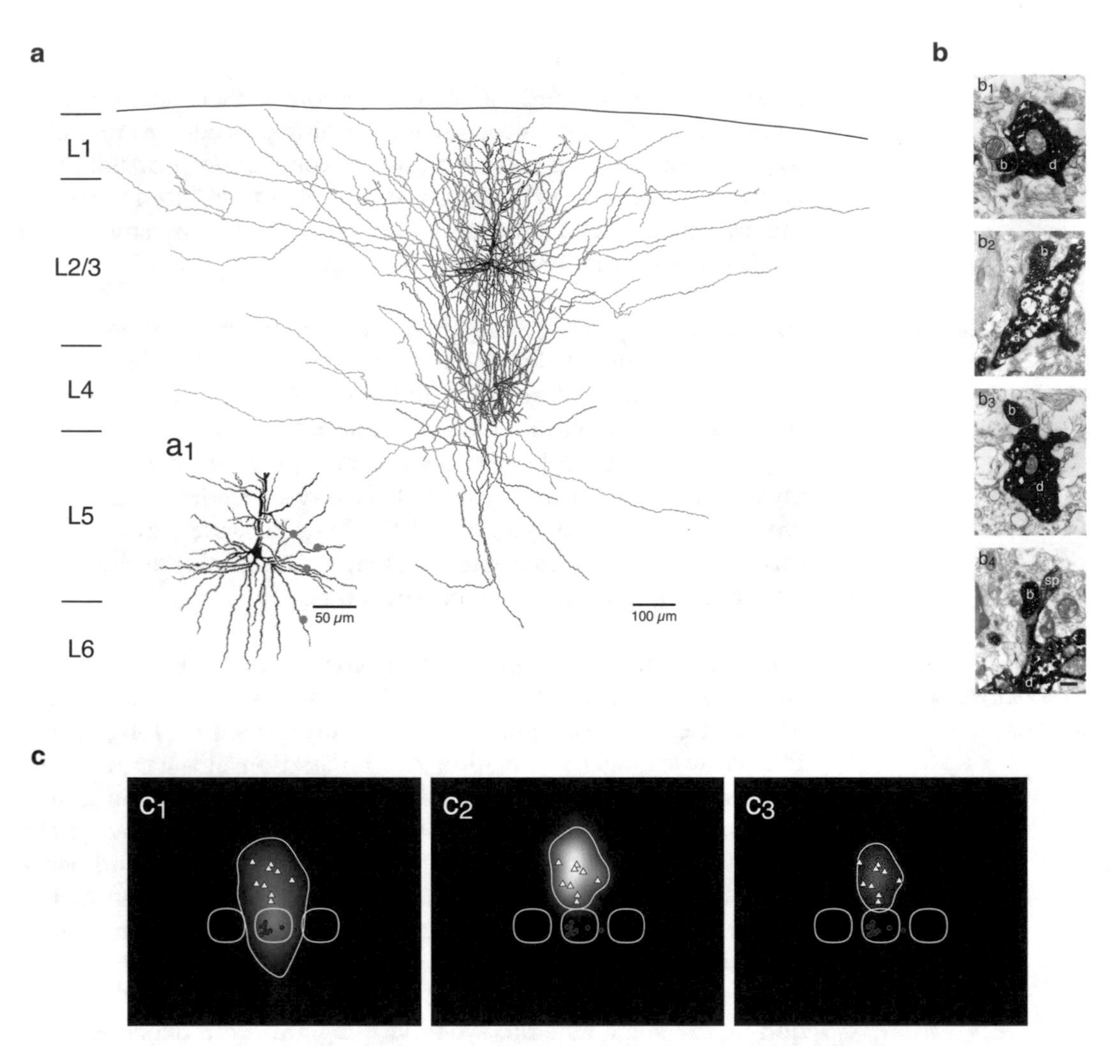

Fig. 3 Structural analysis of neuronal microcircuits with paired recordings. (**a**) Computer-assisted 3D reconstruction of a biocytin-stained, synaptically coupled L4 spiny stellate and a L2/3 pyramidal cell pair. Presynaptic dendrite, *red*; Presynaptic axon, *blue*; postsynaptic dendrite, *black*; postsynaptic axon, *green*. The inset (**a1**) shows the postsynaptic dendrite with the putative synaptic contacts marked by *light blue dots*. (**b**) Electron microscopic images (**b1–b4**) of the four light microscopically identified synaptic contacts shown in subpanel (**a1**). Both pre- and postsynaptic axon and dendrites, respectively, contain the dark precipitate produced by the biocytin-HRP reaction (see 2.4.2.). Presynaptic boutons are marked with "b," postsynaptic dendritic shafts with "d" and the dendritic spine (see subpanel **b4**) with "sp." Axonal boutons are clearly identifiable by their content of transmitter vesicles. The scale bar in subpanel (**b4**) is 0.25 μm and applies also to subpanels (**b1–b3**). (**c**) Density plots of the presynaptic axon (*blue*, **c1**) and the postsynaptic dendrite (*white*, **c2**) of nine different L4 to L2/3 neurone pairs. The predicted innervation domain for this synaptic connection type (*yellow*, **c3**) is calculated as the product of the L4 axonal density and the L2/3 dendritic density. The innervation domain indicates the likelihood of synapse formation between pre- and postsynaptic neurone. In all panels (**c1–c3**) the positions of the presynaptic cell bodies are marked by *red dots*; those of the presynaptic neurones are indicated by *white triangles*

active electrophysiological properties of the pre- and postsynaptic neurone may also contribute to different latencies of synaptic responses.

2.3.2 Synaptic Reliability

An important property of a synaptic connection is its reliability, i.e., whether failure rate and response variability are low. To determine this parameter, the presynaptic neurone is stimulated to elicit EPSPs or IPSPs in the postsynaptic neurone. About 50–200 sweeps are recorded (Fig. 2a, d) in order to determine the mean amplitude of the synaptic response and its variance (=the square of the standard deviation, SD). A measure for this is the coefficient of variation which is calculated as:

$$CV_{PSP} = \sqrt{\left(\sigma^2_{PSP} - \sigma^2_{Noise}\right)}/\mu_{PSP}$$

where $\sigma^2{}_{PSP}$ is the variance of the mean PSP amplitude μ_{PSP} and $\sigma^2{}_{Noise}$ the variance of the membrane potential fluctuation. The variance of the PSP is corrected by subtracting the membrane potential variance (i.e., the membrane potential "noise"). The CV_{PSP} is thus the standard deviation of the unitary PSP amplitude normalized to the average amplitude. For each synaptic connection the CV_{PSP} is determined separately and is a measure for the release probability of a synapse. However, this measure is only indirect and a detailed quantal analysis is required to determine its actual value. For this the CV_{PSP} is measured under conditions with varying release probability, e.g., with variable extracellular free Ca^{2+} concentrations (see [11–13] and chapter by Lanore and Silver in this book).

An additional parameter related to the neurotransmitter release probability at an identified synaptic connection is the short-term synaptic plasticity expressed as the paired pulse behavior. This is calculated by eliciting action potentials in the presynaptic neurone at a fixed interval (e.g., 100 ms, Fig. 2c) and measuring the amplitude of the postsynaptic response. The paired pulse ratio (PPR) is then calculated as the quotient of PSP_2/PSP_1 This is, however, even more indirect than the measurement of the CV_{PSP}. The PSP paired pulse behavior can be either facilitating or depressing so that the second (and the following) stimuli produce larger or smaller PSPs, respectively, than the first one ([27–29]). This is referred to as paired pulse facilitation or paired pulse depression. Different synaptic connections show differential synaptic behavior (e.g., *non-fast-spiking inhibitory interneurones* [30–32]; *hippocampal CA3-CA1 synapses* [33]; *neocortical excitatory synapses* [22, 34, 35]).

The failure rate is another parameter that depends on the release probability at an identified synaptic connection. It describes how frequently a synapse fails to respond to a presynaptic action potential. Only synaptic connections with a low neurotransmitter release probability and/or few synaptic contacts show a significant number of failures. However, despite a relatively low release probability failures may not occur when the number of synaptic contacts for one connection is high. Therefore, the failure rate is a less conclusive indicator of release probability.

2.3.3 Pharmacological Characterization of a Synaptic Connection

To identify the neurotransmitter receptors at an identified synaptic connection pharmacological techniques are used. Here, specific antagonists are used to block the postsynaptic response in full or in part, in case several receptor types coexist at a postsynaptic site. At most excitatory synapses, ionotropic glutamate receptors of the AMPA and NMDA receptor type (GluA and GluN receptors) coexist (e.g., [36]). Because of their markedly distinct ion channel properties, they are the basis of the fast and the slow component of the glutamatergic EPSP. In addition, the so-called kainate-type glutamate receptors (GluK) are also present at some excitatory synapses, mostly during early postnatal brain development (e.g., [37, 38]).

In the neocortex, inhibitory synaptic transmission is mostly mediated by $GABA_A$ receptor channels. Generally, the time course of IPSCs/IPSPs is much slower than that of EPSCs/EPSPs. In addition, some inhibitory connections exhibit IPSPs that have two components, a faster that is mediated by $GABA_A$ receptor channels and a slower that is based on the activation of G Protein-coupled $GABA_B$ receptors (e.g., [39]).

In addition, synaptic connections are also under the influence of neuromodulatory neurotransmitters such as acetylcholine ([14]) as well as dopamine, noradrenaline, adenosine, and many others. The effect of these neuromodulators can be profound and will affect cortical microcircuits in a differential manner because the receptor types and their density differ in pre- and postsynaptic neurones.

2.4 Morphological Characterization of a Synaptic Connection

Up to now, paired recordings from synaptically coupled neurones are the only method to provide a detailed correlation of the functional and structural characteristics of an identified synaptic microcircuit. For a quantitative analysis of the morphological properties of neuronal microcircuits, optimal filling with the chromogenic dye biocytin and careful histochemical processing are of paramount importance. An optimized protocol for these procedures currently used in our laboratory is briefly outlined below (for more detailed descriptions see [9, 40]).

2.4.1 Biocytin Filling

An optimal staining of the neurones in a paired recording experiment requires adequate filling with biocytin. Here, the size of the patch pipette opening is a critical factor. Low resistance patch electrodes (2–4 MΩ, large tip diameter) provide good electrical access and allow a rapid diffusion of biocytin into the cell. However, to obtain well-stained neurones the formation of a GΩ seal after recording and biocytin filling is indispensable, but when the tip diameter is large (and hence the resistance of the patch pipette is low), the removal of the patch pipette and the subsequent formation of a GΩ seal become increasingly more difficult. When the resistance of the patch pipette is lower than 2 MΩ, a GΩ seal will not form, the cell membrane is damaged, and leakage of biocytin out of the cell will ensue. Under these circumstances a quantitative morphological analysis will not be possible.

Pipettes with a smaller tip diameter and ~6–8 MΩ resistances (depending on the size of the neurone cell body) will give better staining results although the dye diffusion time has to be longer. After recording such patch pipettes can be retracted easily and a GΩ seal forms readily ensuring clear staining results for an in-depth morphological analysis. In our hands, recording for about 30 min was sufficient to completely fill even large neurones with the dye.

It is necessary to minimize leakage of biocytin-containing intracellular solution from the pipette into the extracellular space. After histochemical processing this will result in background staining of the surrounding tissue. In addition, deposits of dye precipitates reduce the contrast and increase light diffraction thereby compromising neuronal reconstructions (see below). In some instances dye uptake into the soma of neighboring neurones can also be observed. However, this is clearly distinct from labeling of gap-junction-coupled neurones. In this case, dye coupling between the neurone stained directly via the patch pipette and surrounding electrically coupled neurones is frequently observed but the extracellular space does not contain dark precipitates, which show a marked contrast between extracellular space and labeled cell bodies.

2.4.2 Immunocytochemical Processing

Following the electrophysiological recording and biocytin filling slices are fixated for either light or electron microscopy (see [9, 40]). The fixatives differ for both methods. For light microscopy a relatively weak fixative (4 % paraformaldehyde) is used; this allows additional labeling approaches, i.e., immunofluorescence labeling for intracellular marker proteins (e.g., parvalbumin for fast-spiking basket cell interneurones; [18]). For electron microscopy strong fixatives (2.5 % glutaraldehyde + 1.0 % paraformaldehyde) have to be used to ensure the preservation of subcellular structures and cutting of ultrathin slices.

Because biocytin itself is not staining the cell when it is in solution, an immunohistochemical reaction has to be performed after the fixation (which lasts usually 1–2 days). Biocytin is a carboxamide, the exact chemical name of which is *N_ε-(+)-biotinyl-L-lysine*. The labeling principle lies in its high affinity for avidin (or streptavidin) so that it can be used as a permanent label when using avidinylated visualization reagents. These could be fluorescence markers but they have the disadvantage of bleaching after long-term light exposure which makes detailed morphological reconstructions difficult. A permanent labeling protocol was first introduced by Horikawa and Armstrong ([41]) who used avidin-horseradish-peroxidase-(HRP) to obtain a dark precipitate. Later, a refined and improved protocol was developed for labeling of biocytin-filled neurones: A streptavidin-biotinylated horseradish peroxidase (HRP) reaction catalyzed with diaminobenzidine and started by the addition of H_2O_2 results in the formation of a dark precipitate in the neurone ([42–44]). In our laboratory a modified procedure is used (described in detail in [40] and [9]). Furthermore, to intensify the contrast of the biocytin stain, nickel sulfate ($(NH_4)_2Ni(SO_4)_2$) and cobalt chloride ($CoCl_2$) are added to the solution; in some instances uranyl acetate ($(UO_2(CH_3COO)_2{\cdot}2H_2O)$) is also used. All reactions have to be monitored carefully by eye and stopped once an optimal staining quality is reached to avoid overly dark specimens.

It is also possible to combine electrophysiological recordings with immunofluorescent staining for specific cellular marker proteins (such as parvalbumin, somatostatin, calretinin for different types of cortical interneurones) *and* biocytin-HRP labeling. During recording, the neurone is filled with biocytin and a fluorescent dye so that it is easily discernible from other neurones after paraformaldehyde fixation. In a second step, immunofluorescent staining is performed using a biotinylated primary antibody for the marker protein in question and a secondary antibody coupled to a fluorescent group. The fluorescent wavelength needs to be different from that of the dye with which the neuron is filled during electrophysiological recording. Finally, a biocytin-HRP labeling is made to allow morphological reconstructions of the neurones labeled with the fluorescent group. However, this multiple staining protocol compromises the quality of the biocytin-HRP stain to some extent, making reconstructions of the neuronal morphology less reliable.

After the chromogenic reaction, slices are transferred to slides and embedded in different mounting media such as epoxy or acrylic resins (both of which are not soluble in water) or the water-soluble hydrocolloid mucoadhesive Mowiol®. Mowiol does not require laborious dehydration and has the advantage to be liquid during mounting and forming only a weak permanent bond between slide, tissue, and coverslip. This allows an initial light microscopic examination after which the medium can be removed and replaced by a epoxy resin so that a subsequent electron microscopic analysis can be performed with the *same* tissue sample. In addition, Mowiol does not quench fluorescence labels like most permanent resins do.

On the other hand, Mowiol is inferior to solid embedding media when it comes to realistic 3D morphological reconstructions. This is because in viscous media nonlinear tissue distortions occur particularly in the z-plane of the specimen [45]. Such nonlinear tissue distortions are very difficult if not impossible to correct for because they depend on the depth of the neurone in the slice.

Epoxy and acrylic resins do allow shrinkage correction and will give a more realistic neurone morphology ([40]; see also below). An additional advantage of acrylic resins (in particular Eukitt) is that cytoarchitectonic features such as the cortical lamination and "barrels" in layer 4 of the primary somatosensory cortex are clearly discernible without additional staining. Moreover, the biocytin-labeled structures show a higher contrast when the embedding medium is an acrylic resin: Even small and thin structures such as spine necks or axonal segments between synaptic boutons are more readily visible making high-resolution tracing of these structures significantly easier.

2.4.3 Neuronal Reconstructions and Morphological Analysis

Following the histochemical processing biocytin-labeled pairs of synaptically coupled neurones are examined under the light microscope using a 100× or a 60× oil immersion objective. In order to be able to focus throughout the entire slice thickness, oil immersion objectives with high numerical aperture ($N_A = 1.4$) have to be used. Subsequently, computer-assisted 3D neuronal reconstructions are made (Fig. 3a) using the NEUROLUCIDA® system (Microbrightfield). This is a neuroanatomical analysis system for tracing the projection of dendrites and axons; it is coupled to a microscope with a computer-controled, motorized stage and focus axis.

Tracing is normally done manually because the experience of the reconstructing person makes it easier to distinguish dense and profusely branching dendritic and in particular axonal structures often the pre- and postsynaptic neurones. Their dendrites and axons should be traced at high resolution, i.e., with 0.5–2.0 μm z-length per turn of the focus wheel. A lower resolution can lead to significant errors in the estimation of the dendritic and axonal length and may even introduce spatial errors such as large "steps" in the z-axis of a reconstructed process. All structures that are ambiguous with respect to direction, depth, or branching pattern should be checked carefully using the eyepieces (10×) of the microscope. This is essential for neurones with a dense axonal domain as is the case for interneurones (see, e.g., [18]). Furthermore, frequent alignments in the *x*, *y*, and *z*-dimensions of the neurone need to be made to avoid distortions in the neuronal structure because of error propagation.

In some laboratories half-automatic procedures are used for neuronal tracings. However, reconstructions of more complicated neuronal morphologies and in particular of synaptically coupled pairs of neurones will involve heavy editing and are often just as laborious as manual tracings.

For pairs of synaptically coupled neurones, the next step is to identify possible sites where synaptic contacts could have formed between the pre- and postsynaptic neurone. For this, neurones are inspected at the highest possible magnification. Putative synaptic contacts are sites where a presynaptic axonal bouton (visible as a swelling of the axon) comes in close apposition to a dendritic spine or shaft of the postsynaptic neurone. Potential synaptic contacts are marked in the 3D reconstruction and their dendritic location can be determined. In order to verify synaptic contacts a subsequent electron microscopic (EM) analysis is required (Fig. 3b); only then presynaptic boutons (with synaptic vesicles) and postsynaptic sites can be identified unambiguously. However, we found that only ~10 % of light microscopically identified putative contact could not be verified by EM [11].

After completing the reconstruction a quantitative morphological analysis can be performed using the NEUROEXPLORER® (Microbrightfield) software. It serves to analyze parameters such as, e.g., axonal and dendritic length, degree of collateralization, orientation etc. and may serve to classify neuronal cell types by, e.g., a cluster analysis. In addition, morphological data about the axonal and dendritic domains of the pre- and postsynaptic neurones, respectively, can be further processed to calculate axonal and dendritic "length density" maps (Fig. 3c; [46]). These "density maps" represents the length of an axonal or dendritic segment in voxels of a defined size. In this way, the so-called average innervation domains can be calculated as the product of presynaptic axonal and postsynaptic dendritic density. Such "innervation domains" are maps of the probability of synaptic contact formation for an identified neuronal microcircuit (Fig. 3c3; [21, 46–48]). The higher the presynaptic axonal and postsynaptic dendritic length in a given voxel is the more likely is the formation of a synaptic contact between them.

3 Discussion

Paired recordings from synaptically coupled neurones are a powerful technique to investigate the structure-function relationship of neuronal microcircuits at the subcellular, cellular, and network level. Because it allows the analysis of both the pre- and postsynaptic neurone of a neuronal microcircuit at the same time, it is possible to clearly distinguish the site of action for different experimental interventions which is not possible using extracellular stimulation (be it electrical or optical) of presynaptic neurones. Furthermore, paired recording combined with simultaneous biocytin fillings can identify the morphology of both neurones in a synaptic connection. However, to appreciate properly the results gained from paired recordings, one needs to be aware of some of its shortcomings.

3.1 Issues Concerning Synaptic Connectivity

A major problem of paired and in particular of multiple recording studies is the synaptic connectivity ratio. Many publications provide the so-called connectivity estimates. These estimates are meant to give information about the connection probability of a specific neuronal connection; however, they are subject to several types of errors and misinterpretations.

First and foremost, axonal and to a lesser extent dendritic truncations will result in a substantial underestimation of the true synaptic connectivity (i.e., values of 1 % and lower); in the worst case, the connectivity is reported to be zero although a synaptic connection exists (see, e.g., [20] vs. [10]). This is particularly prominent when the axon projects over several mm to other ipsi- and contralateral cortical areas and even to subcortical structures ([49–51]). A comparison of in vivo and in vitro data suggests that the axonal length of long-range projecting excitatory neurones such as L5 and L2/3 pyramidal cells can be underestimated by a factor of 100×; however, axonal truncations are more likely for distal collaterals than for local ones. Therefore, only local synaptic connections can be studied when using the paired recording technique. Connectivity estimates for such local synaptic connections may approach the true connectivities but become less and less reliable the more distal the presynaptic boutons are. In particular, for GABAergic interneurones with a dense but local axonal domain, the connectivity estimates have been reported to be as high as 70–90 % ([16, 18, 52]).

The detection of synaptic connections is also affected by its physiological properties. A low release probability (and hence a large degree of PSP failures) or a small quantal EPSP or IPSP amplitude (i.e., PSP amplitude at a single synaptic contact) will result in a poor signal-to-noise ratio which in turn decreases the probability of finding a synaptic connection. With respect to the PSP amplitude, increasing the driving force for the ions permeating the ligand-gated synaptic channel may help to improve the detectability of synaptic connections. This is often used for inhibitory synaptic connections were the reversal potential of Cl^- ions is around −90 mV and the resting membrane potential between −70 and −65 mV, resulting in a driving force of only 20–25 mV. This is significantly smaller than that for EPSPs for which the electrochemical gradient is about 70 mV.

For connections with a low release probability, a higher extracellular Ca^{2+} concentration will increase the release probability and hence result in a lower failure rate. In turn, this will improve the likelihood of identifying synaptic connections. However, this approach affects the physiological properties of synaptically coupled neurones so that it is generally not used for paired recordings, except in case of quantal analysis studies ([11–13]).

Another factor that will influence the detection of synaptic connectivity is the location of synaptic contacts. For most synaptic connections, synaptic contacts are relatively close to the soma (i.e., at a dendritic distance <200 μm); therefore, the PSP amplitude and time course are not strongly distorted by dendritic filtering. However, for some synaptically coupled neurone pairs the axonal projection pattern of the presynaptic neurone suggests rather distal synaptic locations, e.g., the dendritic tufts of apical dendrites. Such distal synaptic contacts have been described for both inhibitory and excitatory synaptic connections, e.g., the neurogliaform cell-to-L5B pyramidal cell connection, the Martinotti cell-to-L5B pyramidal cell connection, and the L4 spiny neurone-to-L6A pyramidal cell connection ([53–55]). For these connection types, somatic recordings are affected by substantial dendritic filtering resulting in a strongly attenuated PSP amplitude and a very slow time course (see [55]). This will lower the detection rate for synaptic connections markedly because in somatic recordings, small unitary PSPs at distal synaptic locations will disappear in the electrical noise. Only very strong connections will be detected under these recording conditions. For pyramidal neurones patch-clamp recordings from the apical dendrites will improve the detection ratio of synaptic inputs to the tuft dendrites. Here, the degree of dendritic filtering is smaller and hence the PSP amplitude decrement weaker. However, because this approach is technically challenging it has not been used frequently (but see [53, 54]).

It has been argued that neuronal connectivity is non-random, i.e., that certain types of neurones preferentially target neurones of a distinct type (see, e.g., [56, 57]). This statement, however, needs an in-depth analysis of the 3D axonal and dendritic projection pattern of pre- and postsynaptic neurones, respectively. For this, only neuron pairs with an optimal staining can be used; otherwise the hypothesis of a neuronal connection specificity is not safe.

3.2 Time-Dependent Changes in Synaptic Structure and Function

To date, all experimental techniques used to investigate synaptic connectivity are based on the implicit assumption that connectivity is static, i.e., does not change over time. This is true for studies at the micro-, meso-, and macroscopic scale. In particular pure morphological techniques using fixed tissue (i.e., for light or electron microscopy) provide only a rather static image of the neuronal connectivity in a brain region.

However, in reality both the functional and structural synaptic connectivity is highly dynamic and changes within a few minutes up to weeks or even months and—during aging—years, a feature that cannot be captured by most functional and structural experimental techniques.

Imaging the motility, formation, and elimination of dendritic spines as well as the expression of proteins involved in the formation of a functional postsynaptic site suggest that short-term fluctuations in synaptic connectivity may occur within the range of a few minutes (for a review see [58]). There is also evidence that synaptic contacts are formed or pruned during behavioral states such as sleep ([59, 60] for reviews) suggesting a circadian regulation of synaptic connectivity. Such changes may be related also to the release of neuromodulatory transmitters such as acetylcholine, dopamine, noradrenaline, serotonin, and adenosine. Furthermore, functional and morphological synaptic plasticity during brain development or sensory deprivation has also been demonstrated [61]; it takes place over the course of several days or weeks and is therefore difficult to measure during the course of individual experiments.

References

1. Agmon A, Connors BW (1991) Thalamocortical responses of mouse somatosensory (barrel) cortex in vitro. Neuroscience 41:365–379
2. Land PW, Kandler K (2002) Somatotopic organization of rat thalamocortical slices. J Neurosci Methods 119:15–21
3. Cruikshank SJ, Rose HJ, Metherate R (2002) Auditory thalamocortical synaptic transmission in vitro. J Neurophysiol 87:361–384
4. MacLean JN, Fenstermaker V, Watson BO et al (2006) A visual thalamocortical slice. Nat Methods 3:129–134. doi: 10.1038/nmeth849
5. Smeal RM, Gaspar RC, Keefe KA et al (2007) A rat brain slice preparation for characterizing both thalamostriatal and corticostriatal afferents. J Neurosci Methods 159:224–235. doi:10.1016/j.jneumeth.2006.07.007
6. Llano DA, Slater BJ, Lesicko AM et al (2014) An auditory colliculothalamocortical brain slice preparation in mouse. J Neurophysiol 111:197–207. doi:10.1152/jn.00605.2013
7. Feldmeyer D, Radnikow G (2009) Developmental alterations in the functional properties of excitatory neocortical synapses. J Physiol 587:1889–1896. doi:10.1113/jphysiol.2009.169458
8. Debanne D, Boudkkazi S, Campanac E et al (2008) Paired-recordings from synaptically coupled cortical and hippocampal neurons in acute and cultured brain slices. Nat Protoc 3:1559–1568, doi: nprot.2008.147 10.1038
9. Radnikow G, Günter RH, Marx M et al (2012) Morpho-functional mapping of cortical networks in brain slice preparations using paired electrophysiological recordings. In: Fellin T, Halassa M (eds) Neuromethods: neuronal network analysis. Neuromethods. Springer Protocols, vol 67. Humana Press. An imprint of Springer Science + Business Media, LLC 2011, New York, pp 405–431. doi: 10.1007/7657_2011_14
10. Qi G, Radnikow G, Feldmeyer D (2015) Electrophysiological and morphological characterization of neuronal microcircuits in acute brain slices using paired patch-clamp recordings. J Vis Exp 95:52358. doi:10.3791/52358
11. Silver RA, Lübke J, Sakmann B et al (2003) High-probability uniquantal transmission at excitatory synapses in barrel cortex. Science 302:1981–1984. doi:10.1126/Science.1087160
12. Biró AA, Holderith NB, Nusser Z (2005) Quantal size is independent of the release probability at hippocampal excitatory synapses. J Neurosci 25:223–232. doi: 10.1523/JNEUROSCI.3688-04.2005
13. Biró AA, Holderith NB, Nusser Z (2006) Release probability-dependent scaling of the postsynaptic responses at single hippocampal GABAergic synapses. J Neurosci 26:12487–12496. doi:10.1523/JNEUROSCI.3106-06.2006
14. Eggermann E, Feldmeyer D (2009) Cholinergic filtering in the recurrent excitatory microcircuit of cortical layer 4. Proc Natl Acad Sci U S A 106:11753–11758. doi:10.1073/pnas.0810062106
15. van Aerde KI, Qi G, Feldmeyer D (2015) Cell type-specific effects of adenosine on cortical neurons. Cereb Cortex 25:772–787. doi:10.1093/cercor/bht274
16. Packer AM, McConnell DJ, Fino E et al (2013) Axo-dendritic overlap and laminar projection can explain interneuron connectivity to pyramidal cells. Cereb Cortex 23:2790–2802. doi:10.1093/cercor/bhs210
17. Fino E, Yuste R (2011) Dense inhibitory connectivity in neocortex. Neuron 69:1188–1203. doi:10.1016/j.neuron.2011.02.025
18. Koelbl C, Helmstaedter M, Lübke J et al (2015) A barrel-related interneuron in layer 4 of rat somatosensory cortex with a high intrabarrel connectivity. Cereb Cortex 25:713–725. doi:10.1093/cercor/bht263
19. Holmgren C, Harkany T, Svennenfors B et al (2003) Pyramidal cell communication within local networks in layer 2/3 of rat neocortex. J Physiol 551:139–153
20. Lefort S, Tomm C, Floyd Sarria JC et al (2009) The excitatory neuronal network of the C2 barrel column in mouse primary somatosensory cortex. Neuron 61:301–316. doi:10.1016/j.neuron.2008.12.020

21. Feldmeyer D, Lübke J, Sakmann B (2006) Efficacy and connectivity of intracolumnar pairs of layer 2/3 pyramidal cells in the barrel cortex of juvenile rats. J Physiol 575:583–602. doi:10.1113/jphysiol.2006.105106
22. Feldmeyer D (2012) Excitatory neuronal connectivity in the barrel cortex. Front Neuroanat 6:24. doi:10.3389/fnana.2012.00024
23. Sjöström PJ, Häusser M (2006) A cooperative switch determines the sign of synaptic plasticity in distal dendrites of neocortical pyramidal neurons. Neuron 51:227–238. doi:10.1016/j.neuron.2006.06.017
24. Buchanan KA, Blackman AV, Moreau AW et al (2012) Target-specific expression of presynaptic NMDA receptors in neocortical microcircuits. Neuron 75:451–466. doi:10.1016/j.neuron.2012.06.017
25. Geiger JR, Lübke J, Roth A et al (1997) Submillisecond AMPA receptor-mediated signaling at a principal neuron-interneuron synapse. Neuron 18:1009–1023
26. Feldmeyer D, Egger V, Lübke J et al (1999) Reliable synaptic connections between pairs of excitatory layer 4 neurones within a single "barrel" of developing rat somatosensory cortex. J Physiol 521:169–190
27. Katz B, Miledi R (1968) The role of calcium in neuromuscular facilitation. J Physiol 195:481–492
28. Zucker RS, Regehr WG (2002) Short-term synaptic plasticity. Annu Rev Physiol 64:355–405
29. Tsodyks M, Wu S (2013) Short-term synaptic plasticity. Scholarpedia 8:3153. doi:10.4249/scholarpedia.3153
30. Reyes A, Lujan R, Rozov A et al (1998) Target-cell-specific facilitation and depression in neocortical circuits. Nat Neurosci 1:279–285
31. Beierlein M, Gibson JR, Connors BW (2003) Two dynamically distinct inhibitory networks in layer 4 of the neocortex. J Neurophysiol 90:2987–3000
32. Ascoli GA, Alonso-Nanclares L, Anderson SA et al (2008) Petilla terminology: nomenclature of features of GABAergic interneurons of the cerebral cortex. Nat Rev Neurosci 9:557–568. doi:10.1038/nrn2402
33. Bolshakov VY, Siegelbaum SA (1995) Regulation of hippocampal transmitter release during development and long-term potentiation. Science 269:1730–1734
34. Markram H, Wang Y, Tsodyks M (1998) Differential signaling via the same axon of neocortical pyramidal neurons. Proc Natl Acad Sci U S A 95:5323–5328
35. Tsodyks MV, Markram H (1997) The neural code between neocortical pyramidal neurons depends on neurotransmitter release probability. Proc Natl Acad Sci U S A 94:719–723
36. Silver RA, Traynelis SF, Cull-Candy SG (1992) Rapid-time-course miniature and evoked excitatory currents at cerebellar synapses in situ. Nature 355:163–166
37. Kidd FL, Coumis U, Collingridge GL et al (2002) A presynaptic kainate receptor is involved in regulating the dynamic properties of thalamocortical synapses during development. Neuron 34:635–646
38. Kidd FL, Isaac JT (1999) Developmental and activity-dependent regulation of kainate receptors at thalamocortical synapses. Nature 400:569–573
39. Chittajallu R, Pelkey KA, McBain CJ (2013) Neurogliaform cells dynamically regulate somatosensory integration via synapse-specific modulation. Nat Neurosci 16:13–15. doi:10.1038/nn.3284
40. Marx M, Günter RH, Hucko W et al (2012) Improved biocytin labeling and neuronal 3D reconstruction. Nat Protoc 7:394–407. doi:10.1038/nprot.2011.449
41. Horikawa K, Armstrong WE (1988) A versatile means of intracellular labeling: injection of biocytin and its detection with avidin conjugates. J Neurosci Methods 25:1–11, doi:0165-0270(88)90114-8
42. Adams JC (1992) Biotin amplification of biotin and horseradish peroxidase signals in histochemical stains. J Histochem Cytochem 40:1457–1463
43. Adams JC (1981) Heavy metal intensification of DAB-based HRP reaction product. J Histochem Cytochem 29:775
44. Hsu SM, Raine L, Fanger H (1981) The use of antiavidin antibody and avidin-biotin-peroxidase complex in immunoperoxidase technics. Am J Clin Pathol 75:816–821
45. Egger V, Nevian T, Bruno RM (2008) Subcolumnar dendritic and axonal organization of spiny stellate and star pyramid neurons within a barrel in rat somatosensory cortex. Cereb Cortex 18:876–889. doi:10.1093/cercor/bhm126
46. Lübke J, Roth A, Feldmeyer D et al (2003) Morphometric analysis of the columnar innervation domain of neurons connecting layer 4 and layer 2/3 of juvenile rat barrel cortex. Cereb Cortex 13:1051–1063
47. Feldmeyer D, Roth A, Sakmann B (2005) Monosynaptic connections between pairs of spiny stellate cells in layer 4 and pyramidal cells in layer 5A

indicate that lemniscal and paralemniscal afferent pathways converge in the infragranular somatosensory cortex. J Neurosci 25:3423–3431. doi: 10.1523/JNEUROSCI.5227-04.2005

48. Helmstaedter MN, Feldmeyer D (2010) Axons predict neuronal connectivity within and between cortical columns and serve as primary classifiers of interneurons in a cortical column. In: Feldmeyer D, Lübke JHR (eds) New aspects of axonal structure and function, 1st edn. Springer Science + Business Media, New York, pp 141–155. doi:10.1007/978-1-4419-1676-1_8
49. Narayanan RT, Egger R, Johnson AS et al (2015) Beyond columnar organization: cell type- and target layer-specific principles of horizontal axon projection patterns in rat vibrissal cortex. Cereb Cortex. doi:10.1093/cercor/bhv053
50. Oberlaender M, Boudewijns ZS, Kleele T et al (2011) Three-dimensional axon morphologies of individual layer 5 neurons indicate cell type-specific intracortical pathways for whisker motion and touch. Proc Natl Acad Sci U S A 108:4188–4193. doi:10.1073/pnas.1100647108
51. Pichon F, Nikonenko I, Kraftsik R et al (2012) Intracortical connectivity of layer VI pyramidal neurons in the somatosensory cortex of normal and barrelless mice. Eur J Neurosci 35:855–869. doi:10.1111/j.1460-9568.2012.08011.x
52. Packer AM, Yuste R (2011) Dense, unspecific connectivity of neocortical parvalbumin-positive interneurons: a canonical microcircuit for inhibition? J Neurosci 31:13260–13271. doi:10.1523/JNEUROSCI.3131-11.2011
53. Jiang X, Wang G, Lee AJ et al (2013) The organization of two new cortical interneuronal circuits. Nat Neurosci 16:210–218. doi:10.1038/nn.3305
54. Lee AJ, Wang G, Jiang X et al (2014) Canonical organization of layer 1 neuron-led cortical inhibitory and disinhibitory interneuronal circuits. Cereb Cortex. doi:10.1093/cercor/bhu020
55. Qi G, Feldmeyer D (2015) Dendritic target region-specific formation of synapses between excitatory layer 4 neurons and layer 6 pyramidal cells. Cereb Cortex. doi:10.1093/cercor/bhu334
56. Brown SP, Hestrin S (2009) Intracortical circuits of pyramidal neurons reflect their long-range axonal targets. Nature 457:1133–1136. doi:10.1038/nature07658
57. Brown SP, Hestrin S (2009) Cell-type identity: a key to unlocking the function of neocortical circuits. Curr Opin Neurobiol 19:415–421. doi:10.1016/j.conb.2009.07.011
58. Holtmaat A, Randall J, Cane M (2013) Optical imaging of structural and functional synaptic plasticity in vivo. Eur J Pharmacol 719:128–136. doi:10.1016/j.ejphar.2013.07.020
59. Frank MG (2012) Erasing synapses in sleep: is it time to be SHY? Neural Plast 2012:264378. doi:10.1155/2012/264378
60. Tononi G, Cirelli C (2014) Sleep and the price of plasticity: from synaptic and cellular homeostasis to memory consolidation and integration. Neuron 81:12–34. doi:10.1016/j.neuron.2013.12.025
61. Greenhill SD, Juczewski K, de Haan AM et al (2015) Neurodevelopment. Adult cortical plasticity depends on an early postnatal critical period. Science 349:424–427. doi:10.1126/science.aaa8481

Chapter 10

Extracting Quantal Properties of Transmission at Central Synapses

Frederic Lanore and R. Angus Silver

Abstract

Chemical synapses enable neurons to communicate rapidly, process and filter signals and to store information. However, studying their functional properties is difficult because synaptic connections typically consist of multiple synaptic contacts that release vesicles stochastically and exhibit time-dependent behavior. Moreover, most central synapses are small and inaccessible to direct measurements. Estimation of synaptic properties from postsynaptic currents or potentials is complicated by the presence of nonuniform release probability and nonuniform quantal properties. The presence of multivesicular release and postsynaptic receptor saturation at some synapses can also complicate the interpretation of quantal parameters. Multiple-probability fluctuation analysis (MPFA; also known as variance-mean analysis) is a method that has been developed for estimating synaptic parameters from the variance and mean amplitude of synaptic responses recorded at different release probabilities. This statistical approach, which incorporates nonuniform synaptic properties, has become widely used for studying synaptic transmission. In this chapter, we describe the statistical models used to extract quantal parameters and discuss their interpretation when applying MPFA.

Key words Synapse, Vesicle, Active zone, Release probability, Release site, Quantal analysis, MPFA, Variance-mean analysis

1 Introduction

Fluctuations in the amplitude of evoked end-plate potentials recorded intracellularly from frog muscle fibers bathed in high Mg^{2+}-containing solution [1, 2] together with their similarity to spontaneous end-plate potentials (which was first thought to be noise [3]) lead to the quantum hypothesis—the idea that neurotransmitter is released in discrete multimolecular packets or quanta. The probabilistic nature of this all-or-none process was rigorously tested using a number of different approaches [2]. This work showed that fluctuations in the evoked end-plate potentials could be predicted with Poisson statistics under low probability conditions. Soon after, early electron microscopy studies revealed the

Alon Korngreen (ed.), *Advanced Patch-Clamp Analysis for Neuroscientists*, Neuromethods, vol. 113, DOI 10.1007/978-1-4939-3411-9_10, © Springer Science+Business Media New York 2016

presence of synaptic vesicles in motor nerve terminals [4], providing a structural basis for the “quantum” [5].

Quantal analysis refers to a group of methods that use statistical models to extract the basic functional properties of synapses from postsynaptic responses, which are typically measured at the soma. Quantal analysis can provide insights into the function of synapses and identify the locus of changes in synaptic efficacy. Three quantal parameters are commonly used to determine the properties of synaptic transmission: the first is the maximum number of vesicles that could be released at a synaptic connection by an action potential and is often referred to as the number of independent functional release sites (N), the second is the probability of vesicular release (P), and third, the amplitude of the postsynaptic response following the release of a single vesicle or quantum (Q). The size of the postsynaptic responses and its variability from trial-to-trial are determined by the values of these quantal parameters. Presynaptic modulation is associated with P, while postsynaptic changes are associated with Q. Formation of new contacts (or increasing P from zero at existing release sites) would be associated with a change in N.

Early attempts to apply quantal analysis to central synapses met with mixed success due to difficulties in resolving individual quantal events [6]. Nevertheless, it was recognized that binomial rather than Poisson statistical models were more appropriate given the relatively high release probability and few release sites present [6]. Considerable efforts were made to extract quantal parameters from amplitude distributions in subsequent studies, but this approach was challenging and it was often difficult to determine the reliability of the end results because the amplitude of quantal events often fell below the noise at central synapses [7–11]. Moreover, during this period, there was growing evidence of nonuniform quantal properties [12, 13]. The complications arising from such nonuniform quantal size and nonuniform release probability were highlighted in a simulation study that demonstrated the difficulties of using traditional quantal analysis approaches for studying central synaptic transmission [14].

To overcome these problems a different statistical approach was developed to estimate quantal parameters at central synapses, which has its roots in non-stationary fluctuation analysis of ion channels [15] and synaptic currents [16, 17]. Multiple-probability fluctuation analysis (MPFA) [18] or alternatively variance-mean analysis [19–21] does not require that quantal events be distinguished from the background noise. Moreover, the statistical model underlying MPFA is multinomial and can therefore take into account nonuniformities in P and Q. MPFA has been used to determine the quantal parameters of transmitter release at low frequency [22, 23] and has been extended to short repetitive trains of synaptic responses [24–26]. It has also been combined with analysis of covariance between successive stimuli within trains [27, 28]. MPFA is a valuable tool to quantify the pre- and postsynaptic contributions to short-term plasticity changes [18, 25, 26, 29–33]. This quantal analysis method has also been used to determine whether long-term plasticity changes are expressed pre- or postsynaptically [19, 34, 35].

In this chapter, we describe the statistical basis of MPFA, including the binomial and the multinomial models of synaptic transmission. We also describe how to apply MPFA at central synapses and discuss the interpretation of the quantal parameters with this method.

2 Statistical Models of Synaptic Transmission

2.1 Binomial Model

The binomial model assumes that each vesicle is released independently, that the release is synchronous, that P is uniform across vesicles, and that Q is uniform both at the level of a single release site and across release sites. Under these assumptions the mean peak amplitude of the synaptic current response (I) can be expressed as a function of the number of functional release sites (N), the quantal amplitude (Q), and the probability of vesicular release (P) as follows:

$$I = NPQ \tag{1}$$

and the associated variance can be expressed as:

$$\sigma^2 = NQ^2P(1 - P) \tag{2}$$

The relationship between the variance and the mean synaptic amplitude is therefore:

$$\sigma^2 = IQ - \frac{I^2}{N} \tag{3}$$

These results suggest that if a synaptic connection operates in a simple binomial manner, the relationship between the variance and the mean amplitude of synaptic responses is parabolic. Fitting a function of the form $y = Ax - Bx^2$ (where y is the variance and x is the mean postsynaptic current) to the relationship between the variance and mean current recorded at different release probabilities can provide estimates of N and Q (Fig. 1). P can be then calculated from Eq. (1).

2.2 Multinomial Model

Several studies have shown that the assumptions of uniform release probability and quantal size required for a simple binomial model are not valid for many central synapses [12, 36–38]. These considerations lead to the application of a multinomial model for transmitter release [18, 39].

2.2.1 Nonuniform Quantal Size

In the multinomial model, quantal variability can be accommodated both at the single-site level (intrasite or type I variability;

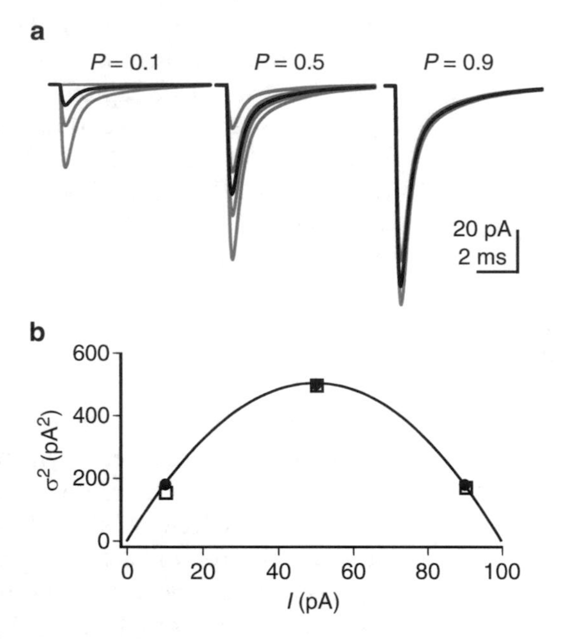

Fig. 1 Binomial model of synaptic transmission. (**a**) Postsynaptic currents for a synapse with $N = 5$, $Q = -20$ pA at $P = 0.1$, 0.5 and 0.9 were simulated stochastically (20 superimposed traces in *grey* and average trace of 200 events in *black*). (**b**) The *dots* represent the theoretical values of variance and mean amplitude at the three different release probabilities tested. The *empty squares* are values from a set of 200 simulated postsynaptic currents. The variance-mean relationship of the theoretical values was fitted using Eq. (3) (*modified from ref* [58] *with permission from Elsevier*)

Fig. 2a) [17, 37] and across sites (intersite or type II variability; Fig. 2b) [37, 40]. Intrasite quantal variability arises from fluctuations in the size of quantal events from an individual release site from trial-to-trial and from fluctuations in their latencies. These sources of type I variance can be defined in terms of their coefficient of variation (CV_{QS} and CV_{QL} respectively). Intersite variability arises from sampling quanta from different release sites, each of which has different mean quantal sizes. This type II variance can be defined in terms of the coefficient of variation CV_{QII}.

Equation (1) can be extended to include nonuniform quantal behavior as follows:

$$I = NP\overline{Q}_P \tag{4}$$

where $\overline{Q}_P$ is the mean quantal size at the time of the peak of the mean synaptic current and is therefore affected by asynchronous release (Fig. 2c, d) (*for details see ref* [41]). The variance is then given by:

$$\sigma^2 = N\overline{Q}_P^2 P(1-P)\left(1 + CV_{QII}^2\right) + N\overline{Q}_P^2 P CV_{QI}^2 \tag{5}$$

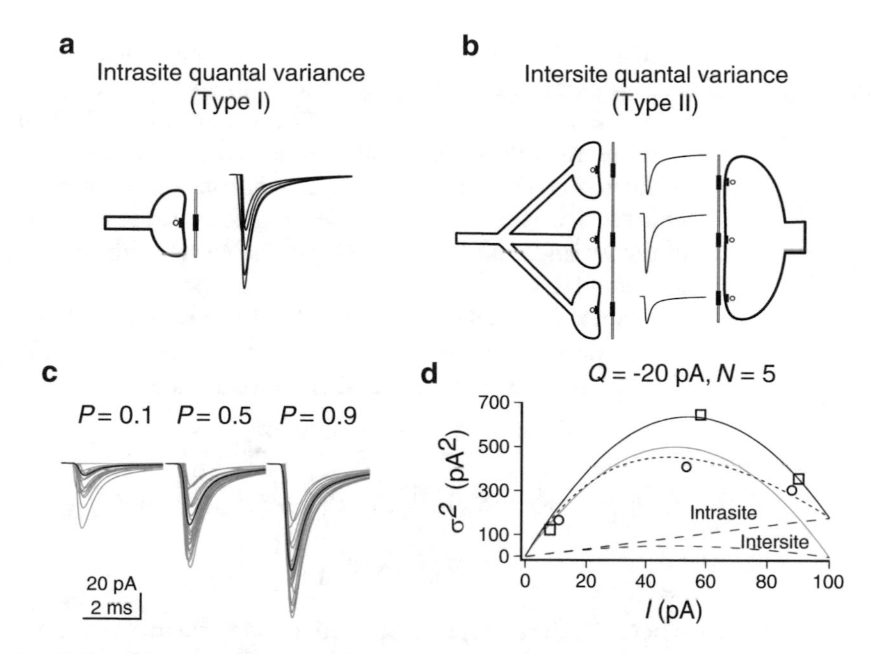

Fig. 2 Multinomial model of synaptic transmission. Intrasite (or type I) (**a**) and intersite (or type II) (**b**) quantal variabilities are illustrated. Intrasite variability arises from a single connection while intersite variability arises from a connection with multiple terminals or from a single synapse containing multiple release sites. (**c**) Postsynaptic currents simulated stochastically for the same parameters as in Fig. 1a except that intra- and intersite variability were introduced in the simulations ($CV_{QI} = 0.3$ and $CV_{QII} = 0.3$). (**d**) Theoretical variance-mean relationship representing nonuniform quantal size (in *black*, fitted with Eq. (7)). The open squares are the values from a set of 200 simulated postsynaptic currents with nonuniformity in quantal size while the release probability was kept uniform. Nonuniform release probabilities were added with an α value equal to 1 and fitted with Eq. (13) (*dotted line*). The *circles* show results of simulation with nonuniform release probability ($\alpha = 1$). The *grey line* represents the simple binomial function as in Fig. 1b. The variance contribution from intrasite and intersite quantal variation is indicated by the *broken lines*

where CV_{QI} combines all the variability observed at a single-site level:

$$CV_{QI} = \sqrt{CV_{QL}^2 + CV_{QS}^2} \tag{6}$$

The relationship between the variance and the mean I is therefore:

$$\sigma^2 = \left(\overline{Q}_P I - \frac{I^2}{N}\right)\left(1 + CV_{QII}^2\right) + \overline{Q}_P I CV_{QI}^2 \tag{7}$$

The sum of the intra- and intersite variability can be defined in terms of the coefficient of variation as follows:

$$CV_{QT} = \sqrt{CV_{QI}^2 + CV_{QII}^2} = \sqrt{CV_{QS}^2 + CV_{QL}^2 + CV_{QII}^2} \tag{8}$$

More details on how the different quantal variances can be estimated are given in Sect. 3.3.

2.2.2 Nonuniform Release Probability

Studies in the spinal cord [12], hippocampal cultures [13, 36, 42], and hippocampal slices [43] have shown that the probability of release is nonuniform across individual synapses. The presence of nonuniform release probability tends to reduce the variance when compared to the uniform case. The impact on the variance of nonuniform P is largest at high P, leading to a wedge-shaped distortion of the variance-mean relationship (Fig. 2d) at high levels of nonuniformity [18, 41]. This behavior can be captured in the multinomial model relatively neatly assuming that there is no correlation between release probability and quantal amplitude across sites [18, 39]. In this case the mean current and the variance are given by:

$$I = N\overline{P}\overline{Q}_{\mathrm{P}} \tag{9}$$

$$\sigma^2 = N\overline{Q}_{\mathrm{P}}^2\overline{P}\left[1 - \overline{P}\left(1 + \mathrm{CV}_{\mathrm{P}}^2\right)\right]\left(1 + \mathrm{CV}_{\mathrm{QII}}^2\right) + N\overline{Q}_{\mathrm{P}}^2\overline{P}\mathrm{CV}_{\mathrm{QI}}^2 \tag{10}$$

where $\overline{P}$ is the mean release probability at the time of the peak of the current and $\mathrm{CV_P}$ represents the coefficient of variation of release probability across sites. The relationship between the variance and the mean amplitude is then:

$$\sigma^2 = \left[\overline{Q}_{\mathrm{P}}I - \frac{I^2}{N}\left(1 + \mathrm{CV}_{\mathrm{P}}^2\right)\right]\left(1 + \mathrm{CV}_{\mathrm{QII}}^2\right) + \overline{Q}_{\mathrm{P}}I\mathrm{CV}_{\mathrm{QI}}^2 \tag{11}$$

Given that P is bounded by 0 and 1, $\mathrm{CV_P}$ changes as a function of $\overline{P}$. $\mathrm{CV_P}$ and how it changes has been modeled using families of beta functions $\beta(\alpha,\beta)$ [41]. These functions mimic the distribution of release probability across release sites and describe how the distribution might change when $\overline{P}$ varies. Furthermore, beta distributions approximate distributions that have been measured in hippocampal synapses in culture [36]. Using this approach $\mathrm{CV_P}$ can be expressed as a function of the mean release probability and a family of beta distributions defined by a single parameter α:

$$\mathrm{CV_P} = \sqrt{\frac{1 - \overline{P}}{\overline{P} + \alpha}} \tag{12}$$

Substituting for $\mathrm{CV_P}$ in Eq. (11) gives:

$$\sigma^2 = \left[\overline{Q}_{\mathrm{P}}I - \frac{\overline{Q}_{\mathrm{P}}I^2(1+\alpha)}{I + N\overline{Q}_{\mathrm{P}}\alpha}\right]\left(1 + \mathrm{CV}_{\mathrm{QII}}^2\right) + \overline{Q}_{\mathrm{P}}I\mathrm{CV}_{\mathrm{QI}}^2 \tag{13}$$

adding only one additional free parameter to the expression. Low α values (<2) indicate nonuniform release probability, while higher values indicate that the probability of release is uniform [41]. However, the beta function is an approximation and α is the least well-constrained parameters in Eq. (13); thus estimates of $\mathrm{CV_P}$

should be considered as a rough indicator of the level of nonuniformity in release probability.

2.3 MPFA During Short-Term Plasticity

MPFA can be extended to examine how quantal parameters change during bursts of activity by determining the quantal parameters from the first evoked postsynaptic current (PSC) in the burst and then analyzing the fluctuations in the subsequent events with CV analysis (Fig. 3) [25]. P can be estimated from the CV of evoked PSCs with the following equation:

$$\mathrm{CV} = \frac{\sigma}{I} = \sqrt{\frac{(1-P)\left(1+\mathrm{CV}_{\mathrm{QII}}^2\right)+\mathrm{CV}_{\mathrm{QI}}^2}{NP}} \tag{14}$$

Changes in $\overline{Q}_{\mathrm{P}}$ can be then calculated from Eq. (4) during the train, assuming N is constant. While simple to implement, this approach does assume that release probability is relatively uniform.

2.4 Interpretation of Quantal Parameters

The multinomial model provides a statistical description of release that incorporates both nonuniform presynaptic and postsynaptic properties. However, other synaptic properties, notably the occupancy of the postsynaptic receptors, affect the interpretation of the

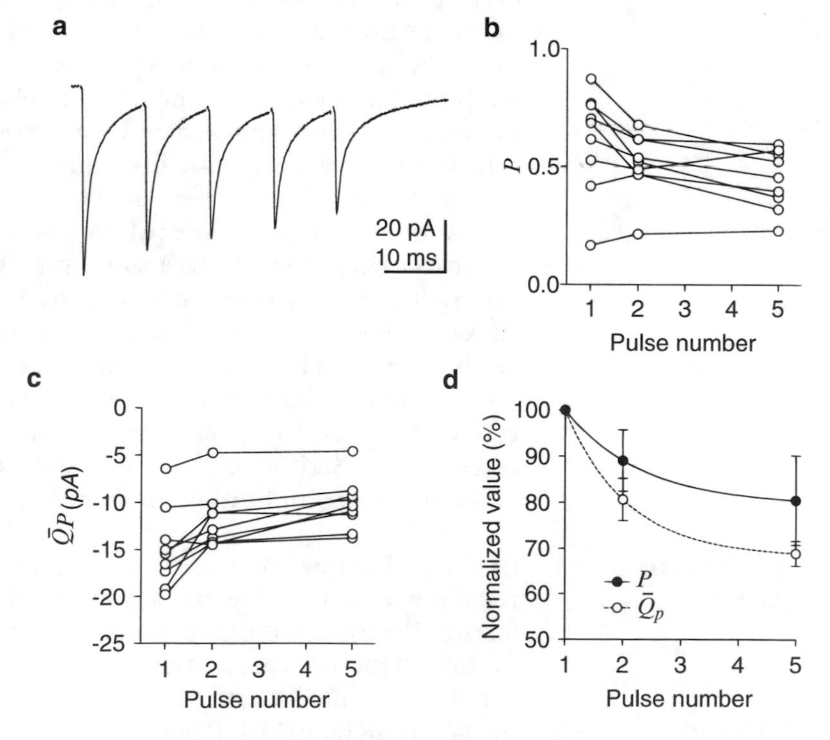

Fig. 3 (**a**) Mean EPSCs of five stimuli at 100 Hz at mossy fiber-granule cell synapse. Changes of P (**b**) and $\overline{Q}_{\mathrm{P}}$ (**c**) during the 100 Hz train of nine different recordings. The graph in (**d**) shows the mean amplitude of P and $\overline{Q}_{\mathrm{P}}$ during the 100 Hz train relative to their initial value, with associated exponential fits (*adapted from ref* [25])

quantal parameters estimated with MPFA. This section describes how to interpret the synaptic parameters obtained with MPFA, discusses potential errors, and outlines some additional tests that can be performed to check that the interpretation is correct.

2.4.1 Interpretation of the Number of Release Sites

When the postsynaptic quantal responses sum linearly, the amplitude of the PSC reflects the number of vesicles released. Under these conditions N is the maximum number of readily release vesicles that could be released by the synaptic connection following an action potential. N therefore corresponds to the number of anatomically distinct synaptic contacts only if a maximum of one vesicle is released per synaptic contact (univesicular release) [22, 30, 41].

In the case of multivesicular release, when multiple vesicles are released at each synaptic contact [31, 44, 45], N will still reflect the total number of functional release sites or equivalently the maximum number of release-ready vesicles that could be released by the synaptic connection following an action potential, if quanta sum linearly. But if the postsynaptic receptors become saturated [46], quantal events no longer sum up because the postsynaptic membrane becomes insensitive to subsequent transmitter release. In the extreme case, where a single vesicle saturates the receptors at a synaptic contact, N would correspond to the number of synaptic contacts where release can occur rather than reflecting the maximum number of vesicles that can be released at those sites (which could be much larger) [17]. Thus, the interpretation of N depends on whether the postsynaptic receptors become saturated or not. Luckily, this can be tested by examining whether the postsynaptic receptor occupancy changes as a function of release probability using rapidly equilibrating low-affinity competitive antagonists (see Sect. 3.4) [22].

At some synapses, spillover of neurotransmitter from neighboring release sites gives rise to a slow current component [47–49]. This spillover-mediated current can introduce a small low variance current component that can lead to an overestimation of N [22]. At cerebellar mossy fiber-granule cell synapses, spillover current can be observed in isolation when direct release fails [47]. If the spillover current and direct release summate linearly, the mean spillover current and its variance can simply be subtracted at the time of the peak of the postsynaptic response at each probability of release [22].

2.4.2 Interpretation of CV_P and $\overline{P}$

As for N, interpretation of $\overline{P}$ also depends on whether quantal responses sum linearly or not. Under conditions when quanta sum linearly, $\overline{P}$ represents the mean probability that a vesicle is released by the time of the mean peak postsynaptic response following an action potential, i.e., the integral of the release rate per functional release site up to that time [50]. $\overline{P}$ can be described in terms of the product of the mean probability of a vesicle being in the release-ready state ($\overline{P}_R$) and the mean probability that a release-ready vesicle will undergo fusion following an action potential ($\overline{P}_F$) [51, 52]:

$$\overline{P} = \overline{P}_{R}\overline{P}_{F} \tag{15}$$

This interpretation is valid for both univesicular and multivesicular release, if the postsynaptic response is linear. On the other hand, if the postsynaptic receptors are saturated by a single vesicle, $\overline{P}$ indicates the mean probability that one or more vesicles have been released at each synaptic contact.

At many synaptic connections, release probability has been found to be reasonably uniform leading to a parabolic variance-mean relationship. When CV_P is large, the variance-mean relationship takes on an increasingly wedge-like shape due to the loss of variance at high $\overline{P}$ [41]. The distribution of release probability and how it changes with $\overline{P}$ can be approximated by a family of beta distributions with the same α value. This approximation is unlikely to give the fine details of P distribution but gives a rough estimate of nonuniformity in P under linear conditions. However, a combination of multivesicular release and receptor saturation could also reduce the variance at high $\overline{P}$. This possibility can be tested for by examining whether receptor occupancy changes with release probability using low-affinity competitive antagonists (see Sect. 3.4) [20, 24, 41]. Low-affinity antagonists can also help prevent the build up of desensitization during trains [53].

2.4.3 Interpretation of the Quantal Size

Estimation of the mean quantal amplitude from the initial slope of the variance-mean relationship is robust to synaptic nonuniformities, because at low $\overline{P}$ the variance is little affected by nonuniform release probability. However, mean quantal amplitude at the time of the peak ($\overline{Q}_P$), estimated from MPFA, does not correspond to the mean peak amplitude of the quantal waveforms unless the vesicular release time course (together with any temporal dispersion arising from axonal delays) is much shorter than the initial decay time course of the PSC. This is because when release is asynchronous, quantal events can occur and partially decay before the time of the peak of the evoked PSC. In cerebellar granule cells, $\overline{Q}_P$ estimated from MPFA is about 20 % lower than rise time-aligned quantal events [22]. Estimation of $\overline{Q}_P$ is also robust in the presence of multiquantal release and receptor saturation because these tend to occur at intermediate to high $\overline{P}$. However, the presence of slow spillover-mediated currents can lead to an underestimate of $\overline{Q}_P$ if it is not corrected for [22].

3 Methods

In the following sections, we describe the experimental and analysis procedures to perform MPFA and explain how to estimate quantal variance and to distinguish between univesicular and multivesicular release at central synapses.

3.1 Experimental Protocol

To perform MPFA, different release probability conditions should be imposed by changing the Ca^{2+} concentration ($[Ca^{2+}]$) and $[Mg^{2+}]$ in the extracellular medium [18, 19, 25]. The divalent cation concentration and thus membrane charge screening can be kept approximately constant by adjusting the ratio of these two ion species [18]. The osmolarity should also be kept constant by adjusting the concentration of glucose and a phosphate-free solution can be used to avoid phosphate precipitation in high $[Ca^{2+}]$. The presence of NMDA receptor antagonists is useful for preventing Ca^{2+}-induced plasticity and high $[Ca^{2+}]$ conditions should be applied at the end of the recording to minimize the impact of plasticity induction.

Measuring synaptic responses under voltage clamp minimizes errors arising from changes in driving force and activation of voltage-gated channels in electrically compact cells such as cerebellar granule cells. However, voltage clamp recordings from the soma are subject to errors in cells where synaptic inputs are located on electrically remote regions of the dendritic tree [54, 55]. Voltageclamp error can be reduced by using capacitance and electrode series resistance compensation functions of the amplifier (but series resistance should be monitored throughout the experiment to ensure it remains constant). While this can improve voltage clamp of the soma and proximal dendrites, it does not compensate for poor voltage control of electrically remote regions of the cell. In cases where synapses are far from the recording site and the "space-clamp" is poor, postsynaptic potentials should be recorded and corrected for the deviation in driving force [56]. However, if synapses are located at different electrotonic locations, dendritic filtering will contribute to type II quantal variance and voltage-dependent dendritic conductances could complicate the variance-mean relationship by shaping postsynaptic potentials as they spread to the soma.

3.2 Data Acquisition

Data acquisition and analysis can be performed with a variety of software applications and hardware configurations. We use NeuroMatic (http://www.neuromatic.thinkrandom.com/), a freely available acquisition and analysis package running under IgorPRO (Wavemetrics).

The sampling frequency of the recorded signal should be at least twofold higher than the highest-frequency component of the signal; otherwise the signal will be distorted by aliasing [57]. Aliasing can be avoided by filtering the data with a low-pass Bessel filter before digitizing and ensuring the cut-off frequency of filter is $<1/3$ of the digitization frequency. Moreover, low-pass filtering the data will reduce the background noise and increase the signal-to-noise ratio. In practice, reliable reconstruction of fast AMPAR-mediated currents can be achieved by low-pass filtering at 10 kHz and digitized at 50–100 kHz [22, 25].

During MPFA experiments, the number of stimulated synapses should be constant. This can be achieved by using paired recordings [56] or minimal extracellular stimulation protocol [18, 23]. In case of extracellular stimulation, changes in excitability or stimulation threshold could occur as a result of the different extracellular $[Ca^{2+}]$ solutions or stimulation frequencies. These should be controlled for by establishing thresholds under the different conditions, monitoring failures and if possible measuring the afferent fiber volley. A minimum of 50 stimuli should be recorded for each release probability condition although more may be required to constrain the fit if few release probability conditions are used. In the case of uniform release probability, a minimum of three different release probability conditions is required and at least four in the nonuniform case.

The higher the frequency the more rapidly the data set can be acquired, but it should not be so high as to induce synaptic depression or facilitation. Time stability of the synaptic responses used to calculate the mean and the variance of the PSC amplitude is essential. Time stability of the amplitude of evoked PSCs can be tested with the Spearman rank order test [18], while amplitude stability across release probability conditions should be checked by returning to a specific condition (e.g., 2 mM $[Ca^{2+}]$) and testing whether the mean PSC amplitude has changed.

3.3 Estimation of Quantal Variance

Unless a simple binomial model is used, quantal variance must be determined. Total quantal variance, CV_{QT}, can be measured from PSCs evoked under conditions of low release probability, when failures of release reach 80–90 %, because the proportion of multiquantal events is negligible regardless of the number of functional release sites [22, 41]. Recording under such low release probability conditions has several advantages: (1) quantal events can be measured from stimulus-aligned PSCs, minimizing the effect of spontaneous currents, allowing a precise measurement of the mean and variance; (2) it minimizes interactions between vesicles at synapses where multivesicular release occurs, and (3) it reduces postsynaptic interactions by reducing the build up of neurotransmitter via spillover.

In another variant of this approach, CV_{QT} can be estimated by evoking PSCs while replacing Ca^{2+} by Sr^{2+} in the extracellular medium in order to induce asynchronous release of quanta [18, 23, 33, 46]. CV_{QT} can also be estimated from the distribution of miniature postsynaptic currents (mPSCs) recorded in TTX. However, this has the serious disadvantage that quantal events arise from all the synapses impinging onto the cell rather than being restricted to an individual synaptic connection.

Once the total quantal variability is determined, the contributions made by intra- and intersite variability can be estimated [41]. For rare cases where there is a single functional release site per synaptic connection, CV_{QI} can be measured directly from the peak of stimulus-aligned PSCs [17]. However, for synaptic connections with multiple functional release sites, direct estimation of CV_{QI} is not possible. In this case CV_{QI} can be inferred from the PSC variance remaining when P is maximal. If quanta release is independent, then the variance remaining at the peak of PSCs should arise purely from intrasite quantal amplitude variability and quantal asynchrony (i.e., $P \rightarrow 1$; $\sigma^2 \approx I\overline{Q}_P CV_{QI}^2$). CV_{QI} can be determined from the remaining variance by dividing it by $I\overline{Q}_P$ and square rooting. CV_{QI} can also be decomposed into CV_{QS} and CV_{QL} using Eq. (6), since CV_{QL} can be determined from the difference in variability in the peak amplitude between stimulus-aligned and rise time-aligned events [22]. Once CV_{QI} is known, CV_{QII} can be calculated from CV_{QT} using Eq. (8). If estimation of CV_{QI} is not possible, assuming that $CV_{QI} = CV_{QII}$ is likely to introduce relatively little error [41].

3.4 Distinguish Between Univesicular and Multivesicular Release

In order to distinguish between univesicular and multivesicular release, PSCs should be recorded in the presence of a rapidly equilibrating low-affinity competitive antagonist (such as γ-DGG or kynurenic acid for AMPA receptors), under low and high probabilities of release [20, 22, 44, 45, 56]. In the case of multivesicular release, the fractional block of the evoked PSC by the competitive antagonist will be less at high release probabilities than low release probabilities, because the postsynaptic AMPA receptors will be exposed to a higher concentration of neurotransmitter, which displaces the antagonist more effectively. Thus, if there is significant dependence of the agonist block on the release probability, this implies the presence of multivesicular release. Moreover, if the $[Ca^{2+}]$-dependence of release changes in the presence of antagonist, this implies that the postsynaptic receptors become saturated at high release probabilities. Under these conditions MPFA should be performed in the presence of a competitive antagonist [46]. However, if the fractional block of the PSC by the antagonist is independent of release probability, this suggests that the postsynaptic receptor occupancy is constant across release probabilities and the interpretation of quantal parameters is straightforward. In this case MPFA should give the same number of functional release sites in the presence or the absence of a competitive antagonist.

3.5 Data Analysis

The peak amplitude of PSCs should be measured from a window (Wp) centered around the maximum of the mean peak PSC and the baseline subtracted using a 1–2 ms window before the stimulus (Wb; Fig. 4). Time stability of the PSC recorded for each condition should be assessed by using statistical tests such as linear regression or the Spearman rank order test. The variance arising from the background noise should be measured from each event using a

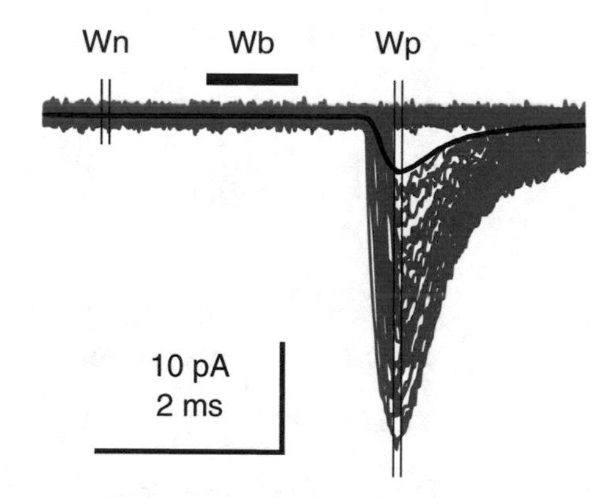

Fig. 4 200 postsynaptic currents were simulated with a multinomial model under a low and uniform probability of release ($P = 0.05$) which gives rise to 70 % failures. 50 stimulus-aligned events are shown in *grey*. The average trace is displayed in *black*. Wb shows the window for the baseline correction and Wn and Wp the windows for the measurement of the mean and variance of the baseline noise and the peak of the postsynaptic current, respectively (*modified from ref* [64] *with permission from Springer*)

window (Wn) that is similar to the one used to measure the amplitude at the time of the peak (e.g., 0.1 ms) (Fig. 4). The two measurement windows (Wn and Wp) should be equidistant to ensure that both measurements pick up similar frequency components of the background noise. The baseline variance can then be calculated and subtracted from the variance of the synaptic amplitude.

$\overline{Q}_P$ and N and their uncertainties can be estimated by performing a weighted fit of the variance-mean relationship to a binomial (Eq. (3)), a multinomial model of release with nonuniform quantal size and uniform release probability (Eq. (7)) or a model that also includes nonuniformity of release probability (Eq. (13)). The initial slope is related to $\overline{Q}_P$, the degree of curvature to P, and the size to N. P can be calculated for each release probability condition from the ratio of the mean current to $NP\overline{Q}_P$ (the maximal possible response). Fits should be accepted using a χ^2 criterion with a significant level of 0.05. P should be greater than 0.6 at high $[Ca^{2+}]$ for an accurate estimate of N [41, 58].

3.6 Weighted Least Squares Method

Conventional least squares fitting (Eqs. (7) and (13)) assumes that the errors estimating the mean PSC amplitude are negligible, but a more sophisticated expression for the weights can be used when the errors in the current cannot be neglected [58]. Errors in the estimation of variance can be determined with unbiased estimators of the variance of the variance that can be obtained using sample moments [58].

A general expression of the variance of the sample variance s^2 is given by:

$$\mathrm{var}\left(s^2\right) = \frac{n}{(n-2)(n-3)}\left[\frac{3(3-2n)(n-1)^2 - n(n-2)(n-3)^2}{(n^2-2n+3)(n-1)^2} m_2^2 + m_4\right] \quad (16)$$

where n is the number of traces acquired and m_2 and m_4 are sample central moments about the mean and can be calculated as $m_r = (1/n)\sum_{i=1}^{n}(X_i - \overline{X})^r$, where m_2 is the variance. Equation (16) is valid even when small sample sizes are used. Maximizing the number of samples improves the estimators' reliability and is therefore anyway desirable. The use of weighted least squares improves the reliability in parameter estimation up to 30 % [58].

4 Discussion

4.1 Summary of Technique

MPFA [18], which is also known as variance-mean analysis [19, 20], is a simple and robust method for estimating the quantal parameters of synaptic connections. MPFA overcomes the technical difficulties encountered with more traditional methods that involved identifying and interpreting peaks in amplitude histograms [14, 59, 60], allowing estimation of quantal parameters at central synaptic connections, where quantal parameter are nonuniform, quantal variance is large, and where the signal-to-noise ratio is often low.

4.2 Limitations of MPFA

While MPFA is straightforward to apply, the accuracy of the results depends on the quality of the experimental data and the particular properties of the synapse under study. Time stable recordings of PSCs at multiple release probabilities are essential. Uncertainty in the values of quantal parameters can be reduced by increasing the number of recordings per release probability condition and the number of different release probabilities recorded, but clearly, for a finite length experiment, there is a trade-off between these two experimental parameters. At many central synapses the assumption that quantal events sum linearly has been borne out. Multiquantal release is not an issue for MPFA if postsynaptic receptor occupancy is in the low-to-intermediate range, but if postsynaptic receptors saturate following quantal release, the postsynaptic responses are no longer linear reporters of vesicular release. Under these conditions it is essential to lower receptor occupancy with rapidly equilibrating low affinity antagonists [20, 24, 41, 53], to bring the postsynaptic response back into a linear regime. But perhaps the biggest limitation for this technique is that good quality voltage clamp is only feasible for synapses close to the soma. While this method can be applied to current clamp conditions, the impact of dendritic filtering and the impact of active dendritic conductances are difficult to control for.

As for all quantal analysis methods, MPFA relies on a statistical model of synaptic transmission. The results therefore depend on the accuracy of the model. Previous quanta analysis methods have been limited by their ability to account for nonuniform quantal parameters and the presence of high quantal variance. While MPFA can account for a wide range of nonuniformities including those arising from release probability and quantal size, it can only estimate the mean values of the quantal parameters and how they change during plasticity. While this is sufficient for many questions, a deeper understanding of the release process, including separation of mean probability of a vesicle being in the release-ready state and the mean probability that a release-ready vesicle will undergo fusion, requires more powerful descriptions of the release process [25–27, 52] that include short-term plasticity [61].

4.3 Comparison of MPFA to Other Methods

Several quantitative methods have been developed for studying synaptic function over the last decade that extends the basic MPFA approach. These include combining variance-mean approaches with covariance analysis [27, 62] and combining deconvolution and variance analysis enabling estimation of both the quantal properties and the time course of the release rate [22, 63]. These approaches together with other more direct methods of measuring release [50] provide valuable additional information on the time course of release. Combining variance-mean analysis and model of short-term plasticity is also effective for investigating the relationship between the quantal properties and short-term plasticity at central synapses [25, 26, 33, 64].

Statistical models based on Bayesian inference have been used to extract information from the fluctuations of the postsynaptic response [65, 66]. This technique assumes that the fluctuations in synaptic signals are described by mixtures of Gaussian distributions. Simulations show that a Bayesian Quantal Analysis (BQA) algorithm can accurately estimate the quantal parameters from a small data set with only two conditions of release probability [66] compared to MPFA which needs at least three release probability conditions [41]. This could be advantageous if recording stability is a problem. The BQA algorithm does not make the distinction between intra- and intersite variability but characterizes their distributions using the "quantal likelihood function." Nevertheless, BQA performed comparably well to MPFA when compared using the same datasets [66].

4.4 Optical Approaches and Future Directions

Most quantitative studies of synaptic connections have used electrophysiological recordings to assess synaptic function. However, optical approaches are potentially very powerful because they can also provide spatial information, thereby enabling the study of individual synaptic contacts [36, 67–70]. Optical approaches have shown, for example, that the probability of neurotransmitter release is nonuniform across release sites [36]. However, using postsynaptic Ca^{2+} changes in spines as an assay of glutamate release is complicated by the presence of voltage-gated channels. The small number of NMDARs activated by a quantum [38] can also complicate interpretation of failures. Nevertheless, optical quantal analysis has been used to probe the locus of expression of short-term plasticity [67, 68]. Presynaptic calcium imaging has also been used to assess release probability at individual synapses [43] as well as changes of probability of release during long-term [69] and short-term plasticity [43, 70]. However, these techniques provide only part of the quantal description of the synapse compared to electrophysiology-based MPFA. Moreover, unlike MPFA, which can be applied to both excitatory and inhibitory synapses, optical quantal analysis can only be applied to synapses that have Ca^{2+}-permeable receptors. However, the development of new fluorescent reporters that sense transmitter directly, like glutamate-sensing fluorescent reporter (iGluSnFR) [71], and high speed 2-photon imaging methods that can monitor synapses distributed in 3D space [72, 73] suggest that several of the limitations of optical quantal analysis could soon be overcome.

5 Conclusions

The quantum hypothesis was proposed to explain the stochastic nature of chemical synaptic transmission. However, the Poisson statistics that describe transmission at the neuromuscular junction under low release conditions are not applicable to most central synapses due to the low number of release sites. Another difficulty at central synapses is that the underlying release probability and quantal size are not uniform. MPFA is a simple approach for determining synaptic properties at synapses with nonuniform quantal parameters. It is robust to noise and large quantal variance, in contrast to more traditional methods that relied on identifying quantal peaks in amplitude histograms. MPFA has been widely used to determine quantal parameters at both glutamatergic [18, 22, 23, 35] and GABAergic synapses [30] and has been applied to trains of responses to determine which quantal parameters change during short-term plasticity [18, 25, 29, 46]. The quantal parameters estimated with MPFA can also be combined with models of short-term plasticity [25, 26, 64], thereby providing a more complete picture of the quantal transmission at central synapses [74].

Acknowledgements

We thank Antoine Valera for comments on the manuscript. FL is supported by an IEF Marie Curie fellowship (FP7) and RAS holds a Wellcome Trust Principal Research Fellowship and an ERC Advanced Grant.

References

1. Fatt P, Katz B (1952) Spontaneous subthreshold activity at motor nerve endings. J Physiol 117:109–128
2. del Castillo J, Katz B (1954) Quantal components of the end-plate potential. J Physiol 124:560–573
3. Fatt P, Katz B (1950) Some observations on biological noise. Nature 166:597–598
4. De Robertis E, Bennett HS (1955) Some features of the submicroscopic morphology of synapses in frog and earthworm. J Biophys Biochem Cytol 1:47
5. Katz B (1969) The release of neural transmitter substances. Liverpool University Press, Liverpool
6. Kuno M (1964) Quantal components of excitatory synaptic potentials in spinal motoneurones. J Physiol 175:81–99
7. Malinow R, Tsien RW (1990) Presynaptic enhancement shown by whole-cell recordings of long-term potentiation in hippocampal slices. Nature 346:177–180
8. Bekkers JM, Stevens CF (1990) Presynaptic mechanism for long-term potentiation in the hippocampus. Nature 346:724–729
9. Larkman A, Stratford K, Jack J (1991) Quantal analysis of excitatory synaptic action and depression in hippocampal slices. Nature 350:344–347
10. Edwards F (1991) Neurobiology. LTP is a long term problem. Nature 350:271–272
11. Kullmann DM, Nicoll RA (1992) Long-term potentiation is associated with increases in quantal content and quantal amplitude. Nature 357:240–244
12. Walmsley B, Edwards FR, Tracey DJ (1988) Nonuniform release probabilities underlie quantal synaptic transmission at a mammalian excitatory central synapse. J Neurophysiol 60:889–908
13. Rosenmund C, Clements JD, Westbrook GL (1993) Nonuniform probability of glutamate release at a hippocampal synapse. Science 262:754–757
14. Walmsley B (1995) Interpretation of "quantal" peaks in distributions of evoked synaptic transmission at central synapses. Proc Biol Sci 261:245–250
15. Sigworth FJ (1980) The variance of sodium current fluctuations at the node of Ranvier. J Physiol 307:97–129
16. Traynelis SF, Silver RA, Cull-Candy SG (1993) Estimated conductance of glutamate receptor channels activated during EPSCs at the cerebellar mossy fiber-granule cell synapse. Neuron 11:279–289
17. Silver RA, Cull-Candy SG, Takahashi T (1996) Non-NMDA glutamate receptor occupancy and open probability at a rat cerebellar synapse with single and multiple release sites. J Physiol 494(Pt 1):231–250
18. Silver RA, Momiyama A, Cull-Candy SG (1998) Locus of frequency-dependent depression identified with multiple-probability fluctuation analysis at rat climbing fibre-Purkinje cell synapses. J Physiol 510(Pt 3):881–902
19. Reid CA, Clements JD (1999) Postsynaptic expression of long-term potentiation in the rat dentate gyrus demonstrated by variance-mean analysis. J Physiol 518(Pt 1):121–130
20. Clements JD, Silver RA (2000) Unveiling synaptic plasticity: a new graphical and analytical approach. Trends Neurosci 23:105–113
21. Clamann HP, Mathis J, Lüscher HR (1989) Variance analysis of excitatory postsynaptic potentials in cat spinal motoneurons during posttetanic potentiation. J Neurophysiol 61:403–416
22. Sargent PB, Saviane C, Nielsen TA et al (2005) Rapid vesicular release, quantal variability, and spillover contribute to the precision and reliability of transmission at a glomerular synapse. J Neurosci 25:8173–8187
23. Lanore F, Labrousse VF, Szabo Z et al (2012) Deficits in morphofunctional maturation of hippocampal mossy fiber synapses in a mouse model of intellectual disability. J Neurosci 32:17882–17893
24. Meyer AC, Neher E, Schneggenburger R (2001) Estimation of quantal size and number of functional active zones at the calyx of held

synapse by nonstationary EPSC variance analysis. J Neurosci 21:7889–7900

25. Saviane C, Silver RA (2006) Fast vesicle reloading and a large pool sustain high bandwidth transmission at a central synapse. Nature 439:983–987
26. Hallermann S, Fejtova A, Schmidt H et al (2010) Bassoon speeds vesicle reloading at a central excitatory synapse. Neuron 68:710–723
27. Scheuss V, Neher E (2001) Estimating synaptic parameters from mean, variance, and covariance in trains of synaptic responses. Biophys J 81:1970–1989
28. Neher E, Sakaba T (2003) Combining deconvolution and fluctuation analysis to determine quantal parameters and release rates. J Neurosci Methods 130:143–157
29. Oleskevich S, Clements J, Walmsley B (2000) Release probability modulates short-term plasticity at a rat giant terminal. J Physiol 524(Pt 2):513–523
30. Biró AA, Holderith NB, Nusser Z (2005) Quantal size is independent of the release probability at hippocampal excitatory synapses. J Neurosci 25:223–232
31. Biró AA, Holderith NB, Nusser Z (2006) Release probability-dependent scaling of the postsynaptic responses at single hippocampal GABAergic synapses. J Neurosci 26:12487–12496
32. Humeau Y, Doussau F, Popoff MR et al (2007) Fast changes in the functional status of release sites during short-term plasticity: involvement of a frequency-dependent bypass of Rac at Aplysia synapses. J Physiol 583:983–1004
33. Valera AM, Doussau F, Poulain B et al (2012) Adaptation of granule cell to purkinje cell synapses to high-frequency transmission. J Neurosci 32:3267–3280
34. Sola E, Prestori F, Rossi P et al (2004) Increased neurotransmitter release during long-term potentiation at mossy fibre-granule cell synapses in rat cerebellum. J Physiol 557:843–861
35. Fourcaudot E, Gambino F, Humeau Y et al (2008) cAMP/PKA signaling and RIM1alpha mediate presynaptic LTP in the lateral amygdala. Proc Natl Acad Sci 105:15130–15135
36. Murthy VN, Sejnowski TJ, Stevens CF (1997) Heterogeneous release properties of visualized individual hippocampal synapses. Neuron 18:599–612
37. Bekkers JM, Richerson GB, Stevens CF (1990) Origin of variability in quantal size in cultured hippocampal neurons and hippocampal slices. Proc Natl Acad Sci 87:5359–5362
38. Silver RA, Traynelis SF, Cull-Candy SG (1992) Rapid-time-course miniature and evoked excitatory currents at cerebellar synapses in situ. Nature 355:163–166
39. Frerking M, Wilson M (1996) Effects of variance in mini amplitude on stimulus-evoked release: a comparison of two models. Biophys J 70:2078–2091
40. Borst JG, Lodder JC, Kits KS (1994) Large amplitude variability of GABAergic IPSCs in melanotrophs from Xenopus laevis: evidence that quantal size differs between synapses. J Neurophysiol 71:639–655
41. Silver RA (2003) Estimation of nonuniform quantal parameters with multiple-probability fluctuation analysis: theory, application and limitations. J Neurosci Methods 130:127–141
42. Branco T, Staras K, Darcy KJ et al (2008) Local dendritic activity sets release probability at hippocampal synapses. Neuron 59:475–485
43. Holderith N, Lörincz A, Katona G et al (2012) Release probability of hippocampal glutamatergic terminals scales with the size of the active zone. Nat Neurosci 15:988–997
44. Wadiche JI, Jahr CE (2001) Multivesicular release at climbing fiber-Purkinje cell synapses. Neuron 32:301–313
45. Christie JM, Jahr CE (2006) Multivesicular release at Schaffer collateral-CA1 hippocampal synapses. J Neurosci 26:210–216
46. Foster KA, Regehr WG (2004) Variance-mean analysis in the presence of a rapid antagonist indicates vesicle depletion underlies depression at the climbing fiber synapse. Neuron 43:119–131
47. DiGregorio DA, Nusser Z, Silver RA (2002) Spillover of glutamate onto synaptic AMPA receptors enhances fast transmission at a cerebellar synapse. Neuron 35:521–533
48. Barbour B, Hausser M (1997) Intersynaptic diffusion of neurotransmitter. Trends Neurosci 20:377–384
49. Sakaba T, Schneggenburger R, Neher E (2002) Estimation of quantal parameters at the calyx of Held synapse. Neurosci Res 44:343–356
50. Minneci F, Kanichay RT, Silver RA (2012) Estimation of the time course of neurotransmitter release at central synapses from the first latency of postsynaptic currents. J Neurosci Methods 205:49–64
51. Vere Jones D (1966) Simple stochastic models for the release of quanta of transmitter from a nerve terminal. Aust J Stat 8:53–63
52. Quastel DM (1997) The binomial model in fluctuation analysis of quantal neurotransmitter release. Biophys J 72:728–753

53. Wong AYC, Graham BP, Billups B et al (2003) Distinguishing between presynaptic and postsynaptic mechanisms of short-term depression during action potential trains. J Neurosci 23:4868–4877
54. Spruston N, Jaffe DB, Williams SH et al (1993) Voltage- and space-clamp errors associated with the measurement of electrotonically remote synaptic events. J Neurophysiol 70:781–802
55. Bar-Yehuda D, Korngreen A (2008) Space-clamp problems when voltage clamping neurons expressing voltage-gated conductances. J Neurophysiol 99:1127–1136
56. Silver RA, Lubke J, Sakmann B et al (2003) High-probability uniquantal transmission at excitatory synapses in barrel cortex. Science 302:1981–1984
57. Nyquist H (1928) Certain topics in telegraph transmission theory. Trans AIEE 47:617–644
58. Saviane C, Silver RA (2006) Errors in the estimation of the variance: implications for multiple-probability fluctuation analysis. J Neurosci Methods 153:250–260
59. Redman S (1990) Quantal analysis of synaptic potentials in neurons of the central nervous system. Physiol Rev 70:165–198
60. Stricker C, Field AC, Redman SJ (1996) Statistical analysis of amplitude fluctuations in EPSCs evoked in rat CA1 pyramidal neurones in vitro. J Physiol 490(Pt 2):419–441
61. Tsodyks M, Pawelzik K, Markram H (1998) Neural networks with dynamic synapses. Neural Comput 10:821–835
62. Scheuss V, Neher E, Schneggenburger R (2002) Separation of presynaptic and postsynaptic contributions to depression by covariance analysis of successive EPSCs at the calyx of held synapse. J Neurosci 22:728–739
63. Sakaba T, Neher E (2001) Quantitative relationship between transmitter release and calcium current at the calyx of held synapse. J Neurosci 21:462–476
64. Saviane C, Silver RA (2007) Estimation of quantal parameters with multiple-probability fluctuation analysis. Methods Mol Biol 403:303–317
65. Turner DA, West M (1993) Bayesian analysis of mixtures applied to post-synaptic potential fluctuations. J Neurosci Methods 47:1–21
66. Bhumbra GS, Beato M (2013) Reliable evaluation of the quantal determinants of synaptic efficacy using Bayesian analysis. J Neurophysiol 109:603–620
67. Oertner TG, Sabatini BL, Nimchinsky EA et al (2002) Facilitation at single synapses probed with optical quantal analysis. Nat Neurosci 5:657–664
68. Yuste R, Majewska A, Cash SS et al (1999) Mechanisms of calcium influx into hippocampal spines: heterogeneity among spines, coincidence detection by NMDA receptors, and optical quantal analysis. J Neurosci 19:1976–1987
69. Emptage NJ, Reid CA, Fine A et al (2003) Optical quantal analysis reveals a presynaptic component of LTP at hippocampal Schaffer-associational synapses. Neuron 38:797–804
70. Sylantyev S, Jensen TP, Ross RA et al (2013) Cannabinoid- and lysophosphatidylinositol-sensitive receptor GPR55 boosts neurotransmitter release at central synapses. Proc Natl Acad Sci 110:5193–5198
71. Marvin JS, Borghuis BG, Tian L et al (2013) An optimized fluorescent probe for visualizing glutamate neurotransmission. Nat Methods 10:162–170
72. Kirkby PA, Srinivas Nadella KMN, Silver RA (2010) A compact acousto-optic lens for 2D and 3D femtosecond based 2-photon microscopy. Opt Express 18:13721–13745
73. Fernández-Alfonso T, Nadella KMNS, Iacaruso MF et al (2014) Monitoring synaptic and neuronal activity in 3D with synthetic and genetic indicators using a compact acousto-optic lens two-photon microscope. J Neurosci Methods 222:69–81
74. Rothman JS, Silver RA (2014) Data-driven modeling of synaptic transmission and integration. Prog Mol Biol Transl Sci 123:305–350

Chapter 11

Acousto-optical Scanning-Based High-Speed 3D Two-Photon Imaging In Vivo

Balázs Rózsa, Gergely Szalay, and Gergely Katona

Abstract

Recording of the concerted activity of neuronal assemblies and the dendritic and axonal signal integration of downstream neurons pose different challenges, preferably a single recording system should perform both operations. We present a three-dimensional (3D), high-resolution, fast, acousto-optic two-photon microscope with random-access and continuous trajectory scanning modes reaching a cubic millimeter scan range (now over $950 \times 950 \times 3000\ \mu m^3$) which can be adapted to imaging different spatial scales. The resolution of the system allows simultaneous functional measurements in many fine neuronal processes, even in dendritic spines within a central core ($>290 \times 290 \times 200\ \mu m^3$) of the total scanned volume. Furthermore, the PSF size remained sufficiently low ($PSF_x < 1.9\ \mu m$, $PSF_z < 7.9\ \mu m$) to target individual neuronal somata in the whole scanning volume for simultaneous measurement of activity from hundreds of cells. The system contains new design concepts: it allows the acoustic frequency chirps in the deflectors to be adjusted dynamically to compensate for astigmatism and optical errors; it physically separates the *z*-dimension focusing and lateral scanning functions to optimize the lateral AO scanning range; it involves a custom angular compensation unit to diminish off-axis angular dispersion introduced by the AO deflectors, and it uses a high-NA, wide-field objective and high-bandwidth custom AO deflectors with large apertures. We demonstrate the use of the microscope at different spatial scales by first showing 3D optical recordings of action potential back propagation and dendritic Ca^{2+} spike forward propagation in long dendritic segments in vitro, at near-microsecond temporal resolution. Second, using the same microscope we show volumetric random-access Ca^{2+} imaging of spontaneous and visual stimulation-evoked activity from hundreds of cortical neurons in the visual cortex *in vivo*. The selection of active neurons in a volume that respond to a given stimulus was aided by the real-time data analysis and the 3D interactive visualization accelerated selection of regions of interest.

Key words Two-photon, Acousto-optical, Dendritic imaging, Network imaging, 3D, Three-dimensional microscopy, Angular dispersion compensation, In vivo imaging, Backpropagating action potentials, Temporal super-resolution, Dendritic spikes, 3D scanning, 3D virtual reality environment

1 Introduction

1.1 The Need for 3D Measurements

The systematic understanding of brain function requires methods that allow neuronal activity to be recorded at different spatial scales in three dimensions (3D) at a high temporal resolution. At the single neuron level, activity is differentially distributed in space

Alon Korngreen (ed.), *Advanced Patch-Clamp Analysis for Neuroscientists*, Neuromethods, vol. 113,
DOI 10.1007/978-1-4939-3411-9_11, © Springer Science+Business Media New York 2016

and time across the dendritic and axonal segments [1–6]. Therefore, in order to understand neuronal signal integration, activity should be simultaneously recorded at many spatial locations within the dendritic and axonal tree of a single neuron. At the neuronal circuits level, closely spaced neurons can have vastly different activity patterns [7]; on the other hand, widely separated cells may belong to the same functional circuit, influencing each other via long axonal processes. Therefore, recording techniques are required that collect information near-simultaneously (in one fast measurement sequence) from many cells of a neuronal population situated in an extensive volume of tissue. Moreover, measurements should be possible on the timescales of dendritic integration and regenerative spike propagation [8], i.e., with sub-millisecond temporal resolution. One last aspect is that these measurements should be performed in as intact as possible neurons on neurons or neuronal networks which are as intact as possible, where neurons are embedded in their original tissue. This poses a challenge: cellular precision must be retained in acute brain preparations where the tissue is at least 100-μm thick or, for in vivo experiments, through millimeters of brain material.

1.2 From Wide-Field Microscopy to Single-Point, Two-Photon Excitation

Light microscopy is important in biological research because it enables living tissue to be observed and suited at a relatively high spatial resolution. This resolution is limited by the wavelength of light (according to Abbe's theory) and does not rival that of electron microscopy, but the scope of electron microscopy is limited when observing living specimens [9]. Other vital imaging technologies, such as MRT (magnetic resonance tomography), PET (positron emission tomography), or X-rays, can neither resolve subcellular structures nor provide high temporal resolution nor the exquisite molecular selectivity that would allow single molecules to be detected in a background of billions of others [10, 11].

Light microscopy inside living tissues is hampered by the degradation of resolution and contrast, caused by absorption and light scattering, which is due to refractive index inhomogeneities present to a varying degree in every tissue. Deeper within the tissue, images become more degraded and high-resolution imaging eventually becomes impossible. A major step toward overcoming this problem was the invention of confocal microscopy [12, 13]. In a confocal microscope, the illumination light is focused on a diffraction-limited spot, and this excites the sample along two cones close to the aimed focal point. Then, using the same objective, the emitted signal photons are focused onto a detector pinhole that rejects all light emitted outside of the focus spot.

The main drawback of confocal microscopy is its wasteful use of both excitation and emission. On the one hand, absorption occurs throughout the specimen, but information is obtained only from a small sub-volume around the focal point. On the other hand, light emitted from the focal point is also rejected by the pinhole if it is scattered by the tissue: this means that tissue scattering has a significant impact on the signal-to-noise ratio (SNR) of the images. This is particularly a major problem in vital fluorescence microscopy, where the limiting factors are usually either photochemical destruction of the fluorophore (photobleaching) or photodynamic damage to the specimen (photodamage). In confocal microscopy, only ballistic photons that are not scattered on their path out of the tissue contribute to the signal, while scattered photons, often the majority, are rejected by the detector aperture, limiting depth penetration of the technology to about 50–80 μm in brain tissue. Excitation needs to be increased in order to compensate for this signal loss which further exacerbates photobleaching and photodamage. All of these problems are general to many other, so-called single photon excitation technologies, among them to the spinning-disc confocal [14] and to the light sheet microscope technologies [15]. Theoretically, these problems can be handled by the use of laser scanning microscopy in combination with two-photon excitation [16, 17].

The concept of two-photon excitation is based on the idea that two photons of low energy can be combined in the same quantum event, resulting in the emission of a fluorescence photon at a higher energy than either one of the two excitatory photons. Although the probability of such absorption is extremely low, its cross section is proportional to the square of the photon flux, making it possible to counterbalance this low initial cross section by using extremely high photon fluxes. Multi-photon absorption is often called nonlinear because the absorption rate is dependent on a higher-than-first-power light intensity. Such high fluxes are only present in the focus of a high numerical aperture lens illuminated by a strong, pulsed near-infrared laser. The possibility of absorbing more than one photon during a single quantum event had been predicted more than 60 years ago [18], but it was confirmed experimentally after the invention of mode-locked lasers with a pulse duration below 1 ps and repetition rates of about 100 MHz. This made two-photon laser scanning microscopy feasible in practice [16].

Mode-locking a Ti:sapphire laser, for example, boosts the two-photon excitation rate by 100,000-fold, compared to continuous-wave (cw) operation at the same average power. On the other hand, the quadratic dependence of the absorption rate on the light intensity gives two-photon microscopy its optical sectioning property because fluorescence is only generated in the vicinity of the geometrical focus where the light intensity is high. While scanning the laser focus in both lateral directions (x and y), fluorescence excitation is limited to the focal plane. No detector pinhole is necessary since—in most cases—no fluorescence is generated outside the focal volume, and all fluorescence photons, whether leaving the sample on scattered or ballistic trajectories, constitute useful signal.

The combination of low phototoxicity enabled by the single-point excitation and the efficient use of fluorescence—even scattered—makes single-point two-photon microscopy a unique tool for observing function deep within the tissue, even resolving structures smaller than 1 mm in brain tissue [19–21]. Its main drawback, however, is speed when compared to camera-based approaches where, instead of the "single-channel" photomultipliers (PMTs), information can potentially be collected from millions of camera pixels simultaneously.

There are several approaches to using multi-site two-photon stimulation in combination with a camera-based detection algorithm to overcome the limitation of single-point illumination. The simplest of these technologies is multi-beam two-photon microscopy [22], where a string of focal spots are generated by breaking the laser beam into parallel beamlets. This causes parallel fluorescent excitation at multiple sites; these are then scanned to illuminate the entire focal plane. Fluorescence is detected by a high-resolution camera after imaging the focal surface onto its sensor chip. This approach makes it possible to image with a higher frame rate than with a scanned single-beam two-photon microscope, but it has one main drawback: imaging florescence onto the camera is sensitive to tissue scattering. As a result, this technology has about the half of the penetration depth of single-point two-photon microscopy and this—especially in the case of in vivo investigations—seriously limits its biological usability.

Recent camera-based approaches can also resolve the 3D structure of a sample by using special light shaping: microlens array [23] or phase mask [24] in front of the camera combined with extended depth of field [25] or holographic [24] illumination. Depth information is obtained from the camera images by computationally reversed transformation, which causes two major limitations. First, there is a strict limit to the number and arrangement of the regions of interest (ROIs) as they should not overlap much on the detector surface. Typically, ROIs situated at the same place but at different depths are imaged to mostly overlapping profiles on the detector surface, making it difficult to distinguish between them. Second, the imaging of the emitted light is sensitive to tissue scattering, causing scattered shapes on the detector surface, challenging the reverse transforming algorithms. For example, Quirin et al. [24] showed penetration depth of about 190 μm (3 × scattering length), which is just enough to reach the first cell layers in mice in vivo, and is much less than the depths achievable by single-point two-photon microscopes.

As we have seen, a single-point illuminated two-photon microscope can reach deep structures with subcellular resolution, which makes this technology the best available tool for functional studies. However, as only one point might be illuminated at a time, special scanning methods need to be developed to sample the important biological features with enough speed to resolve their functionality.

1.3 Why Choose Acousto-optical Scanning for Network and Dendritic Measurements in 3D?

Several novel technologies have been developed to generate 3D readouts of fast population and dendritic activities, including liquid lenses [26], holographic scanning [27], Roller Coaster Scanning [2], piezo-scanning with sinusoidal resonance [28], deformable mirrors, temporal and spatial multiplexing [29], axicon or planar illumination-based imaging [30], fast *z*-scanning based on an axially moving mirror [31], simultaneous multiview light sheet microscopy [15], and optical fiber acousto-optical deflector based 3D scanning [32]. There are three major points why 3D acousto-optical scanning is better than alternative methods:

1. First of all, 3D acousto-optical microscope technology allows single-spot two-photon activation; this permits whole-field detection with high quantum efficiency PMTs. Therefore, photons scattered from deep layers within the tissue can also be collected with a high sensitivity; this is essential for imaging throughout the whole cortex.
2. Moreover, the position of the point spread function (PSF) can be finely adjusted to any spatial coordinates during acousto-optical scanning, with 50–100 nm precision [33]. Therefore, the number of z-planes and coordinates is unlimited, which allows a very precise measurement of neuronal activity by eliminating neuropil contamination.
3. Second, acousto-optical scanning allows flexible selection of regions of interest. Instead of scanning large volumes, we can restrict our measurements only to preselected regions. This subselection increases the product of the measurement speed and the signal collection efficiency by up to several orders of magnitude, as compared to classical raster scanning. More quantitatively, 3D random-access scanning increases the measurement speed and the signal collection efficiency by the following factor:

$$(\text{measurement speed})*(\text{signal collection}) \cong \left(\frac{\text{total image volume}}{\text{volume covered by the scanning points}}\right)$$

where "signal collection" is defined as the number of fluorescent photons detected in a given time interval from a given spatial location, and the "volume covered by the scanning points" is the convolution of the PSF and the preselected scanning points. In a typical in vivo measurement, using a $450 \times 450 \times 650\ \mu m^3$ scanning volume with a 0.47–1.9 μm and 2.49–7.9 μm axial resolution (see Katona et al. [33]) and simultaneously measuring about 100 locations, *random-access scanning will provide a* 2,106,000–46,534-*fold, on average an increase of a six orders of magnitude in the product of measurement speed and signal collection efficiency.* According to the current state of the technology (which now provides an approximately 1 mm^3 scanning volume with high NA objectives), this increase could be even larger, over 1,000,000,000 per locations. *No other currently available 3D scanning method, with high spatial resolution and deep penetration capability, can provide a similar increase in the product of the measurement speed and the signal collection efficiency.*

2 Principles of Acousto-optical Devices

The phrase "acousto-optic" refers to the field of optics that studies the interaction between sound and light waves. In imaging we use acousto-optical devices to diffract laser beams through ultrasonic gratings. The acousto-optical effect is based on a periodic change in the refractive index in the high refractive index medium (usually tellurium dioxide, TeO_2) which is the result of the sound wave induced pressure fluctuation in the crystal. This grating diffracts the light beam just like a normal optical grating, but can be adjusted rapidly.

2.1 Acousto-optical Deflection

Acousto-optical deflectors (AODs) control the optical beam spatially, using ultrasonic waves to diffract the laser beam depending on the acoustic frequency. If we introduce a sine wave at the piezoelectric driver, it will generate an optical deflection in the acousto-optic medium according to the following equation:

$$\Delta\Theta_{\mathrm{d}} = \frac{\lambda}{\nu}\Delta f$$

where λ is the wavelength of the optical beam, ν is the velocity of the acoustic wave, and Δf is the change in the sound frequency (Fig. 1).

In practice, AODs or acousto-optical modulators (AOMs) are used. AOMs modulate only the amplitude of the sound waves, while AODs are able to adjust both the amplitude and frequency.

2.2 Acousto-optical Focusing

Besides deflection, AODs can also be used for fast focal plane shifting [33–36]. If the optical aperture is filled with acoustic waves, the frequency of which increases as a function of time (chirped wave), different portions of the optical beam are deflected in different directions (Fig. 2). Thus a focusing, or alternatively a defocusing, effect occurs, depending on the frequency slope (sweep rate) of the chirped acoustic wave. The focal length of an acousto-optical lens (F) can be calculated from the sweep rate as [34]:

$$F = \frac{\nu^2 T_{\mathrm{scan}}}{2\lambda\Delta f}$$

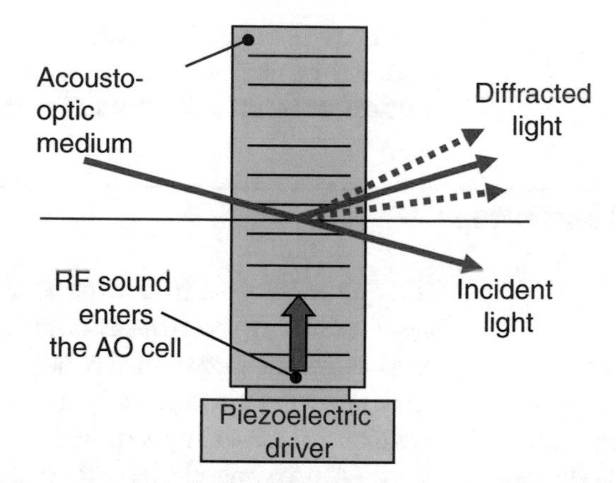

Fig. 1 Operating principle of acousto-optical deflectors. A piezoelectric driver elicits radiofrequency (RF) sound waves due to the externally applied sinusoidal voltage. Sound enters and traverses through the diffracting (TeO_2) medium while interacting with light throughout the aperture. Light is diffracted on the sound wave's refraction index changes as on a steady optical grating providing diffracted light beams whose angle is dependent on the sound wave's frequency

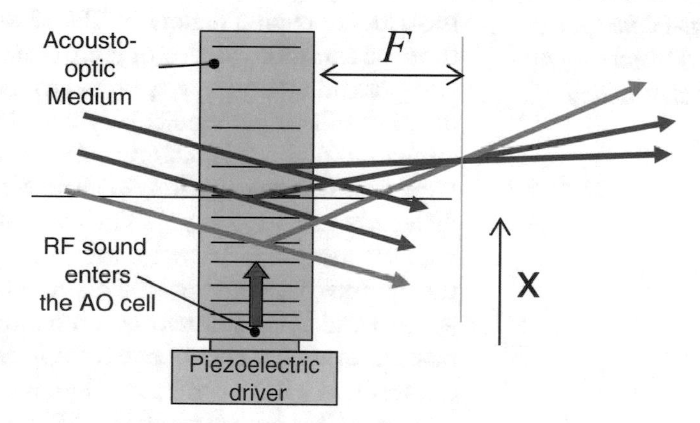

Fig. 2 Acousto-optical focusing. AO deflector arrangement similar to Fig. 1, but here the sound frequency is changing, resulting in a varying grating size along the propagation axis. As a consequence, different parts of the laser beam are diffracted with varying angles creating a focus point whose position can be changed by the parameters of the sound applied to the deflector. *Red* and *orange lines* indicate laser beams and *F* denotes focal distance

where λ is the optical wavelength of the beam, ν is the velocity of the acoustic wave, Δf is the change of the sound frequency, and T_{scan} is the modulation rate of the sound frequency [34].

To keep the focus stable, the frequency gradient should be preserved in the crystal. The frequency should therefore be continuously increased (or decreased) at the piezoelectric driver to preserve the focal distance. This will result in a lateral drift of the focal point, which can be easily compensated by introducing a second acousto-optical deflector with a counter-propagating acousto-optical wave into the optical pathway [34].

3 3D AO Microscope

3.1 Goals

The ultimate goal in neuroscience is to create a 3D random-access laser scanning two-photon microscope simultaneously satisfying two different needs in the largest possible scanning volume. The first need is to record activity across the dendritic tree of a single neuron in 3D at high spatial and temporal resolution in a central core (approximately 290 × 290 × 200 μm^3) of the scanning volume in a way that dendritic spines remain resolvable. The second need is to record in a more extensive volume (now over 950 × 950 × 3000 μm^3 volume in transparent samples with high NA objectives) at high speed but a relatively reduced resolution, in order to capture all activity from a large number of cell bodies in a neuronal population.

3.2 Selection of the Optimal Concept for Fast 3D Scanning by Optical Modeling

Acousto-optical deflectors and lenses provide fast, programmable tools for addressing regions in 3D. The optical grid generated within the acousto-optical deflector is equivalent to a cylindrical lens. In 3D two-photon microscopy, we need to scan points. Therefore, a combination of two perpendicularly oriented cylindrical lenses with the same focal distance is required. As both acousto-optical lenses (x, y) require drift compensation, 3D microscopes need (at least) four AO deflectors (x1, y1, x2, y2). The four AO deflectors can be optically coupled using *afocal projections* (telescopes with two lenses where the distance between the lenses is equal to the sum of each lens' focal length) and air in different combinations (Fig. 3). There are several possible arrangements of passive optical elements and AODs which can result in a 3D microscope (Fig. 3).

A detailed optical model in ZEMAX (and OSLO) is required to find those combinations of all active and passive optical elements in the scanning light path of the microscope which would provide the maximal exploitation of the apertures of all lenses within the 20× objective (numerical aperture, NA = 0.95) at different x-, y-, and z-scanning positions, and therefore result in the smallest PSF in the largest possible scanning range (Fig. 4). According to the optical models, the largest scanning volume can be reached by grouping the deflectors into two functionally distinct groups (z-focusing and lateral scanning) which are coupled together by one afocal projection (see version #5 in Fig. 3) [33], because this arrangement has several optical advantages. For example, limiting the beam divergence settings at the first group of deflectors allows better transmission, and hence larger scanning volume, because the incident angle tolerance of the AODs in the second group is limited.

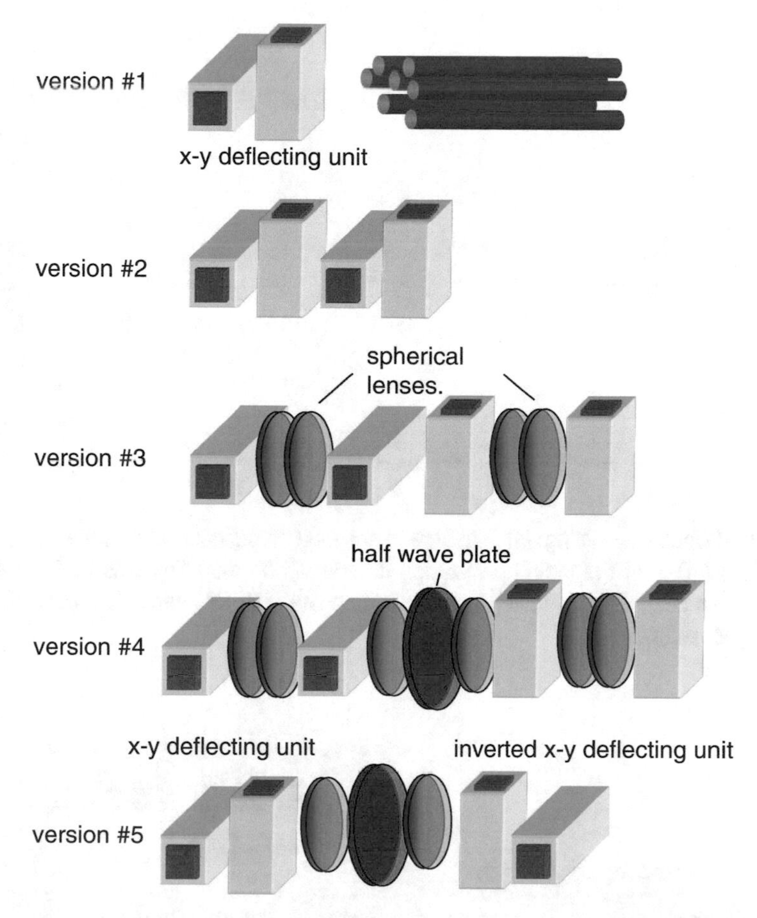

Fig. 3 Different arrangements for 3D AO scanning. Version #1: This was the first realization of a 3D microscope (Rozsa et al., 2003, ISBN 963 372 629 8). The concept uses optical fibers to position the excitation beam in 3D [32]. However, complex mechanical devices are required for each scanning point, which made implementation difficult. Version #2 [35, 37] and version #4 [36, 38–40] have also been described by other laboratories and have recently been used for functional measurements [35, 36]. Versions #2–5 were analyzed by modeling and version #5 was found to be the best [33]. For example, version #5 can provide around two order of magnitude larger scanning volume than other versions

3.3 New Concept for Angular Dispersion Compensation

As AODs are diffractive elements, it is crucial to have a proper compensation for angular dispersion. Angular dispersion is traditionally compensated with prisms [41]. However, 3D scanning requires spatially variable dispersion compensation. The angular dispersion is inherently compensated for following the last (the fourth) AOD in the main optical axis (Fig. 5). However, if we move away from the main optical axis we can find that dispersion increases as a function of distance (Fig. 5). Prisms with increasing apex angle would be required in order to compensate for this

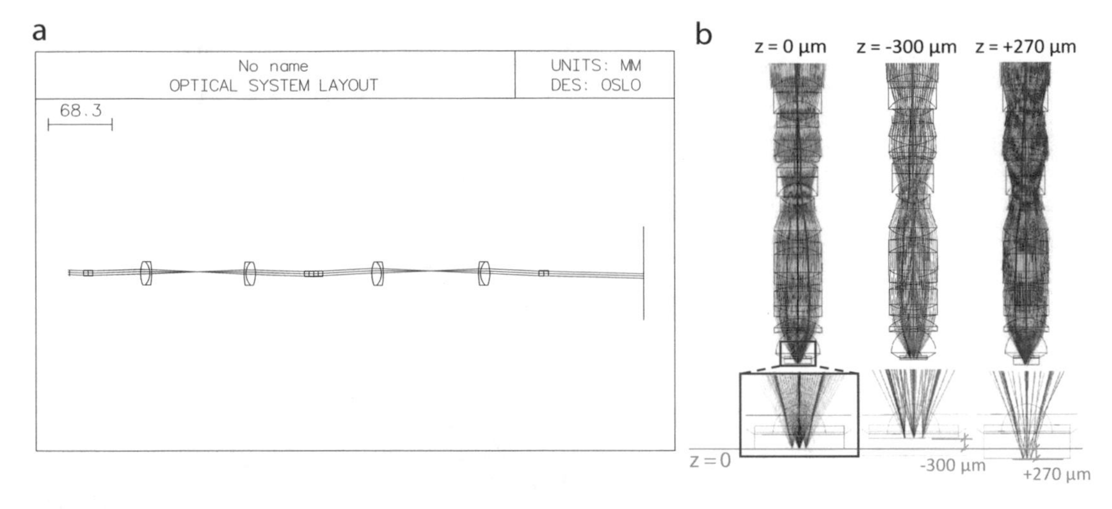

Fig. 4 Optical modeling. (**a**) Print screen showing an optical layout of the 3D microscope's scan head (version #5). (**b**) One of the main optimization criteria during modeling was to fill all the lenses of the objective at variable scanning positions in order to keep the high NA, hence the good PSF, during 3D scanning in the largest volume

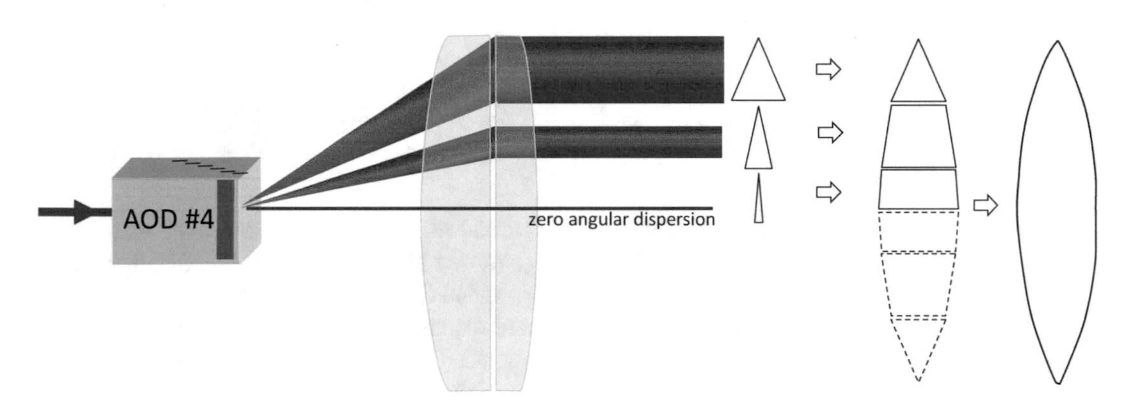

Fig. 5 Concept of angular dispersion compensation

dispersion. However, due to the limited space this theoretical solution does not work. If we make these small prisms wider and "connect" them (in theory, *see* Fig. 5) we can generate a surface which looks like a lens (Fig. 5).

In order to calculate the surface equation of this lens element, we need to go back to the general equation of prisms. The amount of angular dispersion which one prism can compensate for can be matched by properly selecting the incidence angle α_1 and the prism's opening angle α_P. The output angle (α_2) after the prism is given by:

$$\alpha_2(\alpha_1, \alpha_P, \lambda) = a\sin\left(n(\lambda)\sin\left(\alpha_P - a\sin\left(\frac{\sin(\alpha_1(\lambda))}{n(\lambda)}\right)\right)\right),$$

where $n(\lambda)$ is the refractive index. The wavelength-dependent incidence angle can be determined from the rotation angle of the prism β relative to the optical axis, hence the wavelength-dependent angle of propagation of the diffracted beams θ_{def}:

$$\alpha_1(\lambda) = \theta_{\text{def}}(\lambda, f_1, f_2) - \beta,$$

and the total deflection angle of the prism:

$$\Delta\alpha(\alpha_1, \alpha_P, \lambda, f_1, f_2) = \theta_{\text{def}}(\lambda, f_1, f_2) + \beta - a\sin\left(n(\lambda)\sin\left(\alpha_P - a\sin\left(\frac{\sin(\theta_{\text{def}}(\lambda, f_1, f_2) - \beta)}{n(\lambda)}\right)\right)\right) - \alpha_P,$$

The zero angular dispersion requirement after the prism can be expressed as:

$$\frac{\partial[\Delta\alpha(\alpha_1, \alpha_P, \lambda, f_1, f_2)]}{\partial\lambda} = 0$$

The beams deflected at different frequencies in the scanner are spatially separated in the focal plane of the first part of this unit. The different optical wavelength components in each deflected wave are also spatially and angularly dispersed in this plane, and their dispersion also increases as the deflection angle increases, but their spatial dispersion is an order of magnitude less than the separation caused by the acousto-optic deflection. In this case, the angular dispersion increases symmetrically with the distance from this axis, r ($r = \sqrt{x^2 + y^2}$).

The optimal surfaces, however, are not spherical surfaces, but can be expressed as aspheric and conic surfaces with the primary radii given by the equations:

$$c = \frac{\frac{1}{R_1}r^2}{1 + \sqrt{1 - (1 + k_1)\frac{r^2}{R_1}}} + \sum_{n=1:6} a_n r^n$$

where the k_1 conic and a_n aspheric parameters are determined by optical modeling. The value of the radius and the minimum glass thickness at the optic axis is also determined with optimization using the ZEMAX optimization algorithms. Two merit functions are consecutively used, one containing the angular dispersion and aperture in front of the objective, and a second targeting the minimum spot size in the sample plane (the focal plane of the objective). Iteration of these two leads to optimized lens surfaces for the given material and the problem set (see details in Katona et al. [33]).

When comparing a 3D setup using a simple telescope instead of a telescope with the lens element described above, we can obtain a lateral field of view which is ~2-fold larger, with the same focused spot size and dispersion at the edges of this area (*see* Fig. 8 below).

3.4 Construction of the Optical Pathway of a 3D Microscope

According to the optimal arrangement suggested by the optical model [33], a large-aperture (15–17 mm) optical assembly is needed (Fig. 6). A Ti:sapphire laser with automated dispersion compensation provides the laser pulses. The optimal wavelength range is 740–890 nm (but can now be extended to over 1064 nm). A *Faraday isolator* blocks the coherent backreflections (BB8-5I, Electro-Optics Technology). On top of the adjustable pre-chirping provided by the laser source, a fixed four-prism sequence [42] adds a large negative second- and third-order dispersion ($-72{,}000\ \mathrm{fs}^2$ and $-40{,}000\ \mathrm{fs}^3$) in order to pre-compensate for pulse broadening caused by the optical elements in the system (*Proctor and Wise* dispersion compensation unit, Fig. 6) [42]. An automated *beam stabilization unit* (see below) is necessary to precisely stabilize the laser beam in the long optical pathway, and cancel out subtle thermal drift errors. The beam stabilization unit is built from position sensors (*q*, quadrant detectors, PDQ80A, Thorlabs), two backside custom-polished broadband mirrors (BB2-E03; Thorlabs), and motorized mirrors (AG-M100N, Newport), wired in a feedback loop. The positioning feedback loop (U12, LabJack Corporation) is controlled by a program written in LabView (National Instruments). The beam is then magnified by a *beam expander* consisting of two achromatic lenses arranged in a Galilean telescope ($f = -75$ mm, ACN254-075-B, Thorlabs; $f = 200$ mm, NT47319, Edmund Optics; distance = 125.62 mm) to match the large apertures of the first pair of AO deflectors (15 mm). Mirrors, $\lambda/2$ waveplates, and holders were purchased from Thorlabs and Newport. AO deflectors have been custom designed and manufactured at the Budapest University of Technology and Economics. These deflectors form two orthogonal electric cylinder lenses filled with continuously changing frequency ("chirped") acoustic waves [34] and are used for *z*-focusing (*AO z-focusing unit*). Next, laser beams from the *x* and *y* cylindrical lenses are projected to the *x* and *y* AO deflectors (17 mm apertures) of the *2D-AO scanning unit*, respectively, by telecentric projection. Achromatic telecentric relay lenses were purchased from Edmund Optics ($f_{TC} = 150$ mm, NT32-886). The *2D-AO scanning unit* performs lateral scanning and also compensates for the lateral drift of the focal point generated by the cylindrical lenses during *z*-focusing [34] (*drift compensation*). *The angular dispersion compensation unit* optically links the 2D-AO scanning unit and the objective. This unit consists of two achromatic lens doublets, and one aspheric and conic lens element for angular dispersion

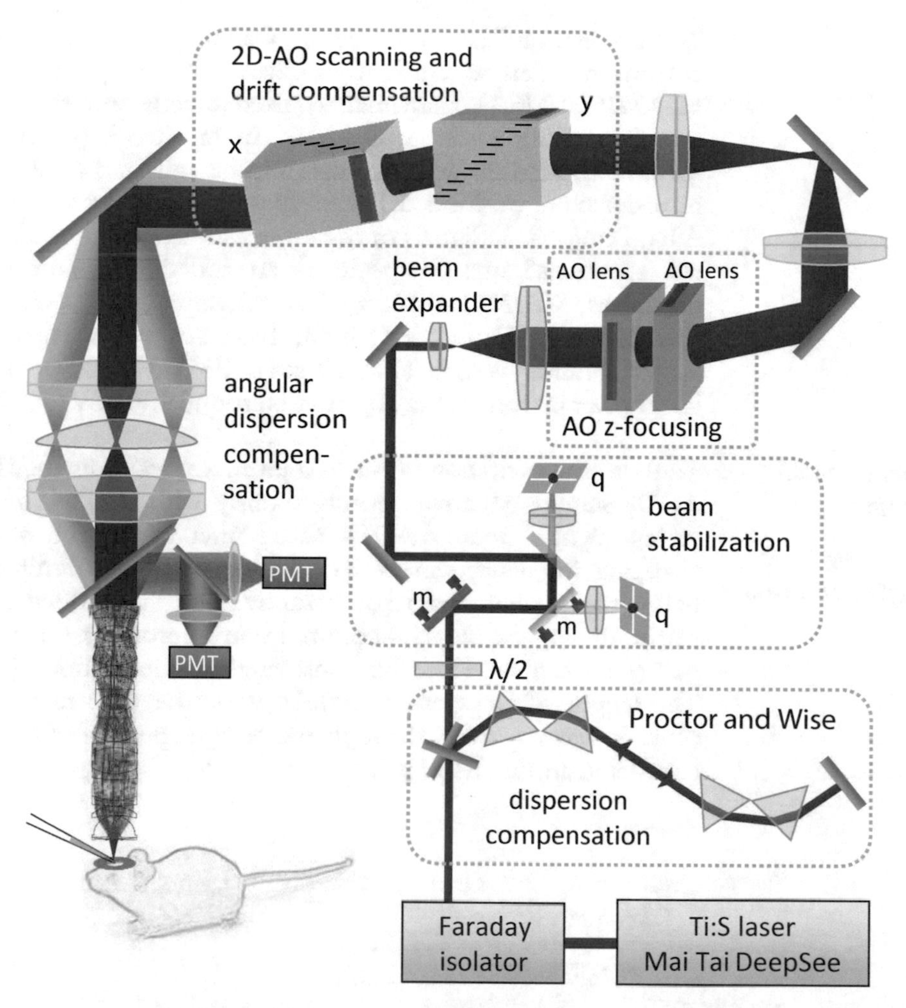

Fig. 6 Design and characterization of the two-photon microscope setup. (**a**) Schematics of the microscope setup. Material-dispersion compensation is adjusted with a four-prism compressor (Proctor and Wise) and a Ti:sapphire laser. A Faraday isolator eliminates coherent backreflections. Motorized mirrors (*m*) stabilize the position of the beam light on the surface of two quadrant detectors (*q*) before the beam expander. Two AO deflectors optimized for diffraction efficiency control the *z*-focusing of the beam (AO *z*-focusing). A two-dimensional AO scanner unit (2D-AO scanning) performs *x–y* scanning and drift compensation during *z*-scanning. A spherical field lens in the second telecentric lens system ($Tc_{3,4}$) provides additional angular dispersion compensation

compensation (Fig. 6). Achromatic scan and tube lenses were chosen from Edmund Optics ($f = 250$ mm, NT45-180) and Olympus ($f = 210$ mm), respectively. Finally, the AO-based 3D scanner system is attached to the top of a galvanometer-based upright two-photon microscope (Femto2D-Alba, Femtonics Ltd.) using custom-designed rails. AO sweeps are generated by the direct digital synthesizer chips (AD9910, 1 GSPS, 14-bit, Analog Devices) integrated into the modular electronics system of the microscope using FPGA elements (Xilinx). Red and green fluorescence are separated by a dichroic filter (39 mm, 700dcxru, Chroma

Technology) and are collected by GaAsP photomultiplier tubes custom-modified to efficiently collect scattered photons (*PMT*, H7422P-40-MOD, Hamamatsu), fixed directly onto the objective arm (Travelling Detector System). In in vitro experiments the forward-emitted fluorescence can also be collected by 2-in. aperture detectors positioned below the condenser lens (Femto2D-Alba, Femtonics). Signals of the same wavelength measured at the epi- and transfluorescent positions are added. The large aperture objectives, XLUMPlanFI20×/0.95 (Olympus, 20×, NA = 0.95) and CFI75 LWD 16XW (Nikon, 16×, NA = 0.8), provide the largest scanning volume (Katona et al. [33]). The maximal output laser power in front of the objective is around 400 mW (at 875 nm).

3.5 Controlling 3D AO Scanning

3.5.1 Generating Driver Signals for 3D AO Scanning

To focus the excitation beam to a given *x,y,z*-coordinate, the four AODs should be driven synchronously with varying frequency voltage signals. Because AODs have limited electrical bandwidths, changing frequency can be maintained only for a limited time, before the frequency has to be abruptly reset. This sudden transient in the driving frequencies results in an improperly formed focal point and can therefore be considered as "dead time" (Fig. 7). The length of this period is defined by the time taken by the acoustic wave to travel through the optical aperture of the crystal (usually in around 5–30 μs).

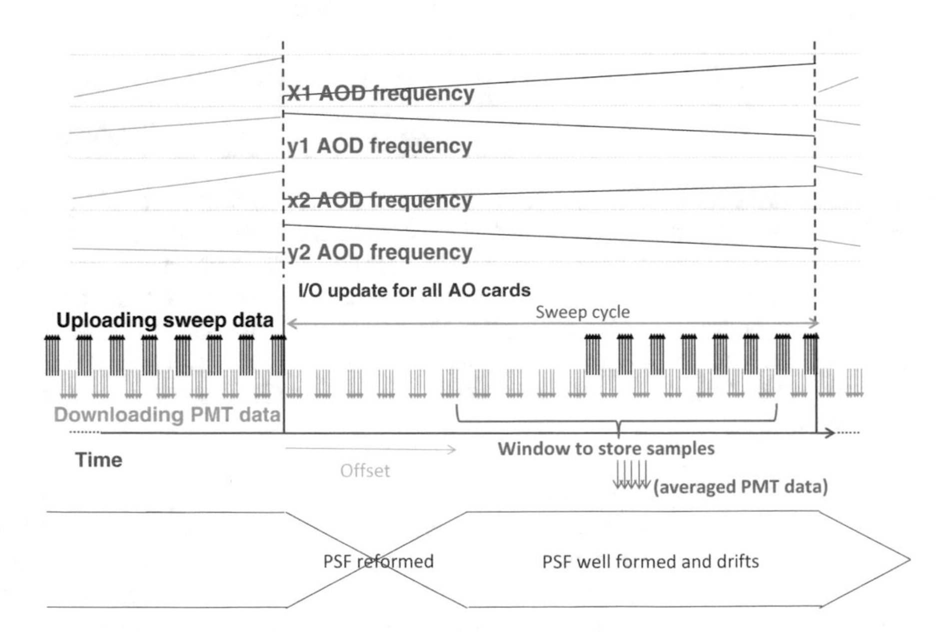

Fig. 7 Driving functions and operation of the four-detector sequence. (*Top*) The four upper traces show the frequency modulation of the sine wave as a function of time at the four acousto-optical deflectors (x1, y1, x2, y2) within one and a half sweep cycle. (*Middle*) PMT data are continuously collected and transferred to the computer. PMT data collected during the "offset" period are eliminated in the analysis process. *Red lines* show the synchronous frequency reset of the driver functions

The position and the movement of the focal point are determined by eight values. Four of them control the starting acoustic frequency of the four AOD drivers, while the other four define the frequency ramp speeds (chirping). All of the eight parameters are updated in every sweep cycle (typical cycle time is 33.6 μs). The starting frequencies and the chirps together define the *xyz* position of the focal spot and its *xy* movement speed (drifting). During every sweep cycle all PMT channels are sampled multiple times. The ratio of the dead time (offset time) and the measurement time (when the focal point has already been formed and is ready for imaging) depends on the *z*-level and is optimal at the nominal focal plane of the objective.

In the simplest case, the goal is to attain a steady focal point with no lateral drifting (random-access scanning mode). In this scanning mode, PMT data are averaged and one value is created for each measurement cycle, corresponding to a single point in *x*,*y*, *z*. Thus, this mode allows random-access 3D point scanning which is ideal for monitoring neuronal network activity. More complex measurement modes are possible when the focal point is allowed to drift, such as during Multiple 3D Trajectory Scanning.

We can also take advantage of the active optical elements to dynamically compensate for optical errors (astigmatism, field curvature, angular dispersion, chromatic aberration) by changing the AOD control parameters: this increases spatial resolution, especially during AO *z*-focusing (at $z \neq 0$ planes by a factor of ~2–3; see "no electric compensation" in Fig. 8).

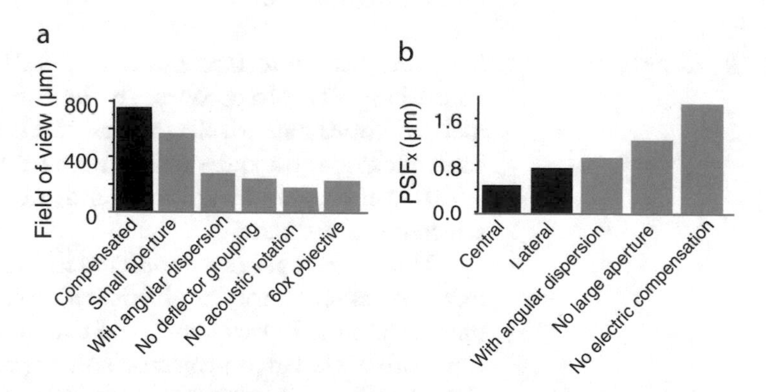

Fig. 8 Characterization of the two-photon microscope setup. (**a**) The maximal field of view (compensated) is shown when both deflector pairs were used for deflection (no deflector grouping) or when optically rotated deflectors (no acoustic rotation), small aperture objectives (60×), no angular dispersion compensation (with angular dispersion), or small aperture acousto-optic deflectors were used (small aperture). (**b**) The compensated PSF size along the *x*-axis (PSF_x) (central) at $(x, y, z) = (150, 150, 100)$ μm coordinates (lateral) or when no angular dispersion compensation (with angular dispersion), no electronic compensation (no electric compensation) or reduced AO apertures were applied (no large apertures)

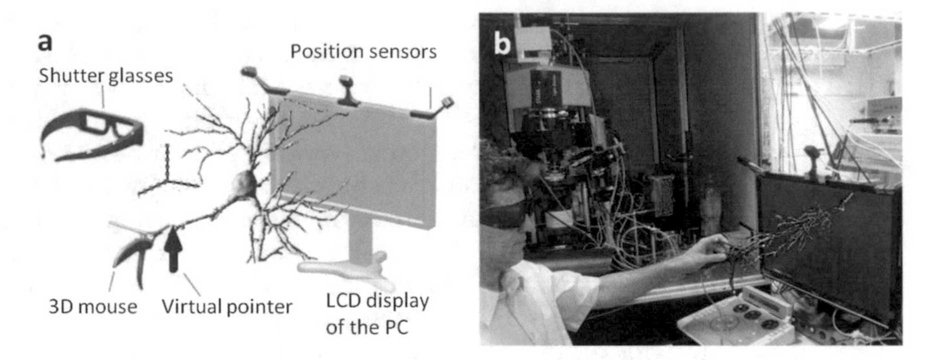

Fig. 9 3D virtual reality environment for 3D two-photon imaging. (**a**) Using the 3D virtual reality environment, the 3D measurement locations can be freely modified or observed from any angle. Head-tracked shutter glasses ensure that the virtual objects maintain a stable, "fixed" virtual position even when viewed from different viewpoints and angles, i.e., the cell's virtual coordinate system is locked in space when the viewer's head position (view angle) changes; however, it can be rotated or shifted by the 3D "bird" mouse. The bird also allows the 3D measurement points to be picked and repositioned in the virtual 3D space of the cell. See also Materials and methods. (**b**) Image of the 3D AO setup and the experimenter using the 3D virtual reality environment (Katona et al. 2012)

The optimal compensation parameters need to be determined in advance for each point of the scanning volume, and loaded into the driver electronics accordingly. During this optimization process the parameters of the chirped sine driver function need to be varied at each of the four deflectors in order to maximize the fluorescence intensity and the sharpness of the fluorescent beads used during this calibration process.

3.5.2 User Interface

Standard user interface modules (GUIs) can also be used for controlling 3D microscopes. In the simplest realization, the optimal "*z*" coordinate is selected first, then using a PC mouse we can easily pick up the *x* and *y* coordinates from the background images, which were taken in advance and are shown according to the selected *z* coordinate.

Thanks to rapid developments in the field of 3D video, we can now implement more sophisticated solutions for controlling 3D measurements. For example, the use of a 3D pointer in combination with a 3D video projector and a head position sensor in a 3D virtual reality environment allows points (or guiding points which determine a 3D object) to be selected for 3D acousto-optical scanning (Fig. 9).

3.6 Advantages of the Detailed Optical Design

One difference between the system described here and previous designs is that the AO deflectors form functionally and physically different groups. The first AO pair is used for *z*-focusing, whereas lateral scanning is performed entirely by the second pair (*2D AO scanning unit*). This arrangement increases the diameter of the lateral scanning range by a factor of about 2.7 (Fig. 8).

Furthermore, not only electronic driver function but also deflector geometry, TeO_2 crystal orientation, and bandwidth are different between deflectors of the *z*-focusing and the 2D AO scanning units. Altogether, these factors increase the diameter of the lateral scanning range to 720 μm using the Olympus 20× objective, and to over 1100 μm when the Nikon 16× objective is used (Fig. 8). Spatial resolution in the whole scanning volume is also increased by the large optical apertures used throughout the system, and about 20 % of this increase is due solely to the use of large AOD apertures (Fig. 8). In contrast to the dominantly *z*-focusing-dependent effect of dynamic error compensation, the angular dispersion compensation unit decreases the PSF in off-axis positions when compared with a simple two-lens telecentric projection (Fig. 8b). These factors which decrease the PSF inherently increase the lateral field of view (Fig. 8a).

3.7 Scanning Modes

Multiple scanning modes have been developed for AO scanning; these can also be used in different combinations as needed by the experimental protocols.

3.7.1 Random-Access Point Scanning

Random-access scanning is one of the most convenient applications for imaging neuronal networks in 3D. In the first step, a reference *z*-stack image is acquired (see later Figs. 19, 20, 21, and 22). In the second step, points can be preselected for fast 3D measurements. Once a region with well-stained cells has been identified, one or more reference structures (typically a brightly red-fluorescing glial cell) can be selected, scanned in 3D, and the 3D origin [(0, 0, 0) coordinate] of the recording defined as the center of this glial cell. In order to compensate for tissue drift, the "origin-glia" cell must be re-scanned regularly during an experiment and moved back into the original position (0,0,0) by moving the microscope table and the objective. Three-dimensional Ca^{2+} responses recorded simultaneously at multiple points can be plotted as curves or, alternatively, responses following spatial normalization can be shown as images (see later Figs. 19, 20, 21, and 22).

3.7.2 Frame Scanning in 3D

One of the basic measurement modes which can be realized with a 3D AO microscope is frame scanning. In this mode one can freely move or rotate arbitrary areas in 3D using the real *z* (by moving the objective) and virtual *z* (by focusing with the AO deflectors at a fixed objective position) for focusing. Since the 3D Frame Scanning mode is relatively slow, this mode can be used to find and preselect ROIs for subsequent fast 3D measurements. Alternatively, multiple areas situated in multiple layers can be selected and measured simultaneously. For example, we can simultaneously measure the basal and apical dendritic regions of pyramidal cells, or image layer II/III and layer V cells simultaneously.

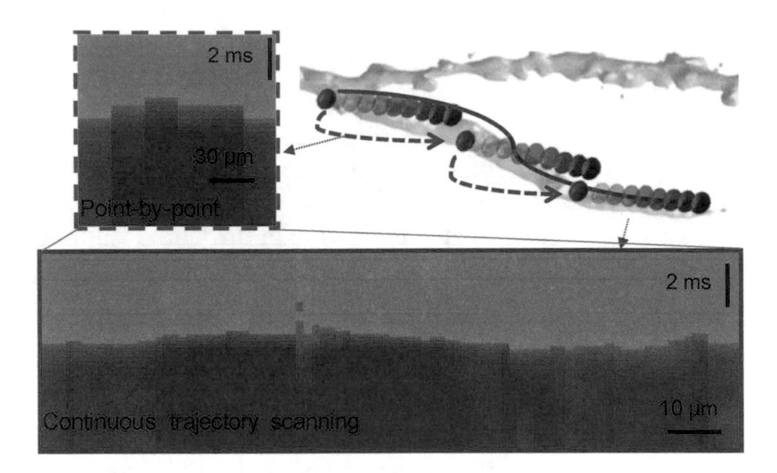

Fig. 10 Comparison of point-by-point and continuous 3D trajectory scanning modes. *Top right*, schema of the scanning modes (*blue*, point-by-point scanning; *green*, continuous scanning). Example of Ca^{2+} responses measured by point-by-point (*top left*) and continuous trajectory scanning modes (*bottom*). Ca^{2+} responses were spatially normalized

3.7.3 Multiple 3D Trajectory Scanning

This mode was developed for simultaneous measurement of multiple neuronal processes. There are two realizations of 3D trajectory scanning. In the Continuous Trajectory Scanning mode, instead of jumping between preselected points (as during by point scanning), the focal point is allowed to drift along neuronal processes (Fig. 10). Information is therefore collected with a much higher spatial discretization within the same time interval.

3.7.4 Volume Imaging

It is possible to capture volumetric data by successively capturing *xy* images at different *z* positions. The jump between the *z*-planes can be obtained by using either the real or the virtual focus. The first will result in better optical quality; therefore it can be used at the end of experiments if only the volume information is required. However, if fast 3D line scanning, 3D frame scanning, or random-access scanning is required, the *z*-stack should be obtained using virtual focusing in order to precisely preserve the coordinates of the cells for the fast 3D AO scanning because the coordinate system of 3D AO microscopes is distorted. Alternatively, a coordinate transformation is needed to correlate the information in the two *z*-stacks.

3.8 Characterizing the Optical Performance of the 3D Microscope

To test the performance of our microscope, the 3D location of fluorescent beads must be imaged to confirm that the microscope can focus and stabilize the PSF at different depths. For example, the stability of fluorescence signals can be measured by repetitively AO-scanning preselected points only, using a short (1.6–16.8 µs) pixel dwell time and a short, 10–20 µs positioning time (typically 16.8 µs, the positioning time is determined by the diameter of the

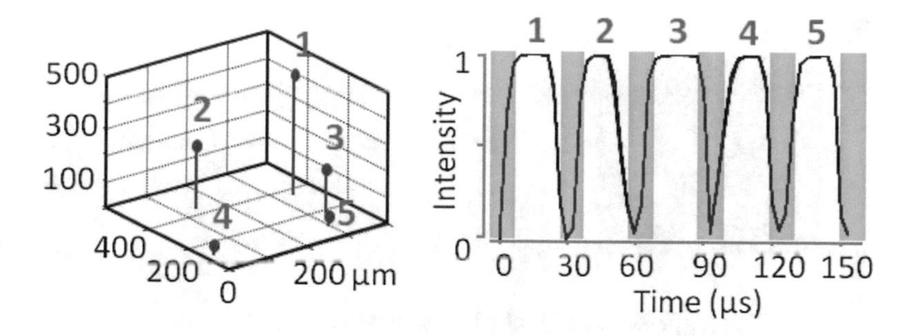

Fig. 11 Stability of 3D AO scanning in random-access mode. (*Left*) Location of the five fluorescent beads (diameter, 6 μm) repetitively scanned in a 3D sample. (*Right*) Five overlaid fluorescence measurements showing dead times (*gray*) and windows of properly formed and positioned PSF (*white*)

optical beam in AODs). These time parameters result in a measurement speed in a range from about 23.8/*N* kHz to 54.3/*N* kHz (where *N* denotes the number of points, typically ranging from 2 to 2000; Fig. 11).

Then the PSF size must be determined at different points in the accessible volume (Fig. 12). The full-width-at-half maximum (FWHM) values of the PSF in the center of the objective are typically slightly larger (for the Olympus 20×, 470 nm, 490 nm, and 2490 nm along the *x*, *y*, and *z* axes, respectively, Fig. 12a) than those compared to the ones measured with a two-dimensional microscope, where the optical pathway is typically more simple.

Although spatial resolution decreases with radial and axial distances from the center of the objective focus (Fig. 12b, c), PSF size remains small (diameter < 0.8 μm, axial length < 3 μm) in the central core of the volume (~290 × 290 × 200 μm^3), allowing the resolution of fine neuronal processes. Moreover, PSF size, being below 1.9 μm diameter and 7.9 μm axial length in an approximately 1400 μm *z* and 700 μm lateral scanning range, allows the optical resolution of individual fluorescent beads whose radius, 6 μm, is smaller than the average diameter of neuronal somata (Fig. 13).

With the help of the AO driver electronics featuring dynamic compensation, the diffraction efficiency inhomogeneities of the AO deflectors and also, partially, the drop of intensity at the sides of the images can be compensate for (Fig. 14).

One of the main advantages of 3D scanning over other 3D technologies is that the wide bandwidth of the AO deflectors allows precise targeting of the center of the focal point (with 51 nm and 140 nm accuracy in the whole volume along radial and axial axes, respectively, which is around 10 % of the optimal PSF size for the system proposed in Fig. 6). Therefore, PSF can be precisely localized in the middle of neurons, avoiding neuropil contamination.

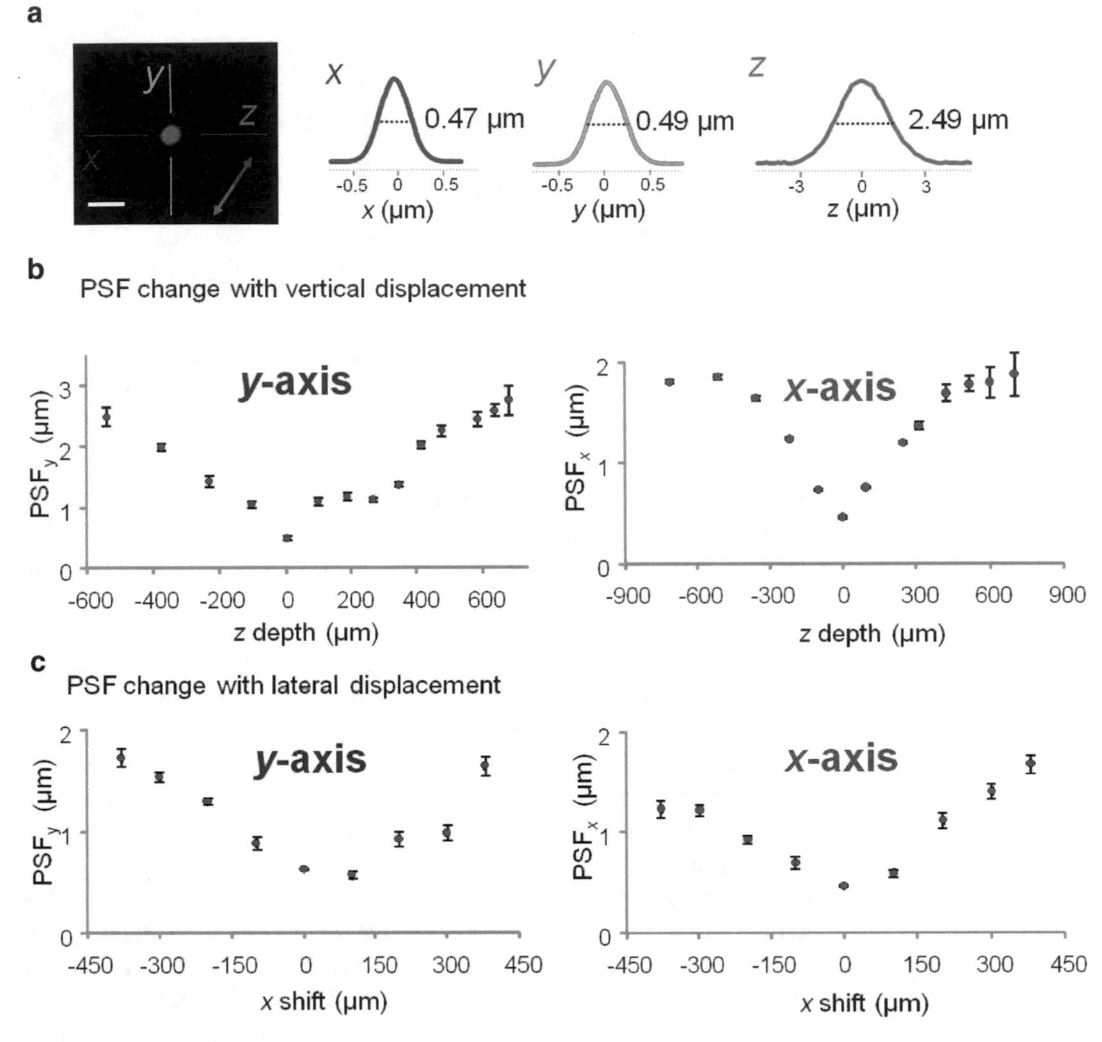

Fig. 12 Size of the PSF in the 3D AO scanning volume. (**a**) A fluorescent bead (0.1 µm) imaged with the optically optimized system utilizing electronic optimization using the 20× objective in the center $(x, y, z) = (0, 0, 0)$ µm coordinates. Corresponding intensity profiles along (x, y, z) axes with full-width-at-half maximum (FWHM) values. Note radial symmetry in center. Scale bar, 1 µm. (**b**) Measured PSF diameters along the *x* and *y* axes are plotted as a function of axial AO *z*-scanning displacement from the nominal object plane of the objective (nominal object plane is also known as front focal plane). (**c**) PSF diameters along the *x* and *y* axes are plotted as a function of lateral displacement from the center of the nominal object. The 20× objective (Olympus XLUMPlanFl20×/0.95) was used for these two-photon measurements whose parameters were also implemented in optical modeling

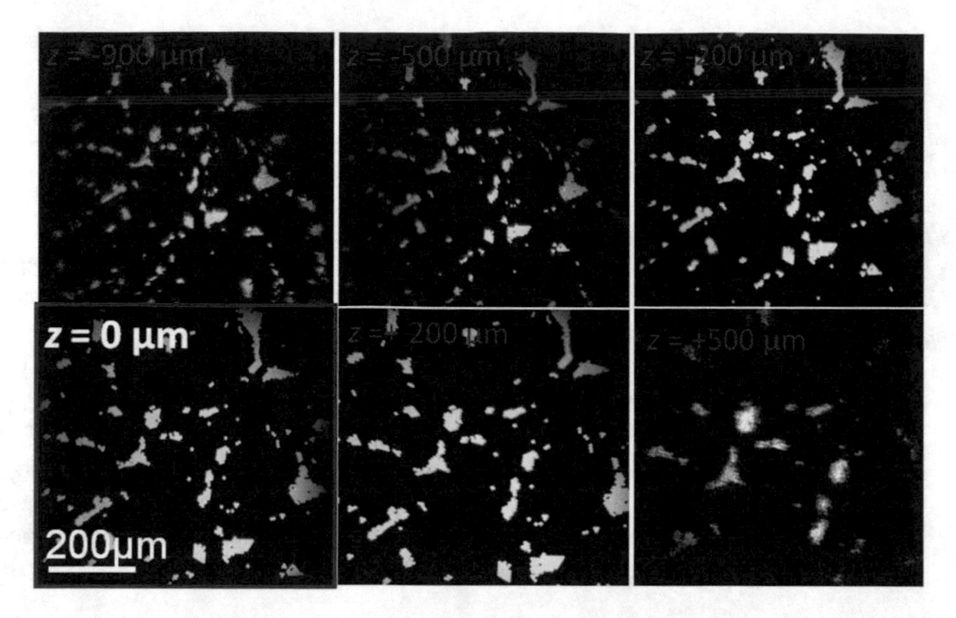

Fig. 13 Fields of view of the 3D microscope system. The analyzed sample consisted of 6-μm-diameter fluorescent beads on a cover glass. AO *z*-focusing was compensated for by refocusing the objective mechanically to keep the fixed sample in focus. The inscriptions show objective shifts along the *z*-axis required for refocusing. The maximal field of view of the system was approximately 700 μm at $z = 0$ μm. $z = 0$ was defined as the nominal object plane. Importantly, lateral displacement of the field of view and change in magnification as a function of AO *z*-focusing need to be compensated for in the acquisition program. Individual beads remained visible in a 1400 μm scanning range

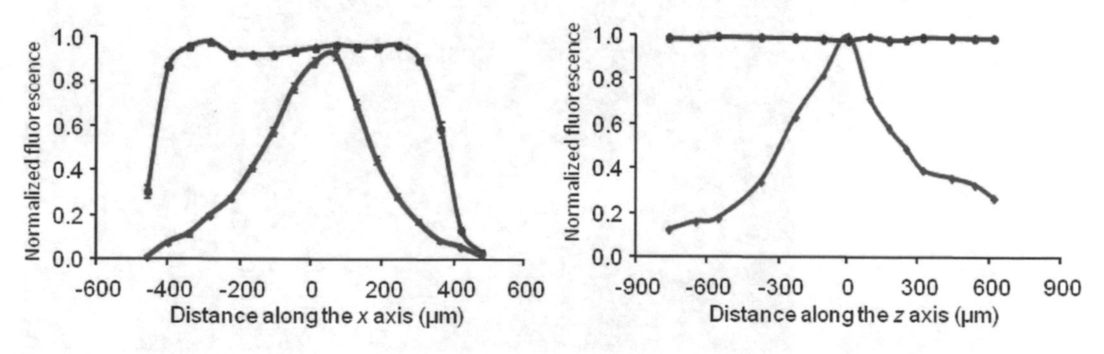

Fig. 14 Maximal FOV of the microscope system with dynamic compensation. Normalized fluorescence intensity as a function of distance from the origin along the *x* and *z* axes with (*red*) and without (*blue*) dynamic compensation. Measurements were performed in a homogeneous fluorescence sample. The fluorescence intensity is affected by the variations in both the excitation intensity and the PSF size (*blue*). In contrast to previous experiments, these technical shortcomings in the driving signals were compensated for by varying both their frequency and power in advance using the dynamic compensation software module (*red*) (from Katona et al. 2012)

In order to properly validate the 3D microscope, the parameters need to be tested in a biological context.

4 In Vitro and In Vivo Testing of the 3D AO Microscope

4.1 3D Random-Access Scanning of Action Potential Propagation

To examine the temporal resolution of a 3D system, imaging one of the fastest regenerative events is the optimal choice. In order to achieve this goal, we can for example measure backpropagating action potentials at multiple 3D locations in a single hippocampal neuron in acute hippocampal mouse brain slices (Fig. 15). Neurons, for example CA1 pyramidal cells, need to be patch-clamped in whole-cell mode and filled with the green fluorescent Ca^{2+} sensor Fluo-5F and the red fluorescent marker Alexa594. The objective must be positioned at the correct depth, such that the center of the region of interest is in focus. The objective is fixed at this position and a reference *z*-stack of images is acquired using 3D AO scanning

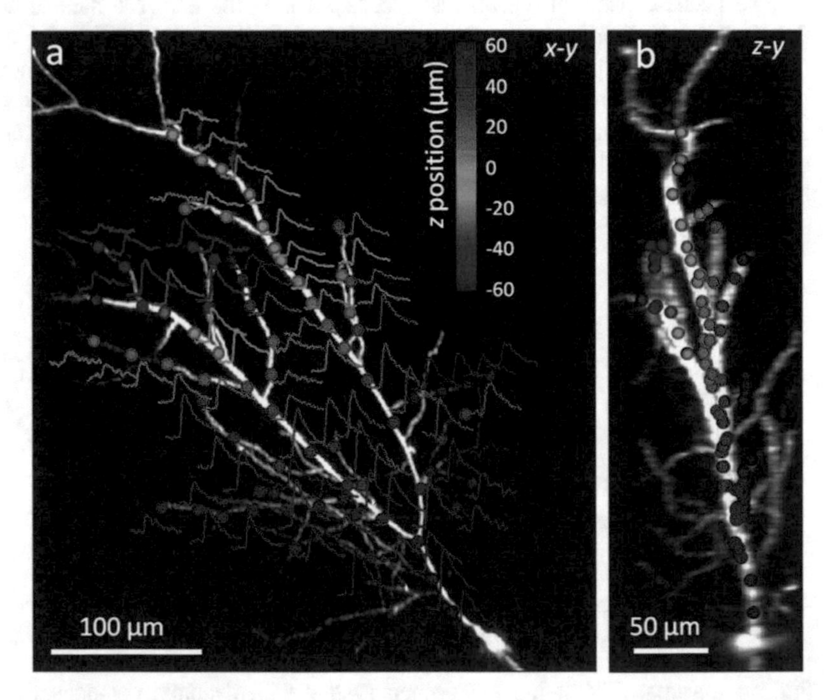

Fig. 15 Three-dimensional measurement of bAPs. (**a**) Maximum intensity *z*-projection of the dendritic arbor of a CA1 pyramidal cell imaged by 3D AO scanning. *Circles* represent the 77 preselected measurement locations for 3D random-access point scanning. *Overlaid curves* show Ca^{2+} transients recorded near-simultaneously in each location, induced by a single bAP (each is the average of five traces). The repetition rate of the measurement was 29.76 kHz (0.39 kHz speed in each location). The *z*-coordinates of the measurement locations are color coded. (**b**) Maximum intensity side projection of the cell with the measurement locations shown in (**a**)

only. Points along the dendritic tree of one single neuron can be selected from the *z*-stack using either a 3D virtual environment (Fig. 9, see also Sect. 3.4.2) or by scrolling through the *z*-stack in a 2D virtual environment [2]. When the cell is held in current-clamp mode, APs can be evoked by somatic current injection while near-simultaneously measuring dendritic Ca^{2+} signals associated with the backpropagating action potentials (bAPs) by repetitively scanning the selected 3D coordinates at 29.76 kHz. When sample drift occurs during measurements, the measurement points need to be repositioned.

Since acute brain slices are about 300-μm thick, only a fraction of the total AO *z*-scanning range can be used. To test whether random-access 3D AO scanning can record bAP-induced Ca^{2+} transients we can, it is possible for example, move the objective focal plane mechanically from +562 μm to −546 μm away from the *z*-center, then use AO *z*-focusing to refocus the recorded dendritic tree (Fig. 16a) and test for AP resolution. Independent of the AO

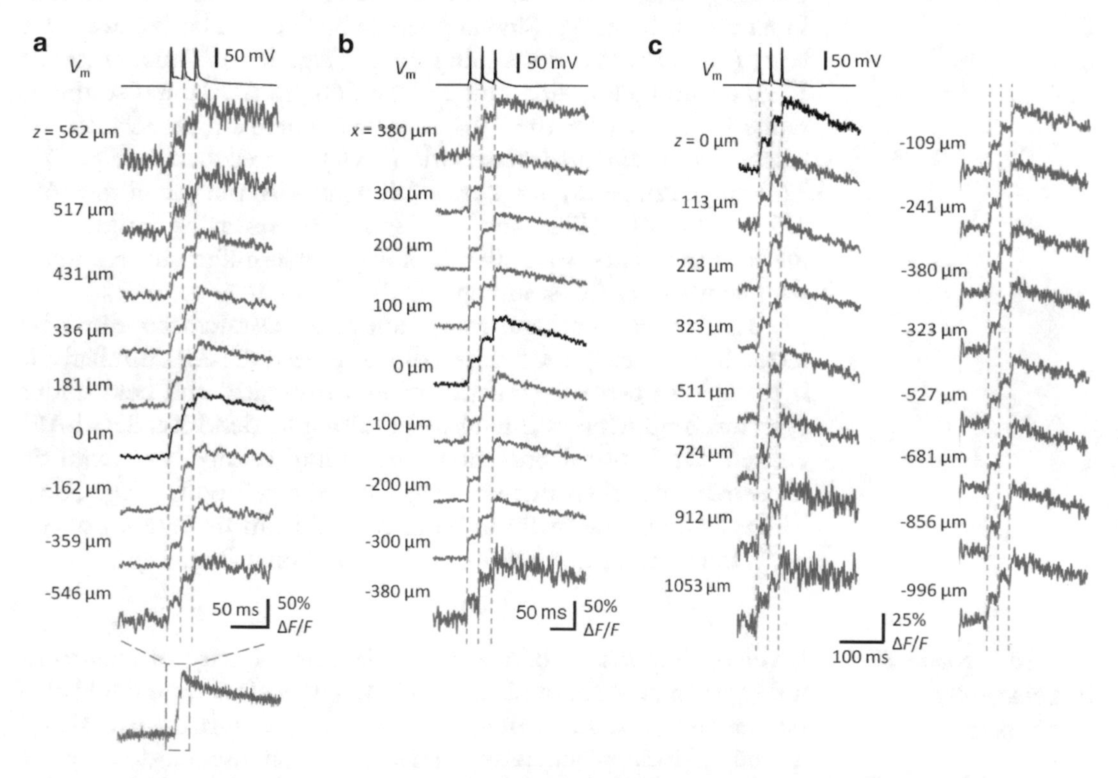

Fig. 16 Resolving bAPs at the sides of the FOV. (**a**) Single traces of three bAP-induced dendritic Ca^{2+} transients measured at the same dendritic point at different AO *z*-focusing settings. To keep the image sharp, the AO *z*-focusing was compensated by mechanical refocusing of the objective. Note the preserved single AP resolution even at the lower and upper limits of the scanning range. Corresponding somatic voltage traces (V_m) are shown on the *top*. (**b**) The same as in (**a**), but transients were recorded while shifting the dendrite along the *x*-axis. (**c**) The same as in (**a**), but single traces were recorded using the 16× Nikon objective. Note that the AO *z*-focusing range with single AP detection level exceeds 2000 μm

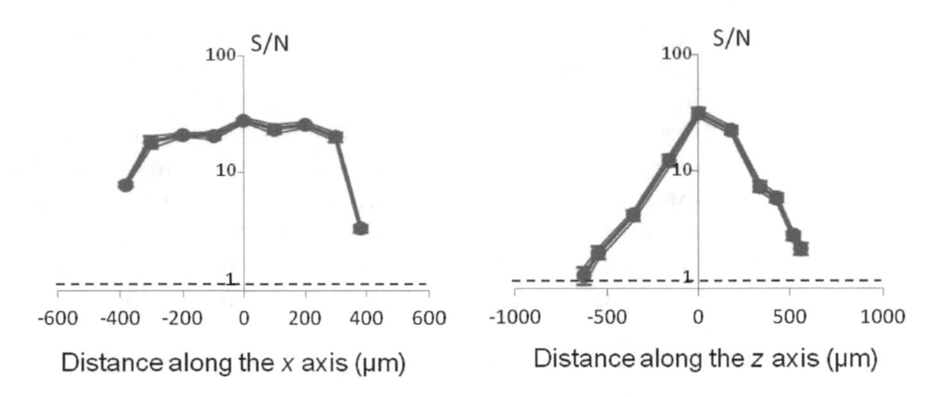

Fig. 17 SNR of bAPs in the field of view. SNR calculated from the pool of transients shown in Fig. 16. Peak of the Gauss-filtered bAP-induced Ca^{2+} transients was divided by the s.d. of the unfiltered 50 ms baseline period (mean ± s.d; $n = 10–50$ transients at each position). Single AP detection level is marked with a *red dashed line*

z-settings, single bAPs could be resolved in Ca^{2+} transients induced by a train of three APs. Similarly, single bAPs could be followed in a large (760 μm) lateral scanning range (Fig. 16b). Thus a total of 1190 μm AO *z*-focusing range and a 760 μm AO lateral scanning range is shown with the 20× objective where single APs can be resolved with signal-to-noise ratio (SNR) greater than 1 (Fig. 17). This range can be extended over 2000 μm with the use of the 16× Nikon objective (Fig. 16c). In summary, resolving single AP-induced transients is possible in a near-cubic-millimeter scanning volume with 3D AO scanning.

The highest temporal resolution of a 3D microscope can be tested by measuring the fast propagation speed of bAPs. Similarly, it is possible to evoke a somatic AP in a pyramidal cell held under current-clamp while scanning in 3D along its dendrite. The bAP-evoked Ca^{2+} transients show increasing latency in dendritic measurements taken further away from the cell body (Fig. 18b). The velocity of the bAP can be calculated from the latency of the Ca^{2+} transients and the distance from the soma (Fig. 18c).

4.2 3D Scanning of Dendritic Spike Propagation

Given the importance of integrating dendritic activity into neuronal information processing [1, 3, 4, 43, 44], the 3D system could also be used to record dendritic spike activity. To measure the extent and spread of local regenerative activity, we can use random-access scanning and detect activity at many discrete points along the 3D trajectory of a dendrite (point-by-point trajectory scanning). As above, we can scroll through a reference *z*-stack of a series of pre-recorded images and select guide points along the length of a dendrite (Sect. 3.4.2). Then, instead of recording from these manually selected points, we can also homogeneously re-sample the trajectory using, for example, piecewise cubic Hermite

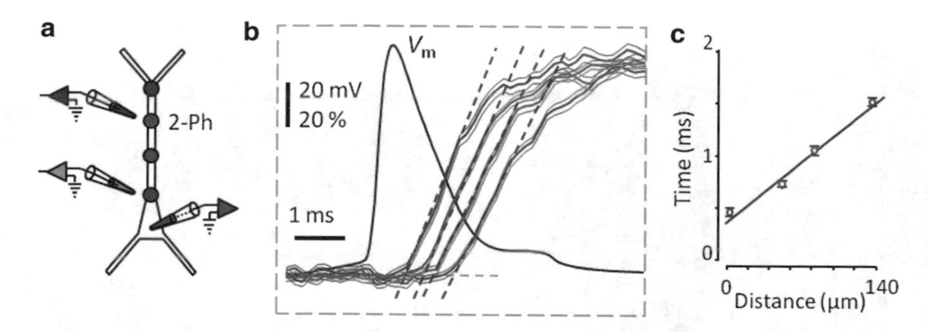

Fig. 18 Measurement of bAP propagation speed. (**a**) Experimental arrangement for signal propagation experiments. Signal propagation speed was measured by simultaneous somatic whole-cell current-clamp (V_m, *black*) and 3D two-photon calcium imaging measurements (*orange*, *pink*, *green*, and *blue*). (**b**) AP peak triggered average of normalized dendritic Ca^{2+} transients induced by APs (mean ± s.e.m.; $n = 54$; *top*). Linear fits (*red dashed lines*) define onset latency times. (**c**) Onset latency times (mean ± s.e.m.; $n = 54$) of Ca^{2+} transients in (**b**) as a function of dendritic distance. Linear fit: average propagation speed. Average bAP propagation speed was 164 ± 13 μm/ms ($n = 9$) at 23 °C

interpolation [2]. This algorithm is better than the conventional spline interpolation because it never overshoots at the guide points. The speed of the 3D measurement depends on the number of recording points along the trajectory.

In the example shown in Fig. 19, a hippocampal CA1 pyramidal cell was filled with Fluo-5F and dendritic Ca^{2+} spikes were induced by focal synaptic stimulation, with an extracellular electrode >400 μm from the cell body. The propagating activity was followed optically at uniformly spaced locations along the main apical dendrite (Fig. 19a). The recorded traces were normalized by calculating $\Delta F/F$ at each point and were projected as a function of time and distance measured along the 3D trajectory line (3D Ca^{2+} responses, Fig. 19b). The Fluo-5F responses revealed dendritic spikes propagating toward the cell body, both when the synaptic stimulation evoked somatic APs (suprathreshold spikes) and when the synaptic stimulation was insufficient to evoke somatic APs (subthreshold spikes, Fig. 19b–e). After a short initiation phase, subthreshold dendritic spikes propagated rapidly and then gradually ceased before reaching the cell body (Fig. 19b). We quantified spike propagation speed by measuring propagation times at the half-maximal amplitude of the transients (Fig. 19c) or at the peak of their first derivatives (Fig. 19d). The average propagation speed was significantly higher for suprathreshold spikes than for subthreshold spikes (subthreshold 81 ± 27 μm/ms, suprathreshold 129 ± 30 μm/ms; 131 %, *t*-test, $P = 0.03$; $n = 7$ cells; Fig. 19e), and both supra- and subthreshold spikes propagated in the direction of the cell body under these conditions (Fig. 19e). Note that, despite the large

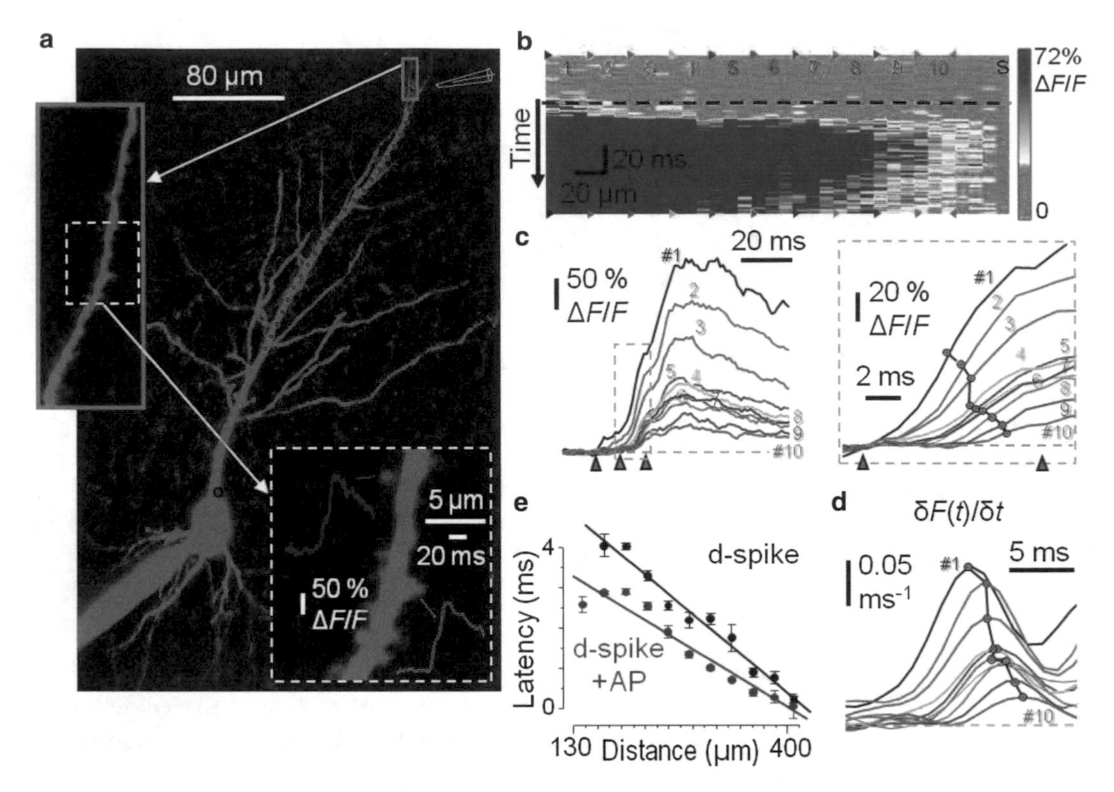

Fig. 19 3D scanning of dendritic Ca^{2+} spike propagation in CA1 pyramidal cells. (**a**) Maximum-intensity projection AO image of a CA1 pyramidal cell. Ca^{2+} transients in dendritic spines (*orange* and *magenta* traces) following induction of dendritic Ca^{2+} spike by focal extracellular stimulation (electrode: *yellow*). *Left* and *bottom insets*, enlarged views. *Purple dots* represent scanning points in a dendrite. (**b**) Spatially normalized and projected Ca^{2+} signals in the dotted dendritic region shown in (**a**) (average of five subthreshold responses). *Black dashed line*, stimulus onset. Column *s*, somatic Ca^{2+} response. (**c**) Ca^{2+} transients derived from the color-coded regions indicated in (**b**). *Right*, Baseline-shifted Ca^{2+} transients measured in the region contained in the *dashed box* in (**a**). *Yellow dots*, onset latency times at the half maximum. (**d**) Onset latency times at the peak of the derivative ($\delta F(t)/\delta t$) of Ca^{2+} transients shown in (**c**). (**e**) Onset latency times as a function of dendritic distance from the soma for somatic subthreshold (*black*) and suprathreshold (*blue*) dendritic Ca^{2+} spikes

total scanning volume (290 × 290 × 200 μm^3), two-photon resolution was preserved (in the range of 0.45–0.75 μm). The total number of measured points along any given trajectory was 23.8–54.3 points per kHz.

4.3 High-Speed In Vivo 3D Imaging of Neuronal Network Activity

To test the performance of the 3D imaging systems in vivo, it is possible, for example, to record Ca^{2+} responses from a population of individual neurons in the visual cortex of adult anesthetized mice. First, we need to inject a mixture of a green Ca^{2+} dye (e.g., OGB-1-AM) to monitor changes in intracellular Ca^{2+} concentrations, and sulforhodamine-101 (SR-101) [45] to selectively label glial cells (Fig. 20). The red fluorescence of SR-101 allows

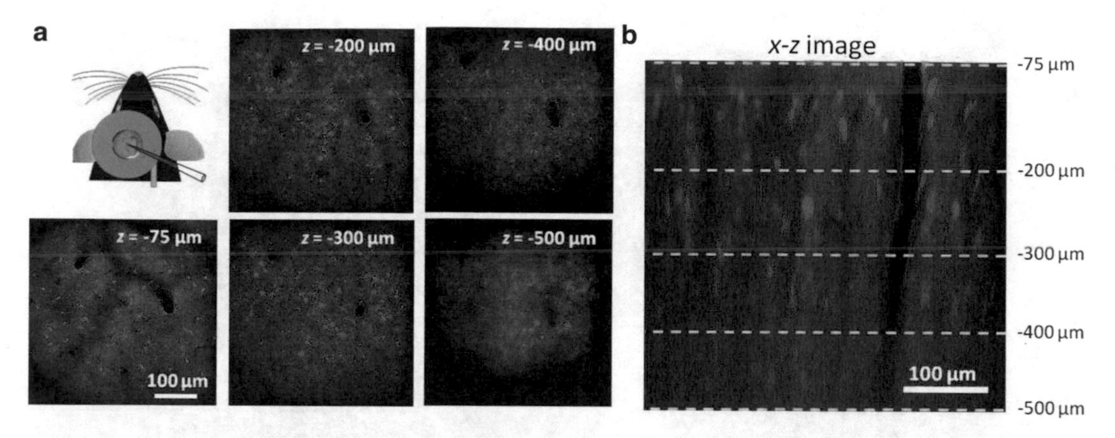

Fig. 20 In vivo image stacks of the neuronal population. (**a**) Representative background-corrected images taken at different depths from the surface of the brain showing neurons (*green*) and glial cells (*red* and *yellow*). *Red* and *green* channel images were overlaid. (*Upper left*) sketch of in vivo experimental arrangement. Staining by bolus loading (OGB-1-AM and SR-101) in mouse V1. (**b**) *x–z* slice taken from the middle of the stack volume *dotted lines* corresponds to the planes in (**a**)

differentiation between neurons and glial cells (green and red, respectively). The maximal power of a Ti:sapphire laser (which is in the range of 3.5–4.5 W) limits the depth of the in vivo recording to a maximum of 500–700 μm from the surface of the cortex (the total imaging volume was about 400 × 400 × 500 μm^3 in Katona et al. [33], and now it is about 600× 600× 650 μm^3 thanks to the ~4-fold increase in transmission efficiency resulting from better AODs).

In the next step, a reference *z*-stack must be recorded. An automated algorithm can identify the neuron and glial cell bodies. When OGB1-AM and SR-101 dyes are bolus-loaded into the animal, cells can be categorized according to their dye content, measured by fluorescence. Neuronal cells can be detected because of their elevated green fluorescence and decreased red fluorescence. The green and red fluorescent channel data of each image can also be normalized in the *z*-stack and then scaled and shifted so that the tenth and 90th percentiles match 0 and 1. Background correction is done by over-smoothing an image and subtracting it from the original, then the red channel data are subtracted from the green, and, finally, each layer of the stack is filtered again (2D Gaussian, $\sigma = 2$ μm). The result is then searched for local maxima, with an adaptive threshold. If two selected locations are closer than a given distance threshold, only one is kept. At the end, the algorithm lists the 3D coordinates of the centers of each neuronal cell body, and these coordinates can be used during random-access scanning.

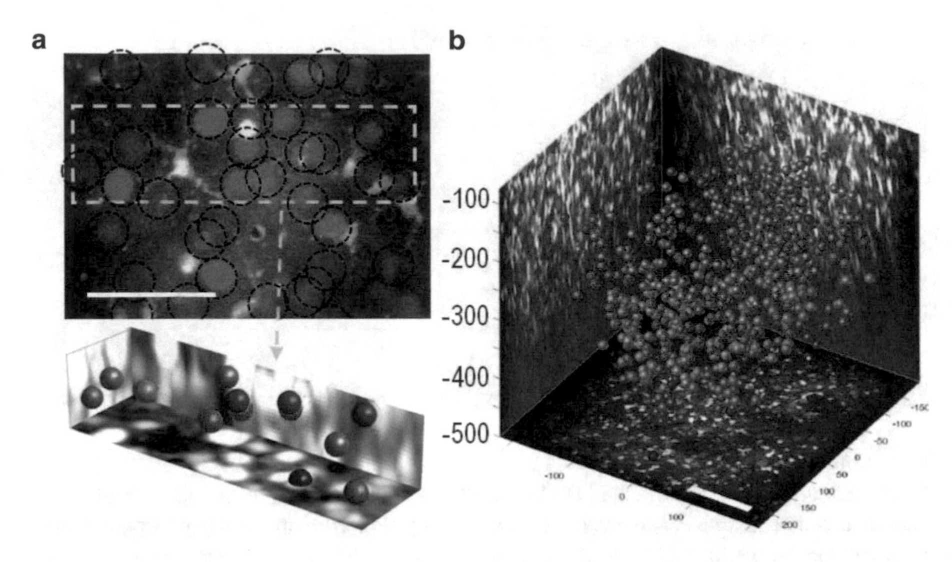

Fig. 21 Automatic localization of neurons in vivo. (**a**) 35 μm *z*-projection of a smaller region of the experiment shown in Fig. 20. *Bottom*, neuronal somata detected with the aid of an algorithm in a sub-volume (shown with projections, neurons in white and glial cells in *black*). Scale bar, 50 μm. (**b**) Maximal intensity side- and *z*-projections of the entire *z*-stack (400 × 400 × 500 μm^3) with autodetected cell locations. *Spheres* are color coded in relation to depth. The detection threshold used here yielded 532 neurons. Scale bar, 100 μm

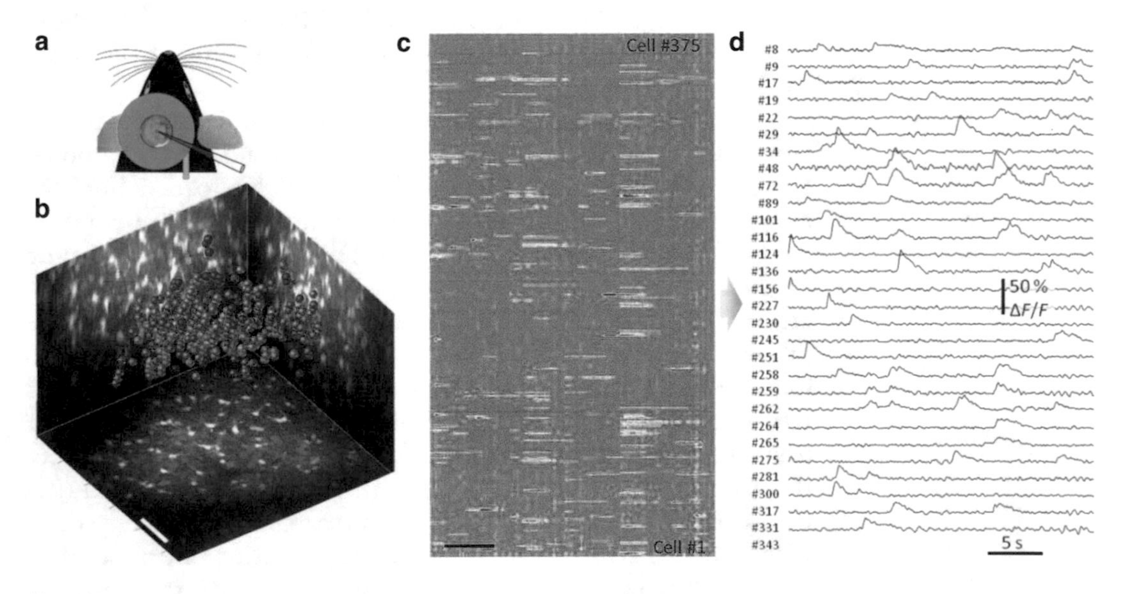

Fig. 22 Spontaneous neuronal network activity in vivo. (**a**) Sketch of in vivo experimental arrangement. (**b**) Maximal intensity side- and *z*-projection image of the entire *z*-stack (280 × 280 × 230 μm^3; bolus loading with OGB-1-AM and SR-101). *Spheres* represent 375 autodetected neuronal locations color coded by depth in the V1 region of the cortex. Scale bars, 50 μm. (**c**) Parallel 3D recording of spontaneous Ca^{2+} responses from the 375 locations. Rows, single cells measured in random-access scanning mode. Scale bar, 5 s. Activity was recorded with ~80 Hz. (**d**) Examples of Ca^{2+} transients showing active neurons from (**c**)

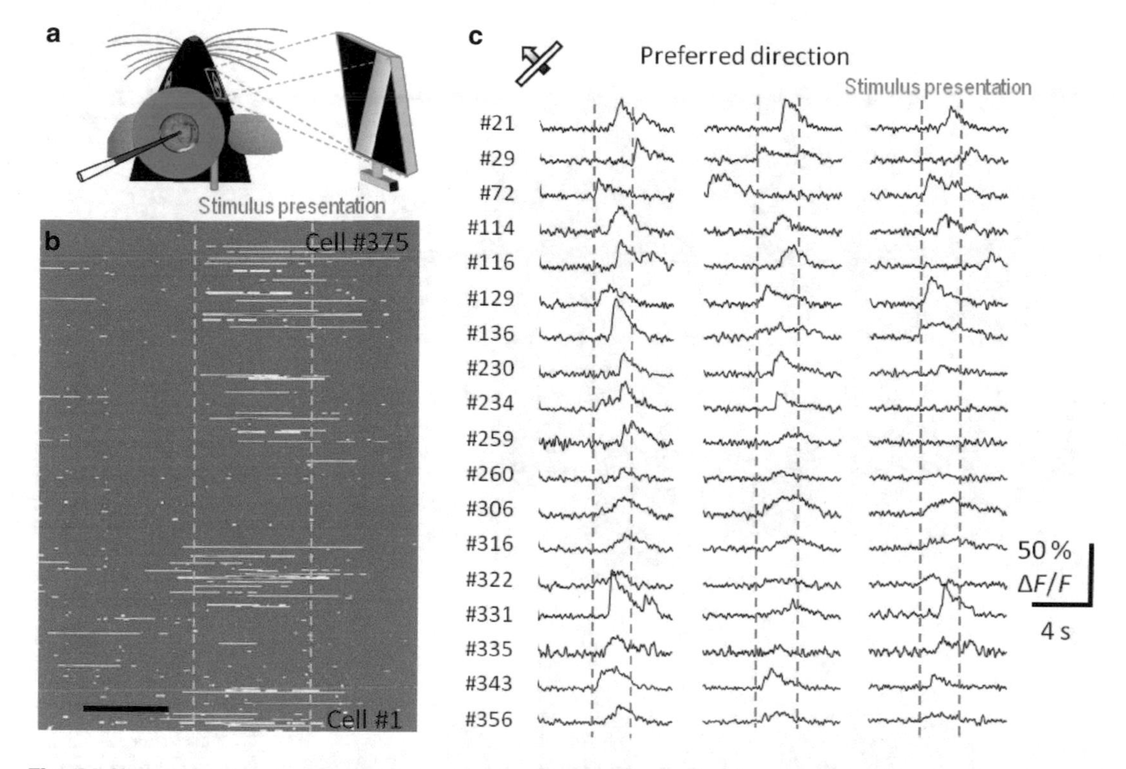

Fig. 23 Neuronal network activity in vivo in response to visual stimuli. (**a**) Sketch of in vivo experimental arrangement. Visual stimulation was induced by moving bars in eight directions at 45° intervals. (**b**) Ca^{2+} responses from the same 375 neuronal locations visible in Fig. 22 (visual stimulation: moving bar at −45°). Rows, single cells from a single 3D measurement. Scale bar, 2 s. (**c**) Examples of Ca^{2+} transients from neurons in **b**, preferentially responding to the −45° bar direction. Bar moved in the visual field during the time periods marked with *dashed lines*

After the selection of the cell bodies (Fig. 22b), the spontaneous activity of each neuron is recorded and the point-by-point background-corrected and normalized fluorescence data are plotted real-time. Data from each cell can also be shown as a single row or column for better visibility (Fig. 22c). The stability of long-term recording can be monitored by measuring the baseline fluorescence.

For neuroscientists, it is more interesting to record neuronal responses following different physiological stimulations in different cortical areas. For example, the mouse could be presented with visual stimuli consisting of movies of a white bar oriented at eight different angles and always moving perpendicular to its orientation (Fig. 23b) and detect responses in the V1 region of the cortex. Visual stimulation with bars oriented at −45° activate a small subpopulation of the measured cells (Fig. 23b and c).

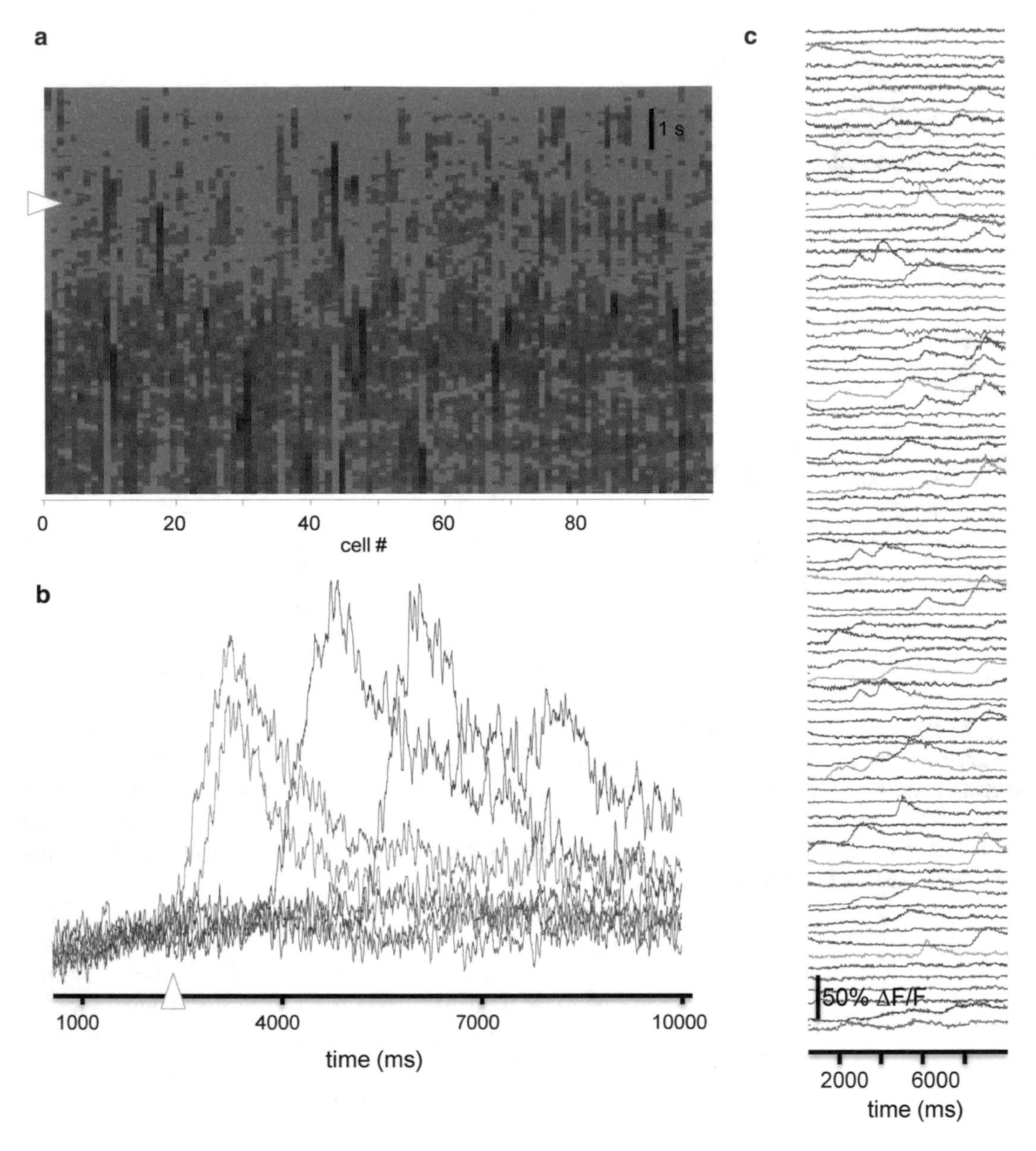

Fig. 24 Neuronal network activity recorded in the V1 region of an awake animal. (**a**) Spatially normalized 3D Ca^{2+} response recorded in 100 genetically labeled neurons. Each column corresponds to a single neuron. Red color indicates high activity. Neurons were labeled using AAV viruses and expressed GCamP6 sensor. (**b**) Exemplified Ca^{2+} responses, raw traces. (**c**) Simultaneously recorded population activity, each trace corresponds to a single neuron

4.4 In Vivo 3D Imaging of Genetically Labeled Neuronal Networks

Similarly to the measurements shown in the previous section, 3D recordings can also be performed on genetically labeled neuronal networks (Fig. 24). The improved capability of the recording technology is mostly because of the recent developments in the field of AOD scanner technology: the novel 3D scan heads provide better transmission, even at longer wavelengths.

5 Future Perspectives

Optically, 3D AO microscopes are now close to the theoretical maximum which can be realized using the currently available high NA objective lenses, but other aspects of AO scanning could still be improved, such as

- develop faster, more specialized scanning algorithms for network measurements
- develop faster AO deflectors
- develop novel correction methods for movement artifacts
- improve lasers and fluorescent dyes to get access to entire thickness of the cortex
- applying the technology to larger FOV objectives
- implement adaptive optics
- overcome the speed of resonant scanning in frame scanning modes
- simplify the system to lower its costs and maintenance efforts
- improve wavelength tunability

Together with these developments, there is a need for systems which can simultaneously scan multiple brain areas in order to study communication between sensory, motor, and higher order areas, and which are able to use photostimulation in combination with 3D scanning in order to map functional connectivity within neuronal networks. Moreover, 3D recordings must be performed in symbiosis with neuronal network and dendritic modeling; 3D recorded data must be explained by neuronal network and dendritic modeling; and, conversely, measurements must be performed according to modeling predictions [44, 46].

References

1. Johnston D, Narayanan R (2008) Active dendrites: colorful wings of the mysterious butterflies. Trends Neurosci 31(6):309–316
2. Katona G et al (2011) Roller Coaster Scanning reveals spontaneous triggering of dendritic spikes in CA1 interneurons. Proc Natl Acad Sci U S A 108(5):2148–2153
3. Losonczy A, Makara JK, Magee JC (2008) Compartmentalized dendritic plasticity and input feature storage in neurons. Nature 452 (7186):436–441
4. Spruston N (2008) Pyramidal neurons: dendritic structure and synaptic integration. Nat Rev Neurosci 9(3):206–221
5. Rozsa B et al (2008) Dendritic nicotinic receptors modulate backpropagating action potentials and long-term plasticity of interneurons. Eur J Neurosci 27(2):364–377
6. Rozsa B et al (2004) Distance-dependent scaling of calcium transients evoked by backpropagating spikes and synaptic activity in dendrites of hippocampal interneurons. J Neurosci 24 (3):661–670
7. Ohki K et al (2005) Functional imaging with cellular resolution reveals precise microarchitecture in visual cortex. Nature 433 (7026):597–603

8. Ariav G, Polsky A, Schiller J (2003) Submillisecond precision of the input-output transformation function mediated by fast sodium dendritic spikes in basal dendrites of CA1 pyramidal neurons. J Neurosci 23 (21):7750–7758
9. Danilatos GD (1991) Review and outline of environmental SEM at present. J Microsc 162 (3):391–402
10. Kherlopian AR et al (2008) A review of imaging techniques for systems biology. BMC Syst Biol 2:74
11. Kerr JN, Denk W (2008) Imaging in vivo: watching the brain in action. Nat Rev Neurosci 9(3):195–205
12. Amos WB, White JG, Fordham M (1987) Use of confocal imaging in the study of biological structures. Appl Opt 26 (16):3239–3243
13. Minsky M (1988) Memoir on inventing the confocal scanning microscope. Scanning 10 (4):128–138
14. Petráň M et al (1968) Tandem-scanning reflected-light microscope. J Opt Soc Am 58 (5):661–664
15. Tomer R et al (2012) Quantitative high-speed imaging of entire developing embryos with simultaneous multiview light-sheet microscopy. Nat Methods 9(7):755–763
16. Denk W, Strickler JH, Webb WW (1990) Two-photon laser scanning fluorescence microscopy. Science 248(4951):73–76
17. Denk W, Svoboda K (1997) Photon upmanship: why multiphoton imaging is more than a gimmick. Neuron 18(3):351–357
18. Goeppert-Mayer M (1931) Ueber Elementarakte mit zwei Quantenspruengen. Ann Phys 9:273
19. Helmchen F, Denk W (2005) Deep tissue two-photon microscopy. Nat Methods 2 (12):932–940
20. Kobat D, Horton NG, Xu C (2011) In vivo two-photon microscopy to 1.6-mm depth in mouse cortex. J Biomed Opt 16(10):106014
21. Theer P, Hasan MT, Denk W (2003) Two-photon imaging to a depth of 1000 micron in living brains by use of a Ti:Al_2O_3 regenerative amplifier. Opt Lett 28(12):1022–1024
22. Niesner R et al (2007) The power of single and multibeam two-photon microscopy for high-resolution and high-speed deep tissue and intravital imaging. Biophys J 93 (7):2519–2529
23. Prevedel R et al (2014) Simultaneous whole-animal 3D imaging of neuronal activity using light-field microscopy. Nat Methods 11 (7):727–730
24. Quirin S et al (2014) Simultaneous imaging of neural activity in three dimensions. Front Neural Circuits 8:29
25. Quirin S, Peterka DS, Yuste R (2013) Instantaneous three-dimensional sensing using spatial light modulator illumination with extended depth of field imaging. Opt Express 21(13):16007–16021
26. Grewe BF et al (2011) Fast two-layer two-photon imaging of neuronal cell populations using an electrically tunable lens. Biomed Opt Express 2(7):2035–2046
27. Nikolenko V et al (2008) SLM microscopy: scanless two-photon imaging and photostimulation with spatial light modulators. Front Neural Circuits 2:5
28. Gobel W, Kampa BM, Helmchen F (2007) Imaging cellular network dynamics in three dimensions using fast 3D laser scanning. Nat Methods 4(1):73–79
29. Cheng A et al (2011) Simultaneous two-photon calcium imaging at different depths with spatiotemporal multiplexing. Nat Methods 8(2):139–142
30. Holekamp TF, Turaga D, Holy TE (2008) Fast three-dimensional fluorescence imaging of activity in neural populations by objective-coupled planar illumination microscopy. Neuron 57(5):661–672
31. Botcherby EJ et al (2012) Aberration-free three-dimensional multiphoton imaging of neuronal activity at kHz rates. Proc Natl Acad Sci U S A 109(8):2919–2924
32. Rozsa B et al (2007) Random access three-dimensional two-photon microscopy. Appl Opt 46(10):1860–1865
33. Katona G et al (2012) Fast two-photon in vivo imaging with three-dimensional random-access scanning in large tissue volumes. Nat Methods 9(2):201–208
34. Kaplan A, Friedman N, Davidson N (2001) Acousto-optic lens with very fast focus scanning. Opt Lett 26(14):1078–1080
35. Fernandez-Alfonso T et al (2014) Monitoring synaptic and neuronal activity in 3D with synthetic and genetic indicators using a compact acousto-optic lens two-photon microscope. J Neurosci Methods 222:69–81
36. Cotton RJ et al (2013) Three-dimensional mapping of microcircuit correlation structure. Front Neural Circuits 7:151
37. Kirkby PA, Srinivas Nadella KM, Silver RA (2010) A compact acousto-optic lens for 2D and 3D femtosecond based 2-photon microscopy. Opt Express 18(13):13721–13745
38. Duemani Reddy G et al (2008) Three-dimensional random access multiphoton

microscopy for functional imaging of neuronal activity. Nat Neurosci 11(6):713–720

39. Iyer V, Hoogland TM, Saggau P (2006) Fast functional imaging of single neurons using random-access multiphoton (RAMP) microscopy. J Neurophysiol 95(1):535–545
40. Reddy GD, Saggau P (2005) Fast three-dimensional laser scanning scheme using acousto-optic deflectors. J Biomed Opt 10 (6):064038
41. Grewe BF et al (2010) High-speed in vivo calcium imaging reveals neuronal network activity with near-millisecond precision. Nat Methods 7(5):399–405
42. Proctor B, Wise F (1992) Quartz prism sequence for reduction of cubic phase in a mode-locked Ti:Al(2)O(3) laser. Opt Lett 17 (18):1295–1297
43. Keren N, Bar-Yehuda D, Korngreen A (2009) Experimentally guided modelling of dendritic excitability in rat neocortical pyramidal neurones. J Physiol 587(Pt 7):1413–1437
44. Almog M, Korngreen A (2014) A quantitative description of dendritic conductances and its application to dendritic excitation in layer 5 pyramidal neurons. J Neurosci 34(1):182–196
45. Nimmerjahn A et al (2004) Sulforhodamine 101 as a specific marker of astroglia in the neocortex in vivo. Nat Methods 1(1):31–37
46. Chiovini B et al (2014) Dendritic spikes induce ripples in parvalbumin interneurons during hippocampal sharp waves. Neuron 82(4):908–924

Chapter 12

Intracellular Voltage-Sensitive Dyes for Studying Dendritic Excitability and Synaptic Integration

Corey D. Acker, Mandakini B. Singh, and Srdjan D. Antic

Abstract

Intracellular voltage-sensitive dyes are used to monitor membrane potential changes from neuronal compartments not readily accessible to glass electrodes, such as basal dendritic segments more than 140 μm away from the cell body. Optical imaging is uniquely suitable to reveal voltage transients occurring simultaneously in two or more dendritic branches, or in two or more locations along the same dendritic branch (simultaneous multi-site recordings). Voltage-sensitive dye recordings can be combined with bath application of drugs that block membrane conductances as well as with focal application of neurotransmitters. The results of dendritic voltage-sensitive dye measurements are naturally incorporated into computational models of neurons with complex dendritic trees. The number of model constraints is notably heightened by a multi-site approach. An interaction between multi-site voltage-sensitive dye recording (wet experiment) and multicompartmental modeling (dry experiment) constitutes one of the most insightful combinations in quantitative neurobiology. This chapter discloses disadvantages associated with voltage-sensitive dyes. It brings useful information for deciding whether voltage-sensitive dye imaging is an appropriate method for your experimental question, and how to determine if a student is ready to work with intracellular voltage-sensitive dyes. Our chapter describes the most important, previously unpublished, practical issues of loading neurons with voltage-sensitive dyes and obtaining fast optical signals (action potentials) from thin dendritic branches using equipment at half price of a standard confocal microscope.

Key words Electrochromism, Chromophores, Hemicyanine, Styryl dyes, Action potential, Dendritic spike, Backpropagation, Plateau potential, Fluorescence and membrane potential

1 Introduction

Our understanding of dendritic physiological properties and dendritic integration depends on our ability to monitor voltage transients in dendritic segments. This chapter is focused on single-photon voltage-sensitive dye recordings from individual neurons in brain slices.

Alon Korngreen (ed.), *Advanced Patch-Clamp Analysis for Neuroscientists*, Neuromethods, vol. 113,
DOI 10.1007/978-1-4939-3411-9_12, © Springer Science+Business Media New York 2016

1.1 A Niche for Dendritic Voltage Imaging

Dendritic physiological properties and dendritic integration are traditionally studied by dendritic patch electrode recordings [1, 2], calcium imaging [3, 4], and voltage imaging [5, 6]. Each method for studying dendritic physiological properties and dendritic integration has its own advantages and disadvantages.

1.1.1 Dendritic Patching

Patch recordings in basal dendrites are currently limited to proximal dendritic segments, less than 140 μm away from the cell body [7]. Action potential (AP) waveforms acquired by dendritic patch electrodes may be distorted by electrode resistance, electrode capacitance, and most acutely by particularly high serial resistance, up to 200 MΩ [7]. Occasional sealing of a high-resistance dendritic patch introduces instability in the recordings manifested by fluctuations in both the amplitude and dynamics of AP transients. By definition, the whole-cell patch electrode recording disrupts (ruptures) dendritic membrane, creates a change in dendritic morphology, change in dendritic capacitance and resistance, as well as it produces a rapid dialysis of the dendritic cytosol resulting in a washout of important ions and molecules.

1.1.2 Dendritic Calcium Imaging

A change in intracellular calcium is an indirect measure of a change in dendritic membrane potential. Intracellular calcium is a complex function of multiple factors including the state of voltage-gated calcium channels, opening or closing of glutamatergic receptor channels, release of calcium from intracellular stores, buffering capacity of dendritic cytoplasm, buffering capacity of the calcium-sensitive dye, and the status of various calcium pumps. The rise in intracellular calcium detected by a calcium-sensitive dye is at least one order of magnitude slower than the rise of voltage detected by an electrode. Obviously, calcium imaging is not ideally suited for the analysis of membrane potential waveforms.

1.1.3 Dendritic Voltage Imaging

For the purpose of studying rapid information processing in central nervous system neurons, one would like to monitor electrical transients occurring in several dendritic branches simultaneously. Simultaneous multi-site recordings of dendritic voltage transients bring three types of useful information. First, such recordings provide critical information about the intrinsic membrane properties of remote dendritic compartments [8, 9]. Second, they provide unique information about the integration of voltage transients in the dendritic tree [10–12]. Third, such measurements can be used to determine the precise locations of excitatory and inhibitory synaptic inputs [13–15]. Recent advances in dendritic voltage imaging hold promise for a wider use of voltage-sensitive dyes in cellular neuroscience [16–18].

1.2 Serious Limitations of Dendritic Voltage Imaging

Dendritic voltage imaging is a difficult experimental procedure characterized by a very low success rate. Experimental failures are often caused by an inherent toxicity of current voltage-sensitive dyes and notoriously low signal-to-noise ratio.

1.2.1 Toxicity

Voltage-sensitive dyes are more toxic than calcium-sensitive dyes or fluorescent tracers such as Rhodamine and Alexa Fluor 594. Overloading neurons with voltage-sensitive dyes causes alterations in neuronal physiological properties, electrical and structural instability of the excitable membrane manifested by broadened APs [6] followed by sustained depolarizations and eventually cell death. Voltage-sensitive dyes are toxic during dye injection procedure, but even more so during the imaging sessions when dye molecules are excited by the incident light.

1.2.2 Low Sensitivity

Voltage-sensitive dyes are characterized by very low sensitivity. Signal amplitudes obtained with voltage-sensitive dyes, expressed as fractional change ($\Delta F/F$), are one or two orders of magnitude lower than signals obtained with calcium-sensitive dyes in the same dendritic segment. In the majority of our experimental measurements, the voltage-sensitive dye signal is buried in noise; it cannot be detected with reasonable averaging procedures (e.g., number of trials less than 9). In some cases, in some dendritic segments, the detected signal is too noisy and therefore it is not useful.

1.2.3 Lack of Calibration

Although the amplitudes of optical and electrical signals obtained from the same location on a neuron are in a strict linear relation [19], the factor that converts $\Delta F/F$ of the optical signal to mV varies from one location to another. Calibration of an optical signal performed at the cell body using an intercellular electrode as a reference must not be applied to a dendrite of the same cell. The reason for this lies in the variable ratio between the concentrations of membrane bilayers in neuronal cytosol (organelles) versus that in the neuronal plasma membrane. If this membrane ratio (organelles/plasmalemma) is not identical in two neuronal locations, then the amplitude of voltage-sensitive dye signal ($\Delta F/F$) will have two different values for the same change in membrane potential [8]. The bottom line is that it is impossible to determine the amplitude of dendritic membrane potential transient in millivolts based solely on the dendritic optical signal and somatic patch electrode recording.

1.3 Is Dendritic Voltage Imaging an Appropriate Method for My Project?

All aspects of this consideration stem from three aforementioned problems, low sensitivity, steep dose-dependent toxicity, and lack of calibration.

An experimental project is currently incompatible with internally applied voltage-sensitive dyes if it requires:

- Determination of absolute amplitude of voltage change in millivolts.

- Monitoring waveforms of miniature synaptic potential (minis) from thin dendritic branches.
- Monitoring waveforms of postsynaptic potentials from thin dendritic branches in single-trial recordings.
- Monitoring spontaneous activity over time scales greater than 1 s. Several seconds of interrupted illumination (dye excitation) are likely to cause irreversible neuronal injury.
- Dendritic imaging in vivo, while theoretically possible, has yet to be achieved.

Dendritic voltage imaging is most successfully used in brain slices to detect:

- Large voltage changes caused by APs, NMDA spikes, and glutamate-mediated plateau potentials [8, 12].
- Evoked voltage transients. Evoked voltage transients (e.g., evoked AP or evoked synaptic potentials) are uniquely suitable for voltage imaging because they allow averaging over trials [8, 13].

1.4 Which Voltage-Sensitive Dye?

1.4.1 Genetically Encoded or Chemical Voltage Indicators?

New fluorescent protein-based voltage sensors are being developed, with the promise of being able to genetically target specific cells in an organism [20, 21]. At present moment, the genetically encoded voltage sensors lack temporal resolution, which is required for determining AP propagation latency and AP half-width. Spike propagation latency and spike half-width are two critical parameters in the data interpretation and analysis portion of many voltage-sensitive dye projects [8, 9, 22].

1.4.2 Extracellular or Intracellular Dye Application?

In order to study voltage transients in thin processes of neurons in the central nervous system, one faces the choice between two means of membrane staining (extracellular or intracellular). In theory, extracellular application of voltage-sensitive dyes could produce signals that can be calibrated at all dendritic loci using a patch pipette at the soma. However, at present, we are still missing a reliable procedure to adequately stain the outer membrane leaf of distal dendrites of individual neurons in brain slices and produce high quality signals; but see a promising development in this direction [23].

1.4.3 Absorption or Fluorescence?

Absorption voltage-sensitive dyes have been used to obtain excellent signals in invertebrate ganglions and brain slices [24]. However, in order to resolve one neurite or one dendrite in voltage-sensitive dye recordings, one must turn away from absorption dyes and use fluorescent indicators injected into the neuronal cytoplasm (intracellular application), as determined by the Zecevic group [5].

1.4.4 Fluorescent Voltage-Sensitive Dyes for Intracellular Application

Rational design methods, based on molecular orbital calculations of the dye chromophores and characterization of their binding in membranes, were employed to engineer dye structures [25]. The general class of dye chromophores called hemicyanine (also referred to as "styryl" dyes) exhibits electrochromism. The most important hemicyanine dyes include: di-4-ANEPPS, di-8-ANEPPS, di-2-ANEPEQ (aka JPW-1114), di-1-ANEPEQ (aka JPW-3028), RH-421, RH-795, ANNINE-6 and ANNINE-6+, di-4-ANEPPDHQ, di-4-ANBDQBS, and di-4-ANBDQPQ, summarized in [17]. Because the electrochromic mechanism is a direct interaction of the electric field with the chromophore and does not require any movement of the dye molecule, all of these dyes provide rapid absorbance and fluorescence responses to membrane potential; they are therefore capable of recording APs [5]. Other mechanisms can give more sensitive voltage responses in specialized applications [26, 27]. However, to date hemicyanine dyes are arguably the most universally used and applicable voltage-sensitive dyes available for recording membrane potential spikes (action potentials) from individual dendritic branches [16, 28–31].

In this chapter we describe the methodological procedures that apply to multi-site voltage-sensitive dye imaging with two "red" dyes JPW-1114 or JPW-3028 excited by green light (~500 nm) and emitting in red spectrum (>600 nm) and one blue dye JPW-4090, excited by red light (~650 nm) and emitting in infrared spectrum (>700 nm). Two-photon voltage imaging and the best dyes used for two-photon excitation were described in [32].

1.5 When Is One Ready to Try Multi-site Dendritic Voltage Imaging?

Dendritic voltage imaging is a difficult experimental procedure characterized by a low success rate and a high rate of frustration. Several fundamental requirements should be met before embarking upon a voltage imaging project.

Requirement 1: An experimenter must be proficient in whole-cell patch electrode recordings. One must be able to keep a cell in a whole-cell configuration for 1 h, while maintaining excellent access resistance.

Requirement 2: At the end of the staining period, an experimenter must be able to pull an outside-out patch without killing the dye-loaded neuron.

Requirement 3: Following a 2-h incubation period, an experimenter must be able to find and re-patch the dye-loaded cell.

Requirement 4: *Calcium Imaging*. Before trying dendritic voltage imaging one should become proficient with calcium-sensitive dye imaging. Towards this goal, perform whole-cell injection of Oregon Green Bapta-1 calcium-sensitive dye [150 μM dissolved in standard intracellular solution]. Load neurons with this dye for at least 45 min. Monitor backpropagation of APs in thin dendritic branches using dendritic calcium imaging. Improve the injection of the calcium-sensitive dye (increase concentration of the dye, decrease access resistance of the patch) and fine-tune optical recording setup (increase illumination intensity; increase the sensitivity of the optical detector, perform temporal averaging and spatial averaging) until a single AP-induced dendritic calcium transient can be detected more than 140 μm away from the cell body. An experimental setup is not ready for dendritic voltage imaging if it cannot detect single AP-associated calcium transients in distal dendritic segments (ref. [8], their figure 7).

1.6 Positive and Practical Aspects of Dendritic Voltage Imaging

1.6.1 Fast Response

The voltage-sensitive dye signal is fast enough to follow changes in membrane potential at sub-millisecond time scale (Fig. 1). This allows for precise measurement of the AP peak latencies and AP half-widths [8, 9, 22] or dendritic spike waveforms [10, 11]. An example of designing experiments based on fast temporal response of a voltage-sensitive dye is shown in Fig. 2. In this example, voltage imaging revealed a clear difference in AP waveform (duration) between an oblique apical dendrite and proximal axon of the same cell (Fig. 2b).

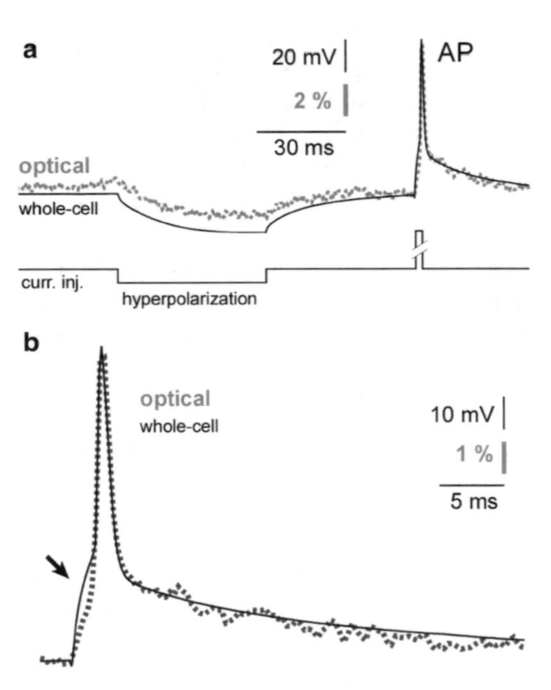

Fig. 1 Fast response time of a voltage-sensitive dye. (**a**) Simultaneous whole-cell (*black*) and optical (*gray*) recordings of the evoked membrane potential changes using voltage-sensitive dye JPW-4090. Optical signal was sampled from the basal dendrite 45 m away from the cell body. Average of six sweeps. (**b**) A segment of trace containing AP is blown up to show temporal correlation between whole-cell (*black*) and optical signal (*gray*). A distortion in electrical recording caused by poorly compensated series resistance (*arrow*) does not appear in the optical signal. Adapted from Zhou et al., 2007

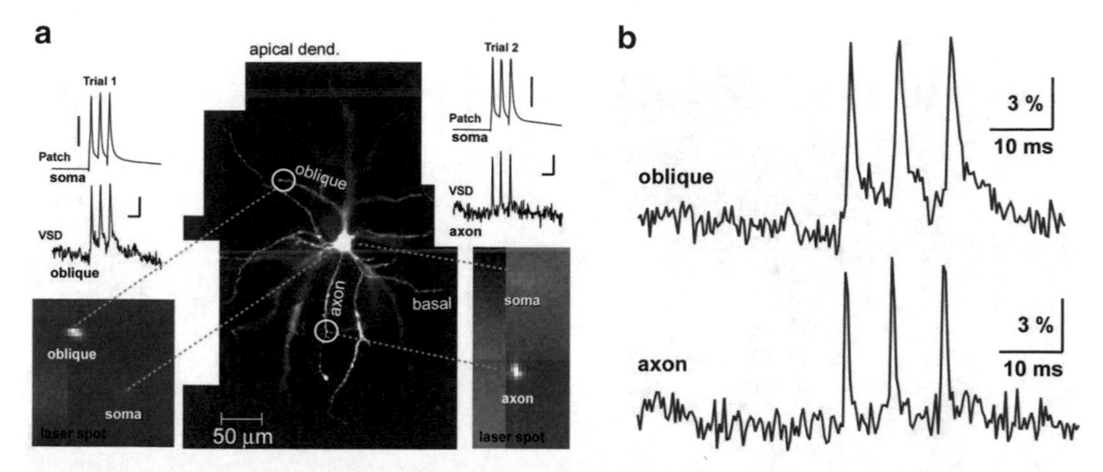

Fig. 2 Optical recordings of action potentials in oblique dendrites and axons of pyramidal cells. (**a**) Central image is a composite microphotograph of a layer 5 pyramidal neuron filled with voltage-sensitive dye JPW-4090. (*Left*) Simultaneous whole-cell (Patch) and optical (VSD) recording of an AP triplet. Optical signal was sampled from the oblique branch at path distance of 128 m from the soma, as shown in the movie frame captured by data acquisition camera at 2.7 kHz frame rate. Only the distal tip of the oblique branch is positioned inside the laser illumination spot. *Dashed gray lines* connect corresponding structures. (*Right*) Same as in the Left except axon is being imaged. Optical signals (VSD) were sampled from a short axonal segment (20 m length) right at the first branch point (path distance from soma=114 m). Optical signal is product of 4 averaging. (**b**) Dendritic (oblique) and axonal (axon) optical signals are aligned on faster time scale to emphasize difference in temporal dynamics of the voltage change in two compartments of the same neuron. Adapted from Zhou et al., 2007

1.6.2 Linear Response

The amplitudes of optical and electrical signals are in strict linear relation [19]. An example of designing experiments based on the linearity of the dye response is shown in Fig. 3. In this example, voltage imaging reveals relative amplitudes of subsequent APs at each location tested (Fig. 3a). At one location (dendrite #2) voltage imaging reveals uniform AP amplitudes during the train of APs (Fig. 3b). Optical recording from another location on the same cell (dendrite #3) shows that the neighboring dendritic branch although similar in shape, length, and diameter experienced a drastic amplitude difference between the first and third AP in the train (Fig. 3c, arrow). These data provided direct experimental evidence for a failure of the first AP in some but not all basal branches.

1.6.3 Compatibility with Standard Experimental Procedures

Voltage-sensitive dye recordings can be easily combined with focal application of pharmacologic agents [33], with focal application of neurotransmitters [29, 34], or bath application of drugs that block membrane conductances (Fig. 4a–c). The fast and linear responses of voltage-sensitive dyes have proven critical for characterizing changes in AP waveforms induced by blocking dendritic sodium (Fig. 4d) or potassium (Fig. 4e) conductances.

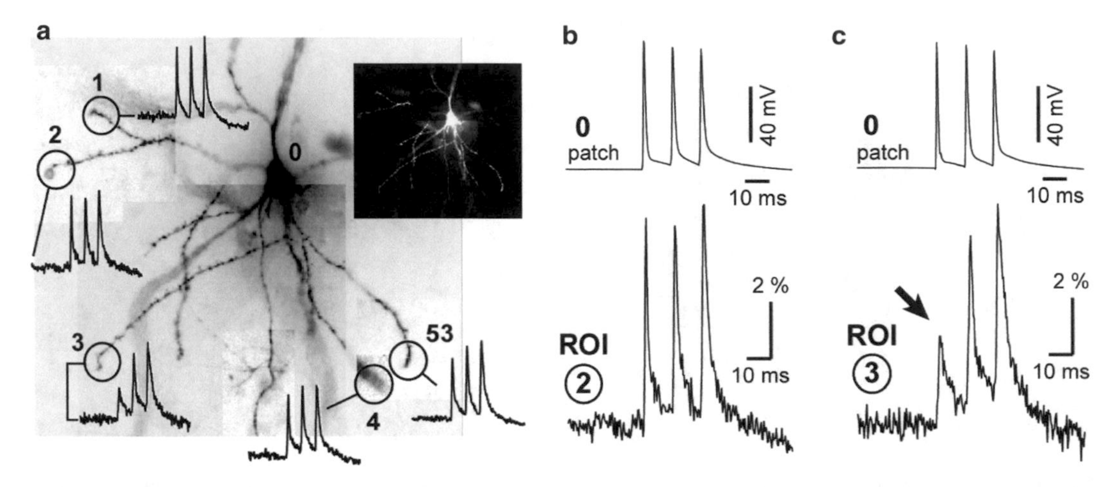

Fig. 3 Voltage transients in tips of basal dendrites. (**a**) Color inverted microphotograph of a neuron filled with JPW3028. Five regions of interest (ROIs 1–5) were explored. *Inset*, same image non-inverted. ROI 5 was re-focused before voltage imaging. (**b**) Simultaneous recording of AP triplets from the soma (patch) and dendritic tip (ROI 2). (**c**) Same as in (**b**) except ROI 3 is shown. Voltage-sensitive dye signal changes linearly with the change in membrane potential, and therefore, relative differences in the optical signal amplitude (ROI 3) reveal exact relative differences in AP amplitudes. *Arrow* marks an action potential which failed to invade the tip of the basal dendrite #3. Adapted from Zhou et al., 2008

1.6.4 Superior Compatibility with Multicompartmental Modeling

The results of dendritic voltage-sensitive dye measurements can be easily incorporated into computational models for the purpose of studying dendritic membrane excitability and dendritic integration [8, 9]. Voltage imaging is uniquely compatible with biophysical models because they provide multiple parameters (signal shape, peak latency) from multiple dendritic loci at the same moment of time. The number of model constraints is notably heightened by a multi-site approach [35]. An interaction between multi-site voltage-sensitive dye recording (wet experiment) and multicompartmental modeling (dry experiment) represents one of the most powerful combinations in quantitative neurobiology [8, 9, 28, 36, 37].

2 Materials

1. Standard extracellular solution (artificial cerebrospinal fluid, ACSF) is used to perfuse brain slices at room temperature during dye loading and during incubation; (in mM) 125 NaCl, 26 $NaHCO_3$, 2.3 KCl, 1.26 KH_2PO_4, 2 $CaCl_2$, 1 $MgSO_4$, and 10 glucose, pH 7.4.
2. Standard gluconate-based intracellular solution is used for the dye-free portion (Fig. 5c, dye free) as well as for the dye-rich portion of the pipette lumen (Fig. 5c, dye); (in mM) 135 potassium gluconate, 2 $MgCl_2$, 3 Na_2-ATP, 10 Na_2-phosphocreatine, 0.3 Na_2-GTP, and 10 Hepes (pH 7.3).

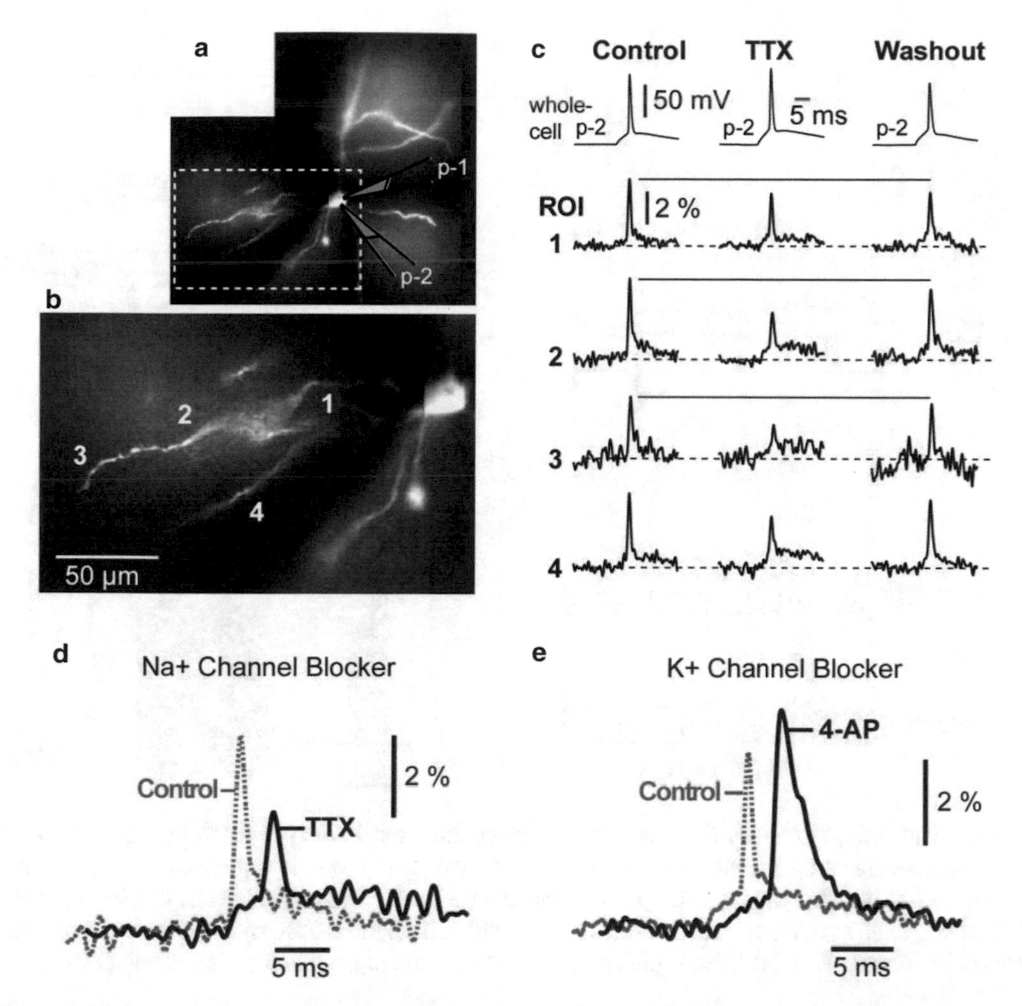

Fig. 4 Dendritic action potential waveforms in the presence of channel blockers. (**a**) Composite photograph of neuron filled with voltage-sensitive dye JPW-1114. (**b**) Blow-up of the *box* shown in (**a**). Four regions of interest (ROIs 1–4) on two basal branches are examined. (**c**) Electrical (whole cell) and optical (ROIs 1–4) signals following natural (Control, Washout) or artificially generated somatic APs (block of sodium channels with tetrodotoxin (TTX), 1 M). The cell body is patched with two electrodes as shown in (**a**). One channel (p-1) is used to voltage-clamp the cell in the shape of a previously recorded AP. The other electrode (p-2) remains in current clamp to monitor membrane potential, necessary to verify the efficacy of the voltage-clamped AP waveform in TTX conditions. Somatic AP amplitude is 109 % of Control in TTX and 88 % of Control in Washout conditions (45-min washout). Spike-triggered averages, $n=4$ sweeps. (**d**) Superposition of optical traces from ROI 2 before (Control) and after blocking sodium channels (TTX). (**e**) Superposition of optical traces from a basal dendrite (different cell) before (Control) and after blocking potassium channels with 4-aminopyridine (4-AP). Adapted from Acker and Antic, 2009

3. Thermo Scientific™ Nalgene™ Syringe Filters—4 mm diameter, nylon membrane, 0.2 µm pore for filtering intracellular solutions.
4. Microcentrifuge (Corning's LSE Mini Microcentrifuge) for filtering intracellular solutions.

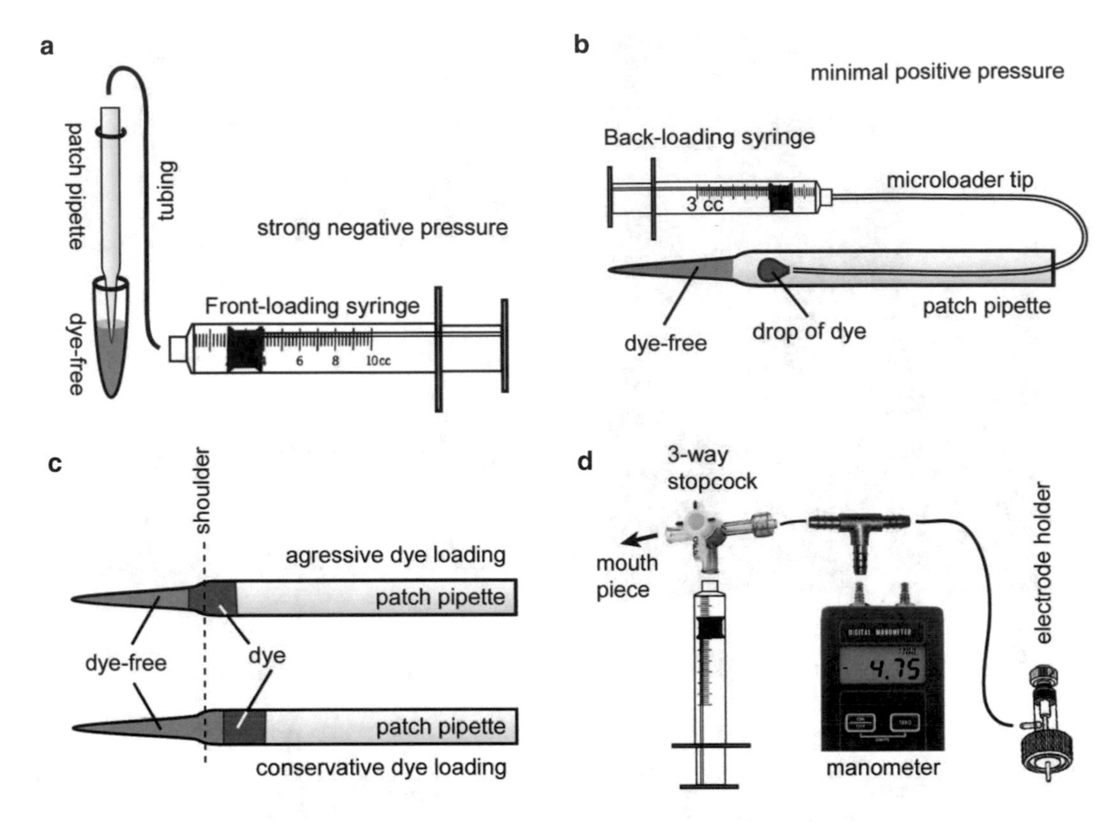

Fig. 5 Loading voltage-sensitive dyes into patch pipettes. (**a**) Front loading of patch pipette with dye-free internal solution drawn up through the tip. A stand holding the micropipette in vertical position is not shown. (**b**) Back loading of patch pipette with dye-rich internal solution. (**c**) Two configurations of a loaded pipette. Note that in the conservative approach the entire tip of the patch pipette is filled with dye-free solution passed the shoulder (*dashed line*). (**d**) Tubing setup for precise monitoring of internal pressure during patching

5. Voltage-sensitive dye JPW-1114, JPW-3028, or JPW-4090 (Molecular Probes, Invitrogen).
6. Microscope. Any patch electrode recording station with an epi-illumination module and two camera ports will do. One camera port hosts a high resolution CCD camera for infrared video microscopy used for patching neurons. The other camera port hosts a low-resolution high-speed camera for voltage imaging. We have successfully used Zeiss Axioskop FS, Olympus BX51WI, and Nikon Eclipse E600. The microscopes should be equipped with either 40× or 63× lenses.
7. Two basic types of epi-illumination are used in our laboratory:
   7a. Wide-field epi-illumination is used for multi-site imaging (Fig. 4) and can be achieved using an arc lamp, metal-halide lamp, laser beam expander, or LEDs. We obtained voltage signals using low-ripple 150 and 250 W arc lamps (OptiQuip Inc., Highland Mills, NY); metal-halide lamp

Lumen 200 (Prior Scientific, Inc., Rockland, MA), and Samba™ 532 nm laser (Cobolt AB, Solna, Sweden).

7b. Laser spot illumination is used for imaging of one location on a dendrite or axon at the time (Fig. 2). Laser spot illumination can be achieved using a laser or laser diode in combination with a beam collimator (ThorLabs, Newton, NJ). The beam collimator is attached to the back port of the microscope in the place of a regular arc lamp.

8. UNIBLITZ® CS35 or CS25 shutters with reflective blades. Shutter driver should be triggered by the data acquisition software.

9. Filter cubes. [Red Dyes] For JPW-1114 and JPW-3028 use an exciter 520 $\pm$ 45 nm; dichroic 570 nm; emission >610 nm. A Chroma filter set for JPW-1114 and JPW-3028 includes exciter D510/60 (480–540 nm bandpass), dichroic 570dcxru and emitter E600lp (600 nm longpass). Alternatively, one can use a 532 nm laser (no exciter). [Blue Dyes] For JPW-4090 use a 658 nm laser diode (no exciter), 700 nm dichroic, and 720 nm longpass emission filter.

10. Camera. NeuroCCD-SMQ, RedShirtImaging, Decatur, GA. This camera allows acquisition of low light intensity (fluorescence) signals at high speed. The resolution of the camera is only 80 by 80 pixels. The camera resolution was sacrificed to achieve fast frame acquisition rates (2 kHz (full frame), 5 kHz (binned 3 $\times$ 3) and 10 kHz (partial frame)). The Neuroplex software (comes with the camera) performs data acquisition and data analysis. Two most versatile features of Neuroplex are: (1) Spatial averaging; and (2) Temporal averaging. Spike-triggered averaging mode uses patch electrode signal, or a referent optical signal from any part of the visual field, to align individual data traces prior to averaging. The spike-triggered averaging mode overcomes problems with signal jitter between subsequent recording trials.

3 Methods

Intracellular application of a styryl voltage-sensitive dye is accomplished by free diffusion of the dye from a patch electrode into the neuronal cell body during a whole-cell recording. Standard procedures for preparation of 300-μm-thick brain slices, standard artificial cerebrospinal fluid, and standard K^+-gluconate-based intracellular solution used for any other whole-cell recordings are also used here for voltage-sensitive dye dendritic imaging.

All experiments aimed at dendritic voltage imaging with internally applied voltage-sensitive dyes comprise four essential maneuvers: (a) Injection of the dye into the neuron; (b) Diffusion of the dye from the cell body into dendrites; (c) Re-patching of the stained neuron; and (d) Optical recordings. A list of individual steps in chronological order follows.

1. Fire-polish glass capillaries prior to pulling patch pipettes. Hold the tip of the capillary above an open flame until the rim of the glass turns red. Once the rim turns red, it takes only 3–4 s to complete the polishing. *See* **Note 1**.
2. Boil the polished glass capillaries in 70 % ethanol for 10 min. Keep capillaries in vertical position inside the beaker. Bubbles of boiled ethanol emerge from the bottom of the beaker and climb through the lumen of the capillary. This continuous flow of boiled ethanol removes impurities and dust from the capillary lumen.
3. Submerge capillaries into acetone (room temperature) for 5 min. Remove from acetone. Gently shake the drops of acetone off the capillaries. It is not necessary to get rid of all acetone drops. Spread capillaries on a sheet of clean aluminum foil. Cover the capillaries with another sheet of aluminum foil and let them dry for 24–48 h. Store washed capillaries in a closed box.
4. Transfer one brain slice to an upright fluorescence microscope configured with imaging equipment. Perfuse brain slice with oxygenated ACSF at room temperature. Note that both the dye loading and the subsequent dye diffusion phase of the experiment are performed at room temperature.
5. Manufacture patch pipettes at 7 MΩ resistance when filled with standard internal solution (dye-free intracellular solution).
6. Voltage-sensitive dyes JPW-1114, JPW-3028, and JPW-4090 are water soluble. Prepare dye stock in standard intracellular solution at 5 mM and store at −20 °C. Prepare final dye-rich solution at concentration range 200–800 μM by dissolving stock in standard intracellular solution. Store aliquots of the final solution at −20 °C. Excellent results can also be achieved with lower dye concentrations (e.g., 30 μm) but longer dye-loading periods [18]. Before loading into patch pipettes both dye-free and dye-rich intracellular solutions should be filtered through syringe filters—4 mm diameter, nylon membrane, 0.2 μm pore. *See* **Note 2**.
7. Fill the patch pipette from the tip with the dye-free intracellular solution by applying negative pressure for a period ranging from 0.5 min to 3 min (Fig. 5a). The time required is established empirically and depends on the parameters of the electrode (e.g., shape and size of the tip) and the intensity of the negative pressure imposed by the front-loading syringe (Fig. 5a). *See* **Note 3**.

8. Backfill the pipette with a drop of intracellular solution containing the voltage-sensitive dye (Fig. 5b). The amount of dye-free solution (Fig. 5c, dye free) will determine the amount of time allowed for maneuvering through brain slice before the dye leaks out and ruins the experiment. The ratio between dye-free and dye-rich solutions will determine the rate of dye diffusion into the cell. For strong-fast dye loading use small dye-free/dye-rich ratio (Fig. 5c, *aggressive*). For more relaxed and more gradual dye-loading use large free/rich ratio (Fig. 5c, *conservative*).

   This system of front loading the patch pipette with dye-free solution prevents leakage of the voltage-sensitive dye into the brain slice tissue before the patch pipette is attached to the neuron. *See* **Note 4**.

9. The dye-loaded patch pipette (Fig. 5c) cannot be stored. It must be used immediately to prevent mixing of dye-free and dye-rich solutions. Insert pipette in the pipette holder (Fig. 5d). Use a digital manometer to set a positive pressure inside the patch pipette before patching (Fig. 5d). For 7 MΩ patch pipettes use a positive pressure of ~8 mbar before the pipette is submerged in the recording chamber. Increase the positive pressure to ~25 mbar just before inserting the patch pipette into a brain slice. *See* **Note 5**.

10. Upon establishing a whole-cell configuration test if the voltage-sensitive dye leaked in the surrounding tissue using very low excitation light intensity. *See* **Note 6**.

11. Keep the whole-cell configuration for 45–60 min to allow for the dye to diffuse from pipette into cytoplasm.

12. During the dye-loading process/period, the cell body may be exposed to a large dye concentration. It is necessary to prevent overloading of the cell since the dye's toxic effects become apparent at large concentrations. *See* **Note 7**.

13. Detach the patch electrode from the cell by forming an outside-out patch. To form an outside-out patch, switch to voltage clamp. Apply a seal test (the same test used for establishing whole-cell). Slowly withdraw the patch pipette away from the cell body. Watch as the seal test current slowly decreases (resistance increases) until a gigaohm seal is formed. At this point pull freely the patch pipette out of the recording chamber. Discard the loading pipette and examine the integrity of the cell under the infrared video-microscopy monitor.

14. If the cell body is still visible in infrared video microscopy, switch to fluorescence microscopy. Now examine the cell same body using very low excitation light intensity.

    If the surrounding brain slice tissue shows no signs of background fluorescence, proceed with the incubation phase. If background fluorescence is present, abandon this brain slice. *See* **Note 6**.

15. Incubate the slice for 90–180 min at room temperature. Incubation is necessary to allow the diffusion of the lipophilic voltage-sensitive dye into distal processes.

    Re-patching of the stained cell is performed at physiological temperature (e.g., 35 °C) with patch electrodes filled with dye-free intracellular solution.

16. After the incubation period is over, use low excitation light intensity and fluorescence video imaging to locate the stained cell body in the visual field. Use infrared DIC to patch the cell. If possible, reduce the intensity of infrared light and increase the video camera gain. This will produce a grainy (low quality) image during re-patching, but it may preserve the cell from photodynamic damage.

17. Focusing and positioning of the dendritic tree prior to a voltage imaging sweep should be performed using the fluorescence video imaging at very low excitation light intensity and a video camera at very high gain. It is important to limit all types of light exposure to a minimum. *See* **Note 8**.

18. The subsequent voltage imaging is performed at high illumination intensity. Illumination intensities necessary for single-photon voltage imaging typically exceed those used in calcium imaging experiments by more than threefold.

19. Ideally, one wishes to limit exposure to strong excitation light to a minimum. Note that the majority of scientific projects utilizing voltage-sensitive dyes were centered on AP backpropagation. In each of these projects, opening of the shutter was limited to ~100 ms per sweep [8, 9].

20. Wait time. Following a 100 ms exposure (one sweep), it is a good idea to pause recordings for at least 1 min. The logic behind inserting long pauses between voltage imaging sweeps is based on the hypothesis that under certain conditions the excitation of dye molecules in the plasma membrane may lead to formation of free radicals that cause structural or functional damage. Long pauses allow natural mechanisms to clear some of the free radicals from the cell.

21. Sampling rate. Fast sampling rates drastically disrupt signal-to-noise ratio, because they reduce the number of collected photons per each sample. One can use the NeuroCCD camera to sample a portion of a chip at 10 kHz. This sampling rate requires high power excitation (typically using a 532 nm laser), which causes fast photodynamic damage. We usually record AP waveforms at 2 kHz (or 2.7 kHz) and then we interpolate (oversample) the AP waveforms from 2 to 20 kHz using

MATLAB (interp, $l = 4$, alpha $= 0.5$, $r = 10$) or other custom-made software [8, 9].

22. The quality of voltage-sensitive dye signal is typically low (poor signal-to-noise ratio). When designing voltage imaging experiments, one should choose biological signals that can be averaged (e.g., evoked AP or an evoked EPSP). Typically, one would average up to nine sweeps. The signal-to-noise ratio improves by the square root of the number of sweeps combined in the temporal average. *See* **Note 9**.
23. Spatial averaging. In addition to the temporal averaging one is encouraged to perform spatial averaging of voltage-sensitive dye signals over the neuronal compartment of interest. *See* **Note 10**.
24. Bleach subtraction. One of the significant artifacts in optical imaging is bleaching, a decrease of the optical signal while the biological signal remains unchanged. *See* **Note 11**. Bleaching correction is performed by fitting an exponential function through a recorded trace and then subtracting the function from the trace (Neuroplex, *Exponential Subtract*). Bleach correction can also be performed by recording traces of the same length (same number of points) at the end of the experiment (stimulus turned off) and subtracting them from experimental traces. *See* **Note 12**.
25. Filtering. Final conditioning of the voltage traces is performed using digital filters available in Neuroplex, MATLAB, and other software. A high-pass filter is used to remove slow, vibration-related oscillations. A low-pass filter is used to remove high-frequency noise, while observing the Nyquist criterion. A free seminar on conditioning optical traces using the Neuroplex software is given by Dr. Larry B. Cohen every year at the RedShirtImaging Workshop, a satellite to the Annual Society for Neuroscience Meeting.

4 Notes

1. Fire-polishing of the electrode glass eliminates sharp edges that can damage silver wire during insertion of the patch electrode into an electrode holder. Sharp edges of the capillary glass can scrape the chips of silver chloride, thus introducing particles into the intracellular solution. Foreign particles block the flow of intracellular solution through the mouth of the patch pipette. Clogging of the pipette tip precludes the formation of the gigaohm seal, as well as it disrupts the intracellular injection of a dye.

2. Filtering of dye-free and dye-rich internal solutions is performed in a microcentrifuge at slow speed. Insert a 4-mm filter into an open 0.5 or 0.7 mL standard plastic tube before placing the tube into microcentrifuge. Load 50 μL of the solution into the filter lumen using a pipettor. It takes 2–3 min of spinning to push a 50 μL of solution through the filter into the tube lumen. One 0.5 mL plastic tube can only take three 50 μL loads, because the tip of the filter is inserted deep into the lumen of the plastic tube.
3. An inexperienced voltage-sensitive dye user should fill the entire tip of the dye-loading patch pipette with dye-free solution passed the shoulder (Fig. 5c, shoulder). This allows plenty of time for maneuvering pipettes through brains slices and making a gigaohm seal.
4. This system of loading the patch pipette prevents leakage of the voltage-sensitive dye into the brain slice tissue before the patch pipette is attached to the neuron. The volume of dye-free intracellular solution in the tip of the patch pipette must be optimized empirically. If too little dye-free volume in the tip, the dye will leak from the pipette during the formation of a gigaohm seal (Fig. 5c, aggressive). If too much dye-free volume in the tip of a dye-loading patch pipette, the staining of the cell membrane with voltage-sensitive dye will be slow and inefficient (Fig. 5c, conservative).
5. It is a good practice to measure and control positive pressure while approaching a target cell with a dye-loading patch pipette. Having consistent pressure levels and achieving consistent time spent in the recording chamber prior to establishment of a gigaohm seal, from experiment to experiment, allows one to fine-tune and improve the efficacy of dye loading.
6. If voltage-sensitive dye leaks out of the loading pipette before the gigaohm seal was accomplished, the dye will bind strongly to the neighboring membranes and to the extracellular side of the target cell. Voltage-sensitive dyes cannot be washed out from the neuropil in a reasonable amount of time. The "leak" dye disrupts voltage signals in two ways: (1) by increasing the "F" component in $\Delta F/F$; and (2) by decreasing the "ΔF" component in $\Delta F/F$. Voltage-sensitive dye in the outer leaf of the plasma membrane (extracellular side) produces voltage signal of opposite sign from the internally bound voltage-sensitive dye. Two signals cancel each other causing a severe deterioration of the signal-to-noise ratio.
7. One practical strategy for preventing a dye overload is constant monitoring of evoked AP doublets during the dye injection period. Every 2–4 min inject a depolarizing current pulse (duration 50 ms) to evoke 2 APs. Compare the first recording

established right upon whole-cell break-in against the current recording. If the amplitude of the second AP in the current recording was decreased by 30–40 % since the break-in, it is time to pull out the dye-loading pipette and leave the cell to incubate. Following a 2-h incubation period and successful re-patch, the concentration of voltage-sensitive dye in the cell body is reduced, which is manifested by the recovery of the second AP amplitude. Note that neurons injured by voltage-sensitive dye loading can recover after incubation and re-patch [6].

8. It is a common mistake to search for dendrites using high illumination intensity. In experiments with neurons injected with voltage-sensitive dyes, it is paramount to limit the light exposure to a minimum. Use neutral density filters in both the epi-illumination and transmitted light paths.
9. Although the Neuroplex recording software can perform online real-time averaging including spike-triggered averaging to eliminate sweep-to-sweep temporal jitter of APs, always save all sweeps for offline analysis and averaging. Some sweeps may be noisier than others, exclude noisy sweeps from the average.
10. A neuronal compartment of interest (e.g., dendrite) projects on many camera detectors. Using Neuroplex or other software, select and combine optical signals from several neighboring detectors receiving light from the same compartment. With 40× magnification one camera pixel typically collects light from 5 μm × 5 μm in the object filed. We normally average 12 pixels (4 × 3 pixel quadrant) from a given dendritic segment. The signal-to-noise ratio improves by the square root of the number of pixels combined in the spatial average. One can combine temporal and spatial averaging in the same experiment. A simple numerical example shows that using nine sweeps for temporal averaging and selecting an area containing nine detectors for spatial averaging would improve the signal-to-noise ratio nine times ($\sqrt{9} \times \sqrt{9} = 9$), compared to the output of a single detector in a single sweep.
11. Bleaching should always be minimized by reducing the excitation intensity. However, the absence of any bleaching indicates that excitation power needs to be increased to improve signal and signal-to-noise ratio.
12. Any noise in the bleach trace will appear in the final product after subtraction if directly subtracted from the experimental trace. To reduce noise in "bleach" traces one should average 4–9 sweeps at the end of experiment, in the absence of any stimulus.

Acknowledgments

We are grateful to Leslie Loew for comments. Supported by institutional Health Center Research Advisory Council (HCRAC) grant and NIH U01 grant to SDA.

References

1. Stuart GJ, Sakmann B (1994) Active propagation of somatic action potentials into neocortical pyramidal cell dendrites. Nature 367 (6458):69–72
2. Korngreen A, Sakmann B (2000) Voltage-gated K+ channels in layer 5 neocortical pyramidal neurones from young rats: subtypes and gradients. J Physiol 3:621–639
3. Miyakawa H, Ross WN, Jaffe D et al (1992) Synaptically activated increases in Ca2+ concentration in hippocampal CA1 pyramidal cells are primarily due to voltage-gated Ca2+ channels. Neuron 9(6):1163–1173
4. Schiller J, Helmchen F, Sakmann B (1995) Spatial profile of dendritic calcium transients evoked by action potentials in rat neocortical pyramidal neurones. J Physiol 487(Pt 3):583–600
5. Antic S, Zecevic D (1995) Optical signals from neurons with internally applied voltage-sensitive dyes. J Neurosci 15(2):1392–1405
6. Antic S, Major G, Zecevic D (1999) Fast optical recordings of membrane potential changes from dendrites of pyramidal neurons. J Neurophysiol 82(3):1615–1621
7. Nevian T, Larkum ME, Polsky A et al (2007) Properties of basal dendrites of layer 5 pyramidal neurons: a direct patch-clamp recording study. Nat Neurosci 10(2):206–214
8. Antic SD (2003) Action potentials in basal and oblique dendrites of rat neocortical pyramidal neurons. J Physiol 550(1):35–50
9. Acker CD, Antic SD (2009) Quantitative assessment of the distributions of membrane conductances involved in action potential backpropagation along basal dendrites. J Neurophysiol 101(3):1524–1541
10. Milojkovic BA, Radojicic MS, Antic SD (2005) A strict correlation between dendritic and somatic plateau depolarizations in the rat prefrontal cortex pyramidal neurons. J Neurosci 25(15):3940–3951
11. Milojkovic BA, Radojicic MS, Goldman-Rakic PS et al (2004) Burst generation in rat pyramidal neurones by regenerative potentials elicited in a restricted part of the basilar dendritic tree. J Physiol 558(Pt 1):193–211
12. Milojkovic BA, Zhou WL, Antic SD (2007) Voltage and calcium transients in basal dendrites of the rat prefrontal cortex. J Physiol 585(2):447–468
13. Djurisic M, Antic S, Chen WR et al (2004) Voltage imaging from dendrites of mitral cells: EPSP attenuation and spike trigger zones. J Neurosci 24(30):6703–6714
14. Canepari M, Willadt S, Zecevic D et al (2010) Imaging inhibitory synaptic potentials using voltage sensitive dyes. Biophys J 98(9):2032–2040
15. Willadt S, Nenniger M, Vogt KE (2013) Hippocampal feedforward inhibition focuses excitatory synaptic signals into distinct dendritic compartments. PLoS One 8(11):e80984
16. Canepari M, Popovic M, Vogt K et al (2010) Imaging submillisecond membrane potential changes from individual regions of single axons, dendrites and spines. In: Canepari M, Zecevic D (eds) Membrane potential imaging in the nervous system: methods and applications. Springer Science + Business Media, LLC., New York, NY
17. Yan P, Acker CD, Zhou WL et al (2012) Palette of fluorinated voltage-sensitive hemicyanine dyes. Proc Natl Acad Sci U S A 109 (50):20443–20448
18. Rowan MJ, Tranquil E, Christie JM (2014) Distinct Kv channel subtypes contribute to differences in spike signaling properties in the axon initial segment and presynaptic boutons of cerebellar interneurons. J Neurosci 34 (19):6611–6623
19. Cohen LB, Salzberg BM, Grinvald A (1978) Optical methods for monitoring neuron activity. Annu Rev Neurosci 1:171–182
20. Akemann W, Mutoh H, Perron A et al (2010) Imaging brain electric signals with genetically targeted voltage-sensitive fluorescent proteins. Nat Methods 7(8):643–649
21. Baker BJ, Jin L, Han Z et al (2012) Genetically encoded fluorescent voltage sensors using the voltage-sensing domain of Nematostella and Danio phosphatases exhibit fast kinetics. J Neurosci Methods 208(2):190–196
22. Zhou WL, Yan P, Wuskell JP et al (2008) Dynamics of action potential backpropagation

in basal dendrites of prefrontal cortical pyramidal neurons. Eur J Neurosci 27(4):1–14

23. Aseyev N, Roshchin M, Ierusalimsky VN et al (2012) Biolistic delivery of voltage-sensitive dyes for fast recording of membrane potential changes in individual neurons in rat brain slices. J Neurosci Methods 212(1):17–27
24. Wu JY, Lam YW, Falk CX et al (1998) Voltage-sensitive dyes for monitoring multineuronal activity in the intact central nervous system. Histochem J 30(3):169–187
25. Loew LM, Scully S, Simpson L et al (1979) Evidence for a charge-shift electrochromic mechanism in a probe of membrane potential. Nature 281(5731):497–499
26. Fink AE, Bender KJ, Trussell LO et al (2012) Two-photon compatibility and single-voxel, single-trial detection of subthreshold neuronal activity by a two-component optical voltage sensor. PLoS One 7(8):e41434
27. Miller EW, Lin JY, Frady EP et al (2012) Optically monitoring voltage in neurons by photo-induced electron transfer through molecular wires. Proc Natl Acad Sci U S A 109 (6):2114–2119
28. Popovic MA, Gao X, Carnevale NT et al (2014) Cortical dendritic spine heads are not electrically isolated by the spine neck from membrane potential signals in parent dendrites. Cereb Cortex 24(2):385–395
29. Acker CD, Yan P, Loew LM (2011) Single-voxel recording of voltage transients in dendritic spines. Biophys J 101(2):L11–L13
30. Kampa BM, Stuart GJ (2006) Calcium spikes in basal dendrites of layer 5 pyramidal neurons during action potential bursts. J Neurosci 26 (28):7424–7432
31. Holthoff KP, Zecevic DP, Konnerth A (2010) Rapid time-course of action potentials in spines and remote dendrites of mouse visual cortex neurons. J Physiol 588:1085
32. Acker CD, Loew LM (2013) Characterization of voltage-sensitive dyes in living cells using two-photon excitation. Methods Mol Biol 995:147–160
33. Zhou WL, Antic SD (2012) Rapid dopaminergic and GABAergic modulation of calcium and voltage transients in dendrites of prefrontal cortex pyramidal neurons. J Physiol 590(Pt 16):3891–3911
34. Milojkovic BA, Wuskell JP, Loew LM et al (2005) Initiation of sodium spikelets in basal dendrites of neocortical pyramidal neurons. J Membr Biol 208(2):155–169
35. Keren N, Peled N, Korngreen A (2005) Constraining compartmental models using multiple voltage recordings and genetic algorithms. J Neurophysiol 94(6):3730–3742
36. Meyer E, Muller CO, Fromherz P (1997) Cable properties of dendrites in hippocampal neurons of the rat mapped by a voltage-sensitive dye. Eur J Neurosci 9(4):778–785
37. Prinz AA, Fromherz P (2003) Effect of neuritic cables on conductance estimates for remote electrical synapses. J Neurophysiol 89 (4):2215–2224

Chapter 13

Modeling the Kinetic Mechanisms of Voltage-Gated Ion Channels

Autoosa Salari, Marco A. Navarro, and Lorin S. Milescu

Abstract

From neurons to networks, the kinetic properties of voltage-gated ion channels determine specific patterns of activity. In this chapter, we discuss how experimental data can be obtained and analyzed to formulate kinetic mechanisms and estimate parameters, and how these kinetic models can be tested in live neurons using dynamic clamp. First, we introduce the Markov formalism, as applied to modeling ion channel mechanisms, and the quantitative properties of single-channel and macroscopic currents obtained in voltage-clamp experiments. Then, we discuss how to design optimal voltage-clamp protocols and how to handle experimental artifacts. Next, we review the theoretical and practical aspects of data fitting, explaining how to define and calculate the goodness of fit, how to formulate model parameters and constraints, and how to search for optimal parameters. Finally, we discuss the technical requirements for dynamic-clamp experiments and illustrate the power of this experimental-computational approach with an example.

Key words Voltage-gated ion channels, Kinetic mechanism, Markov model, Dynamic clamp, Voltage clamp, Sodium channels

1 Introduction

Ion channels are the molecular building blocks of cellular excitability, forming highly specific and efficient pores in the membrane. Gated by various types of stimuli (chemical ligands, electricity, mechanical force, temperature, or light), ion channels form a superfamily of transmembrane proteins that underlie a vast number of physiological and pathological events [1]. Within this superfamily, voltage-gated ion channels [2] play a uniquely important role: they detect changes in membrane potential using specialized voltage-sensing structures [3], and further modify the membrane potential by allowing ions to flow through the lipid bilayer. To perform this function, the channel molecule undergoes conformational transitions within a set of conducting and non-conducting states, governed by specific kinetic mechanisms [4–6].

Alon Korngreen (ed.), *Advanced Patch-Clamp Analysis for Neuroscientists*, Neuromethods, vol. 113,
DOI 10.1007/978-1-4939-3411-9_13, © Springer Science+Business Media New York 2016

The relationship between membrane voltage and ionic current is simple and can be derived from basic principles [7]. Electrically, a patch of membrane is equivalent to a capacitor (the lipid bilayer) connected in parallel with a variable conductance (the ion channels, swinging between closed and open states), which is connected in series with a battery (the electrochemical potential of permeant ions). If no external current is injected in this circuit, the current flowing through the conductance and the current charging the capacitor sum to zero. Thus, ignoring spatial effects, the change in the membrane potential vs. time is proportional to the net ionic current flowing through the membrane, as described by the differential equation:

$$C\frac{\mathrm{d}V}{\mathrm{d}t} = -I, \tag{1}$$

where C is membrane capacitance, V is membrane potential, $C \times \mathrm{d}V/\mathrm{d}t$ is the capacitive current that charges the membrane, I is the ionic current flowing through the membrane, and t is time.

All the ion channels within the membrane contribute to I, which is the algebraic sum of all the single-channel currents. Thus, any individual ion channel that *opens* or *closes* will cause an immediate and finite change in the net current I, unless V happens to be equal to the reversal potential for that channel. From this perspective, a closing of a channel is as significant as an opening. In turn, this change in the current modifies the rate at which the membrane potential changes over time. Then, as V evolves in time, the driving forces for the permeating ions and the kinetic properties of voltage-gated channels will also change. These changes will again modify I, closing the causal loop between membrane potential and ionic current.

Because they both *sense* and *control* membrane potential, voltage-gated ion channels play a key role in action potential generation and propagation, in neurons and other excitable cells [8]. Neurons, in particular, spend considerable amounts of chemical energy to create and maintain the electrochemical gradients necessary for action potentials to work [9], and thus to establish communication within the nervous system. Different types of neurons display unique patterns of cellular excitability [10] and assemble into brain circuits with distinct network properties [11]. The firing properties of individual neurons and neuronal circuits, and ultimately the function of the entire nervous system, are largely determined by the kinetic properties of voltage-gated ion channels [12–15]. Considering the rich variety of excitable behavior at cellular and system levels, it's not surprising that voltage-gated ion channels have their own impressive repertoire of molecular properties [16]. Understanding these properties, particularly the dynamics of state transitions and their voltage sensitivity, is key to understanding how neurons and circuits work.

1.1 Target Audience and Expectations

The aim of this chapter is to guide the reader through the most important aspects of modeling and testing the kinetic mechanisms of voltage-gated ion channels. We are focused on deriving biophysically realistic models from macroscopic currents obtained in whole-cell voltage-clamp experiments [17] and testing these models in live neurons using dynamic clamp [18]. The reader is expected to have a basic understanding of ion channel and membrane biophysics, and some experience with electrophysiology experiments. We tried to keep the discussion general, without relying on a specific computer program, hoping that the readers will be able to take the basic principles learned here and implement them in their preferred software. Nevertheless, for some of the examples presented here we have used a version of the QuB software (www.qub.buffalo.edu), as developed and maintained by our lab (http://milesculabs.biology.missouri.edu/QuB).

2 Ion Channel Models

Modeling ion channel kinetics is fun. However, when taken beyond exponential curve fitting for time constants, or sigmoid fitting of conductance curves, modeling becomes quite challenging. More experimental data and more sophisticated computational algorithms are necessary, and results are not so easy to interpret. Whether this effort is worthwhile depends on the specific goals of the investigator. For example, one may want to find a model that can be used as a computational building block in large-scale simulations of neuronal networks. For this application, simplified phenomenological models will compute faster and would probably work just as well [19]. However, one could set a more mechanistically oriented goal, where the biophysical knowledge available on a particular ion channel is assembled into a detailed computational model [20], which is then tested and refined against new experimental data, and then further used to quantitatively test various hypotheses.

Starting with the seminal work of Hodgkin and Huxley [21, 22], most ion channel models fall somewhere in the range defined by these two examples. Although phenomenological models that simply describe the data are useful, the ultimate goal would be to quantitatively understand how the ion channel works at the molecular level and how it interacts with its environment at the cellular level. A biophysically realistic model must agree with existing theory and experimental data [23–30], but it should also remain computationally tractable. Above all, keep in mind that "all models are wrong but some are useful" [31].

2.1 Kinetic Mechanisms

First, a kinetic mechanism is defined by a set of possible conformational states. Although in principle a protein can assume a continuum of structural conformations, statistically, the molecule will reside most of the time in a relatively small subset of high-occupancy states. The time spent *continuously* in a given state—the "lifetime"—is a random quantity with an exponential probability distribution [32]. For voltage-gated ion channels, the high-occupancy states are the various conformations that correspond to functional and structural elements, such as resting or activated voltage sensors, closed or open pore, inactivated or non-inactivated channel, etc. [3]. Other states may be characterized by more subtle or less understood conformational changes. A state can be identified experimentally if it is associated with a measurable change in properties (e.g., conductance, fluorescence), or it can be inferred statistically from the data.

A kinetic mechanism is further defined by a set of allowed transitions between states. Powered by thermal energy or other sources, the channel undergoes conformational changes at random times. Which state is next is also a random event, with the average frequency of a given transition being inversely related to the energy barrier separating the two states. Transition frequencies are quantified by rate constants. According to rate theory [33], a voltage-dependent rate constant, k_{ij}, corresponding to the transition from state i to state j, has the following expression:

$$k_{ij} = k_{ij}^0 \times e^{k_{ij}^1 \times V}, \quad (2)$$

where V is the membrane potential and k_{ij}^0 is the rate at zero membrane depolarization. k_{ij}^1 is a factor that indicates how sensitive the rate constant is to the membrane potential, as follows:

$$k_{ij}^1 = \left(\delta_{ij}^0 \times z_{ij} \times F\right)/(R \times T), \quad (3)$$

where z_{ij} is the electrical charge moving over the fraction δ_{ij} of the electric field, F is Faraday's constant, R is the gas constant, and T is the absolute temperature [34]. k_{ij}^1 is zero for voltage-insensitive rates, while k_{ij}^0 is zero for non-allowed transitions. Together, the set of possible states and the set of possible transitions describe the topology of a kinetic mechanism. The rate constants and their voltage dependence define the kinetic parameters of the mechanism.

2.2 Markov Formalism

The mathematical properties of a kinetic mechanism—finite set of discrete conformations, exponentially distributed lifetimes, random conformational changes—are beautifully captured by Markov models. Originally developed for stochastic processes, the Markov formalism can be directly applied to ion channels [35], by mapping each known or hypothesized conformation of the channel into a state of the Markov model. The rate constants associated with a

Markov model can be compactly expressed as a rate matrix **Q**, of dimension $N_S \times N_S$, where N_S is the total number of states. The **Q** matrix has each off-diagonal element, q_{ij}, equal to the rate constant k_{ij}, and each diagonal element, q_{ii}, equal to the negative sum of the off-diagonal elements of row i, so that the sum of each row of **Q** is zero. If a transition is not allowed between states i and j, q_{ij} is zero.

The state of the model as a function of time can be conveniently expressed as a probability vector, **P**. At any time t, each element of **P** represents the occupancy of that state, or the fraction of channels that reside in that state. Under stationary conditions, the average fraction of the total time spent by the channel in each state can be calculated as an equilibrium state occupancy. For an ensemble of channels, the average number of channels residing in state i at equilibrium is equal to $p_i \times N_C$, where p_i is the equilibrium occupancy of state i and N_C is the total channel count.

When conditions change (e.g., when a voltage step is applied in a voltage-clamp experiment), the energy landscape of the channel changes as well. All the voltage-sensitive rate constants take different values, and thus the rate matrix **Q** will change as well. As a result, the equilibrium state occupancies will also be different. For an ensemble of channels, if a state becomes less likely to be occupied under the new conditions, the fraction of channels residing in that state will decrease over time, at a rate that depends on the average lifetime of that state. The same behavior would be observed from repeated trials of a single channel. However, in a single trial, the channel will simply continue its stochastic behavior, just with different transition frequencies.

The process of relaxation towards a new state of equilibrium is described by the ordinary differential equation (ODE):

$$\frac{d\mathbf{P}}{dt} = \mathbf{P} \times \mathbf{Q}. \tag{4}$$

The state occupancies corresponding to equilibrium, $\mathbf{P}_{eq}$, can be obtained by setting the time derivative of **P** equal to zero and solving the resulting algebraic equation:

$$\frac{d\mathbf{P}_{eq}}{dt} = \mathbf{P}_{eq} \times \mathbf{Q} = 0. \tag{5}$$

When conditions are stationary and the rate matrix **Q** is constant, the differential equation has a simple analytical solution:

$$\mathbf{P}_t = \mathbf{P}_0 \times e^{\mathbf{Q} \times t}, \tag{6}$$

where $\mathbf{P}_t$ and $\mathbf{P}_0$ are the state occupancies at an arbitrary time t and at time zero, respectively. The exponential of $\mathbf{Q} \times t$ is another matrix, **A**, that contains the conditional state transition probabilities. Each element of $\mathbf{A}$, a_{ij}, is the conditional probability that the channel will be in state j at time t, given that it was in state i at time zero. No assumption is made about what other transitions would have occurred in that time interval. The transition probability matrix $\mathbf{A}$ for a given time t can be calculated numerically using the spectral expansion method [35], as follows:

$$\mathbf{A}_t = \mathrm{e}^{\mathbf{Q}\times t} = \sum_k \mathbf{B}_k \times \mathrm{e}^{\lambda_k \times t}, \tag{7}$$

where the $\mathbf{B}_k$ values are the spectral matrices derived from the eigenvectors of $\mathbf{Q}$, and the λ_k values are the eigenvalues of $\mathbf{Q}$, with λ_0 always equal to zero.

The $\mathbf{B}_k$ and λ_k values can be calculated easily with a numerical library or with specialized software, such as Matlab or QuB. For analysis of macroscopic currents, it is convenient to calculate the transition probability matrix $\mathbf{A}_{\delta t}$ that corresponds to the data sampling interval, δt. Then, the state occupancies can be calculated recursively, starting with some initial solution, using a simple vector–matrix multiplication:

$$\mathbf{P}_{t+\delta t} = \mathbf{P}_t \times \mathbf{A}_{\delta t}. \tag{8}$$

In summary, the Markov formalism has the outstanding convenience of encapsulating all the properties of a kinetic mechanism, as well as the state of the channel, in a few matrices and vectors. The same mathematical and computational operations will apply to any ion channel model, regardless of its topology (how many states, which transitions are allowed) and kinetic properties (rate constant values). Furthermore, Markov models can be used at both single molecule [36–42] and macroscopic levels [43–48].

2.3 Hodgkin-Huxley-Type Models

The ion channel models originally proposed by Hodgkin and Huxley [21] can also be formulated as Markov models, as they explicitly represent the closed, open, or inactivated states of the channel. While they were empirical at the time of their discovery, HH models remain to this day reasonably realistic. Their main limitation—but also their power, depending on the application—resides in making some strongly simplifying assumptions about the channel, which are simply outdated now (e.g., equal and independent "activation particles," or independent activation and inactivation processes). However, one should keep in mind that HH models are in disagreement with biophysical theory when their rate constants do not follow the Eyring rate theory [33], but instead are formulated as arbitrary functions of voltage. While their limited number of states and transitions would inherently reduce their ability to explain experimental data, HH models can gain more flexibility through these arbitrary rate functions.

3 Experimental Data

3.1 What Is in the Data?

A good way to understand the experimental data is to run simulations. Let's consider the simple ion channel model shown in Fig. 1a. For illustration purposes, this model is a very crude approximation of a voltage-gated sodium (Nav) channel, featuring closed,

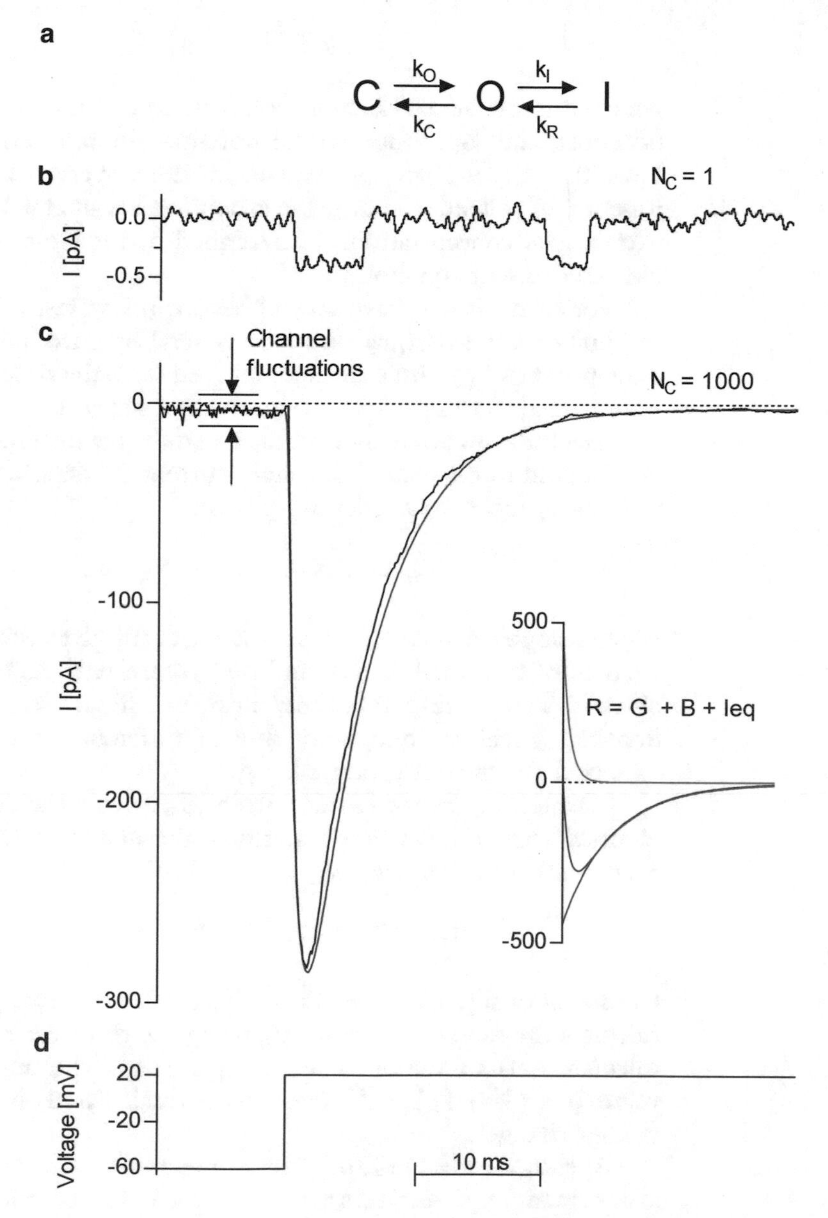

Fig. 1 From model to data. A simple ion channel model (**a**) was used to simulate single-channel (**b**) and macroscopic (**c**) currents in response to a voltage step (**d**). The macroscopic current was simulated with an ensemble of 1000 channels, either deterministically (*black trace*) or stochastically (*red trace*). The *inset* shows a fit of the stochastic macroscopic current (*red*) with a two-exponential function. The individual exponential components of the fit line are also shown (*green* and *blue*)

open, and inactivated states. A single-channel stochastic simulation of a voltage-clamp recording is shown in Fig. 1b, where a noisy signal randomly jumps between zero and a tiny negative current. The noise in the trace is mostly caused by instrumentation, though the open state has its own intrinsic fluctuation in current [49]. The average single-channel current corresponding to the open state can be calculated as follows:

$$i = g \times (V - V_{\mathrm{R}}), \tag{9}$$

where g is the single-channel conductance, V is the membrane potential, and V_{R} is the reversal potential for the permeant ion. Note that Eq. 9 is an approximation: the current is a nonlinear function of voltage when the permeant ion has unequal intra- and extracellular concentrations, as described by the Goldman-Hodgkin-Katz current equation [1].

For channels that have several conducting states, we can make the unitary current equation more general by introducing a conductance vector $\mathbf{g}$, with each element g_{i} equal to the conductance of state i, or equal to zero for non-conducting states. The dot product between the state occupancy vector $\mathbf{P}$ and the conductance vector $\mathbf{g}$ can be used to calculate the unitary current for an arbitrary set of state occupancies, as a function of time:

$$i_t = (\mathbf{P}_t \cdot \mathbf{g}) \times (V - V_{\mathrm{R}}). \tag{10}$$

When a single-channel trace is simulated, at any given time only one element of $\mathbf{P}$ is equal to one, and the rest are zero. As the channel changes state during the simulation, a different element of $\mathbf{P}$ becomes equal to one, and thus a different conductance is "selected" by the dot product $\mathbf{P} \cdot \mathbf{g}$.

To calculate the total ionic current, I_t, given by an ensemble of identical channels, we simply multiply the unitary current by the total number of channels, N_{C}:

$$I_t = (\mathbf{P}_t \cdot \mathbf{g}) \times (V - V_{\mathrm{R}}) \times N_{\mathrm{C}}. \tag{11}$$

Computationally, I_t can be efficiently calculated in two steps: first, calculate the state occupancies $\mathbf{P}_t$, using the recursive Eq. 8; then, calculate I_t as a function of $\mathbf{P}_t$, using Eq. 11. The time-invariant vector $\mathbf{g} \times (V - V_{\mathrm{R}}) \times N_{\mathrm{C}}$ needs to be recalculated only when the voltage changes.

As shown earlier in Eq. 6, for a time interval where conditions are constant (e.g., during a voltage step), $\mathbf{P}_t$ can be calculated as a function of some initial state occupancies, $\mathbf{P}_0$. For a typical voltage-clamp protocol, the $\mathbf{P}_0$ at the beginning of a sweep can be calculated as the equilibrium occupancies corresponding to the holding membrane potential. For this calculation to be accurate, the holding voltage should be maintained long enough to allow channels to reach equilibrium. If the protocol consists of a sequence of voltage step commands, the $\mathbf{P}_0$ of one step can be calculated as being equal to the $\mathbf{P}_t$ at the end of the previous step. This idea could also be applied to protocols where the command voltage varies continuously (e.g., during a "ramp"). In this case, a continuously varying episode can be approximated with a sequence of discrete steps of constant voltage. At the limit, each of these steps is as short as one acquisition sample.

Although very compact, Eq. 11 is not easy to interpret. To clarify its properties, we first replace $\mathbf{P}_t$ with its solution as a function of $\mathbf{P}_0$:

$$I_t = \left(\left(\mathbf{P}_0 \times \mathrm{e}^{\mathbf{Q}\times t}\right) \cdot \mathbf{g}\right) \times (V - V_\mathrm{R}) \times N_\mathrm{C}. \tag{12}$$

Then, we replace $\mathrm{e}^{\mathbf{Q}\times t}$ with its spectral expansion:

$$I_t = \left(\left(\mathbf{P}_0 \times \left(\sum_{k=0}^{N_\mathrm{S}-1} \mathbf{B}_k \times \mathrm{e}^{\lambda_k \times t}\right)\right) \cdot \mathbf{g}\right) \times (V - V_\mathrm{R}) \times N_\mathrm{C}. \tag{13}$$

We rearrange the terms and obtain:

$$I_t = \sum_{k=0}^{N_\mathrm{S}-1} \left(\left(\left(\mathbf{P}_0 \times \mathbf{B}_k\right) \cdot \mathbf{g}\right) \times (V - V_\mathrm{R}) \times N_\mathrm{C} \times \mathrm{e}^{\lambda_k \times t}\right). \tag{14}$$

In the sum above, the term corresponding to $k = 0$ is a constant, because λ_0 is always equal to zero. That term is actually the current that would be generated when channels reached equilibrium under those conditions. The explanation is that all eigenvalues are negative, except λ_0, which makes all terms in Eq. 14 become vanishingly small when t is sufficiently large, with the exception of the λ_0 term, which remains constant:

$$I_{t\to\infty} = \left(\left(\mathbf{P}_0 \times \mathbf{B}_0\right) \cdot \mathbf{g}\right) \times (V - V_\mathrm{R}) \times N_\mathrm{C}. \tag{15}$$

Since channels are at equilibrium when t is sufficiently large, one can recognize that the vector $(\mathbf{P}_0 \times \mathbf{B}_0)$ must be equal to $\mathbf{P}_\mathrm{eq}$. Therefore, the current flowing at equilibrium has the expression:

$$I_\mathrm{eq} = \left(\mathbf{P}_\mathrm{eq} \cdot \mathbf{g}\right) \times (V - V_\mathrm{R}) \times N_\mathrm{C}, \tag{16}$$

where:

$$\mathbf{P}_\mathrm{eq} = \mathbf{P}_0 \cdot \mathbf{B}_0. \tag{17}$$

In the equation above, $\mathbf{P}_0$ can be any arbitrary probability vector. With these results, the macroscopic current can be written as:

$$I_t = I_{\text{eq}} + \sum_{k=1}^{N_S-1} \left(I_k \times e^{\lambda_k \times t}\right), \tag{18}$$

where I_k is a scalar quantity with dimension of current:

$$I_k = ((\mathbf{P}_0 \times \mathbf{B}_k) \cdot \mathbf{g}) \times (V - V_R) \times N_C. \tag{19}$$

The eigenvalues, λ_k, can be replaced with time constants, τ_k, obtaining the final current equation:

$$I_t = I_{\text{eq}} + \sum_{k=1}^{N_S-1} \left(I_k \times e^{-t/\tau_k}\right), \tag{20}$$

where $\tau_k = -1/\lambda_k$. The macroscopic current described by Eq. 20 as a function of time is a sum of $N_S - 1$ exponentials, plus a constant term. Each exponential component is parameterized by a time constant τ_k and an amplitude I_k.

These results are general: any voltage-gated ion channel that has N_S high-occupancy conformations will in principle generate a macroscopic current with $N_S - 1$ exponentials, when subjected to a step change in membrane potential. This is illustrated in Fig. 1c for the simple three-state model: I_t (the red trace) has the expected profile of rise (activation) followed by decay (inactivation). This time course is a sum of two simple exponential components that vanish to zero with different time constants. In this particular case, I_{eq} is almost zero.

In the above equations, the macroscopic current I_t was calculated as a deterministic function of some initial conditions $\mathbf{P}_0$. However, one should keep in mind that I_t is the sum of many unitary currents, each generated by an individual ion channel that makes random transitions between states. These stochastic events at the single-channel level will make the macroscopic current a stochastic process as well. Therefore, the state occupancy at time t is a random quantity, characterized by a probability distribution [44, 46, 47, 50]. The state at time t can be statistically predicted from some known previous state, but the uncertainty of the prediction increases with the time from the reference point. In contrast, the initial state of a deterministic process can predict any future state with equal precision. The difference between stochastic and deterministic processes is illustrated in Fig. 1c, where the trajectory of the stochastically simulated macroscopic current (black trace) consistently diverges from the deterministically calculated current (red trace).

3.2 Protocol Design

As discussed above, an ion channel kinetic mechanism is fully characterized by its number of states, connectivity matrix of allowed state transitions, and rate constants quantifying transition frequency and voltage dependence. This information is encoded in single-channel or macroscopic voltage-clamp recordings as a stochastically fluctuating current, mixed with noise and artifacts. In single-channel recordings, the mean value of the current randomly jumps between two (or more) levels, corresponding to molecular transitions between conducting and non-conducting channel conformations. For example, in the single-channel trace shown in Fig. 1b, there happens to be four conductance changes over 50 ms. A channel with faster kinetics would result in more transitions per second, or, equivalently, in shorter average lifetimes in each state. Furthermore, a channel with greater voltage sensitivity would exhibit transition frequencies that change more substantially with voltage. Overall, the statistical properties of single-channel data can be analyzed with a variety of mathematical methods and computational algorithms, to extract the kinetic mechanism of the underlying ion channel [36–42, 51]

A macroscopic current is also a stochastic sequence of events, where individual channels randomly change state. Thus, stochastic fluctuations in the macroscopic current [52, 53] are also a potential source of information that could be used to extract the kinetic mechanism [44, 46, 47, 50]. However, these fluctuations are more difficult to separate from experimental noise and artifacts. First, depending on the recording technique, the experimental preparation, and the noise levels of the recording system, a change in the conductance state of a single channel may be very difficult or impossible to detect experimentally. Second, the frequency of transitions in the overall state of the ensemble is proportional to the total number of channels, and it may exceed the bandwidth of the recording system. For example, if the average single-channel transition frequency is 10 s^{-1}, an ensemble of 10,000 channels would exhibit 100,000 transitions per second, while the recording bandwidth may be smaller by an order of magnitude. Thus, information encoded in the magnitude and frequency of stochastic current fluctuations may be lost.

A generally more reliable source of information is the mean value of the macroscopic current, as a function of time and voltage. Even in this case, although the mean value can be easily extracted from noisy data, decoding the kinetic mechanism is far from being trivial. The main difficulty lies in the ambiguous relationship between the exponential parameters describing the macroscopic current (time constants τ_k and amplitudes I_k) and the kinetic parameters of the channel (rate constant factors k_{ij}^0 and k_{ij}^1). Following a step change in conditions, the overall state of the ensemble relaxes exponentially towards a new equilibrium. For a channel with N_S states, this relaxation process is quantified by a set of $2 \times N_S - 1$

parameters, as described by Eq. 20: $N_S - 1$ time constants τ_k, $N_S - 1$ amplitudes I_k, and the equilibrium current I_{eq}. Every one of these exponential parameters, including I_{eq}, is a mathematical function of all the rate constant parameters, and implicitly a function of membrane potential. Thus, while calculating the exponential parameters from the kinetic mechanism is straightforward, the inverse calculation is not. Furthermore, this also implies that no more than a maximum of $2 \times N_S - 1$ kinetic parameters can be extracted from a macroscopic current generated in response to a single voltage step. In reality, kinetic mechanisms may have more parameters than that. For example, the three-state model in Fig. 1a has eight kinetic parameters but only five exponential parameters. Even with this unrealistically simple model, it is clear that the kinetic parameters of the model cannot be unequivocally determined from the mean value of the macroscopic current, unless the voltage-clamp protocol is expanded to more than one voltage step.

A second difficulty is related to the theoretical and experimental observability of all the exponential components, given the limited resolution of the recording system. The idea is that, although each pair of exponential parameters (τ_k, I_k) depends on all the rate constants, fast or slow exponential components will be influenced most by similarly fast or slow rates, respectively. Then, if a certain exponential component is weakly represented in the data, some of the kinetic parameters will also be weakly determined. The contribution of an exponential component to the data, given a set of kinetic and conductance properties, depends on two factors. First, the amplitudes I_k depend on the initial state occupancies $\mathbf{P}_0$. Thus, depending on the voltage-clamp protocol, some components may have very small, or even zero, amplitude, and can be undetectable relative to the experimental resolution. Overall, a change in state occupancy is accompanied by a change in current only if the total occupancy of the conducting states changes. If this fraction doesn't change, or if the change is small relative to the resolution of the recording system, the mean current value will remain approximately constant, even though the properties of the stochastic fluctuations may change. Second, an exponential component can be observed experimentally only if the bandwidth of the recording system is adequate. Thus, very fast exponentials may be distorted or filtered out, while very slow components may not be detected in short protocols. A property worth remembering is that these exponential components vanish in order, from the smallest to the largest time constant. As a result, the fastest components will be affected the most by experimental artifacts associated with abrupt changes in the command voltage.

In conclusion, voltage-clamp protocols must be designed carefully and optimized to minimize these issues. Overall, the most important practical recommendation we can make is to design and apply as many types of stimuli as feasible, to force the channel to visit as many states as possible, which should result in well-observed exponential components and well-determined kinetic parameters. Ultimately, designing a good set of stimulation protocols is an iterative process, without a priori solutions. It may well happen that applying yet another protocol exposes a new behavior of the channel, which then needs to be investigated with new or refined stimuli.

An example of a typical set of voltage-clamp protocols is given in Fig. 2, as applied to recording whole-cell Nav currents from mammalian neurons in brain slices [54]. A minimum of four protocols is necessary to investigate the kinetic properties of Nav channels, as illustrated in Fig. 2a–d. Each of these voltage-clamp protocols forces the channel on a different state trajectory, thus exposing a different set of kinetic properties. For example, the protocol in Fig. 2a starts the channel in a state of deactivation and takes it through activation, opening, and inactivation. Several exponential components are well defined in the data, particularly the two time constants of inactivation. In contrast, the protocol in Fig. 2c starts the channel in a state of deactivation as well, but the channel is taken directly into inactivation, without opening. Two time constants of inactivation can also be detected in the data, but the exponential components have lower amplitude and thus are slightly less well defined.

With some of these protocols, the raw data can be used directly to determine the kinetic parameters (e.g., the time course of activation and inactivation in Fig. 2a). With others, the raw data are first processed to extract some empirical measure of state occupancy, which is then used to estimate kinetic parameters. Examples are the (pseudo) steady-state activation and inactivation in Fig. 2e, and the time course of recovery from inactivation and the subthreshold inactivation in Fig. 2f. Generally, the raw data are used directly when an exponential time course is experimentally observable in the macroscopic current. For example, when the channel activates, opens, and then inactivates (Fig. 2a). When state changes are not associated with changes in conductance, information is obtained from two-pulse protocols. For example, when the channel inactivates at membrane potentials where it cannot activate and open (Fig. 2c). In this case, the peak of the current is used as an empirical measure for the total occupancy of non-inactivated states available to generate current upon activation.

3.3 Experimental Artifacts

The data recorded in voltage-clamp experiments do not contain just the current of interest but are contaminated by a variety of artifacts, including other currents active in the preparation, experimental noise, voltage-clamp errors, etc. [17]. All of these artifacts will negatively affect fitting algorithms and can result in a distorted model. Although artifacts cannot be eliminated, they can be reduced to acceptable levels. Thus, the effects of random

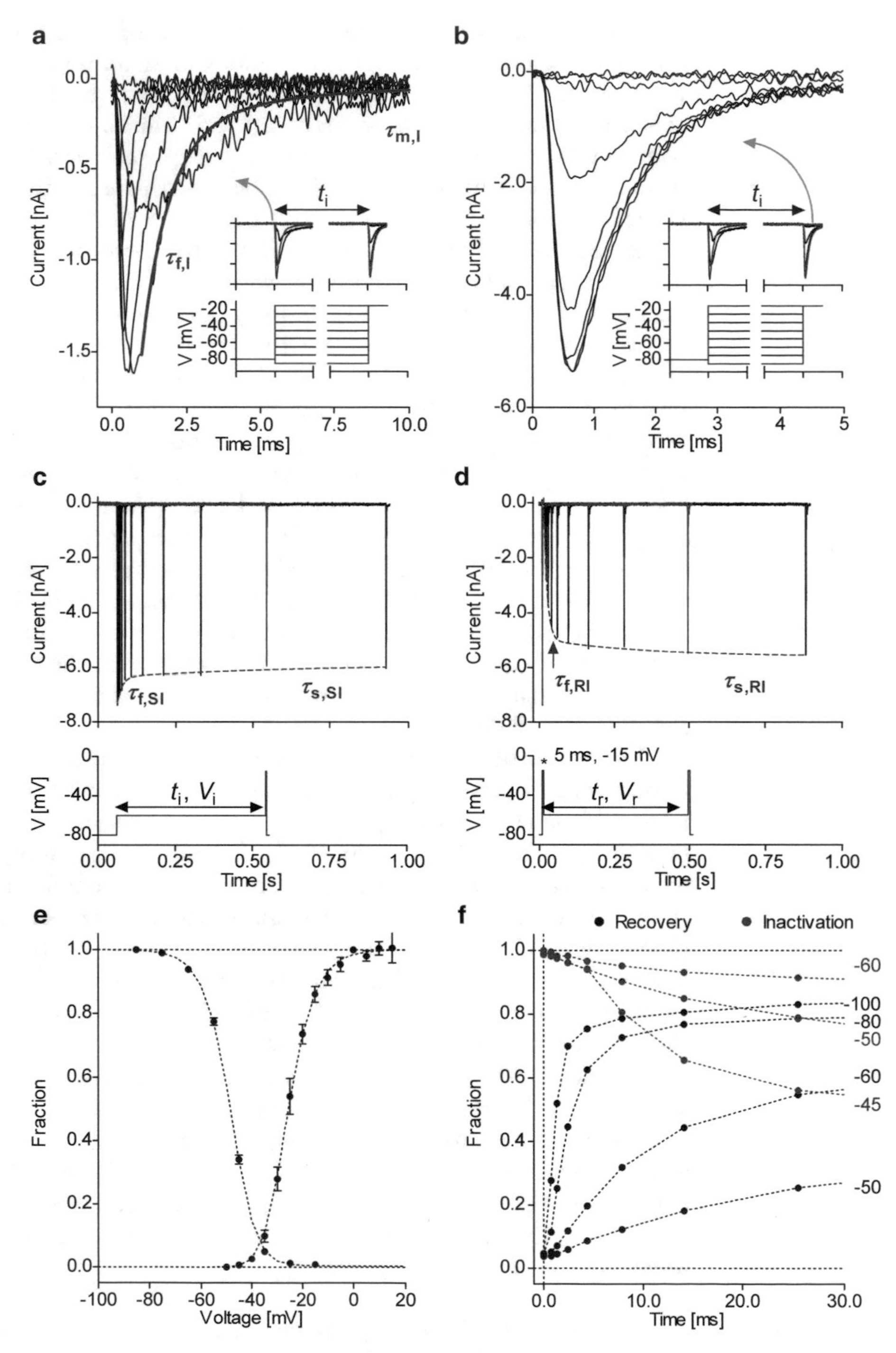

Fig. 2 Designing voltage-clamp protocols for Na^+ currents. To gather information about the kinetic mechanism, the channels are forced to make transitions between different sets of states, as follows: deactivated to open to inactivated (**a**), deactivated to inactivated to open (**b**), non-inactivated to inactivated (**c**), and inactivated to non-inactivated (**d**). Raw data are further processed to extract state occupancies as a function of time and voltage (**e** and **f**). Adapted from [54]

measurement noise, which lowers the precision of parameter estimates, can always be reduced by collecting more data, and generally are not an issue. Deterministic power line interference (50 or 60 Hz, or harmonics) can be easily removed online or offline. Uncompensated brief transients that occur when the command voltage changes abruptly can simply be excluded from the fit, provided that they don't overlap with significant channel activity. However, longer transients must be somehow separated from the signal. Voltage-clamp errors caused by incomplete compensation of the series resistance could be significant. However, the actual voltage at the membrane can be either measured directly in some techniques (e.g., two-electrode voltage clamp) or calculated from the measured series resistance and the recorded current. Then, a corrected version of the command voltage protocol can be constructed and used in the data fitting procedure. As explained above, an arbitrary voltage waveform can be approximated with a sequence of constant voltage steps. Another artifact is imperfect space clamp, which can occur when recording from neurons in vivo or in brain slices [55]. In this case, the current recorded from the soma can be contaminated with action potentials backpropagating from the axon [56], which usually escapes voltage-clamp control. Space-clamp errors can be reduced with a simple technique that selectively inactivates axonal sodium channels and thus makes the axon a passive compartment [57]. Finally, the bandwidth of the recording system should be sufficiently wide for the kinetics of the ion channel investigated. Initially, the cutoff frequency of the low-pass filter and the sampling rate of the digitizer should be set as high as possible to identify the fastest time constant present in the data. Then, acquisition parameters can be set to the values recommended by Nyquist's sampling theorem [58]. In many cases, the fastest time constant corresponds to channel activation or deactivation.

Even when these artifacts cannot be eliminated, in principle they can be parameterized and included into the fitting algorithm. Unfortunately, contamination with other ionic currents that are active in the preparation cannot be easily encoded in the algorithm. These currents are generally unknown quantities and cannot be compensated computationally. The ideal solution is to isolate the current of interest pharmacologically, with a very specific blocker. If this is available, the same protocols can be repeated under control conditions and with the blocker applied, and the two sets of data can be subtracted from each other, giving the current of interest. This subtraction not only eliminates all other currents, including leak, but will also remove capacitive transients. However, not all channels can be completely isolated by pharmacological subtraction. The solution is to reduce all other currents as much as possible. Some currents can be blocked pharmacologically, while others can be rendered inactive by exploiting their kinetic properties, particularly voltage dependence. Furthermore, the background leak currents and possibly other currents left unblocked can be subtracted using the P/n technique [59], which will also remove capacitive transients. However, one should be aware that the P/n method relies on the assumption that leak currents are linear with voltage and thus cannot subtract voltage-sensitive currents.

When designing protocols to isolate the current of interest, pharmacologically or via the P/n technique, one should keep in mind that these procedures will extend the total acquisition time, and recording parameters may change over time. For example, the seal resistance may deteriorate, causing an increase in leak, the level of solution in the bath may change and alter pipette capacitance and transients, and series resistance may fluctuate and change the amplitude of recorded currents. Generally, currents may run down over time. All these changes will distort the subtracted current. One other problem with subtraction methods is that uncompensated series resistance errors depend on the total current flowing. Then, if the total current takes significantly different values under control versus pharmacological block or P/n conditions, the actual voltage at the membrane will also differ. As a result, the subtracted current will be contaminated with some leftover current. Thus, even if a good blocker is available for the channel of interest, reducing all the other currents pharmacologically is still recommended. A similar artifact occurs when the current of interest is functionally coupled with other currents (e.g., Ca^{2+} – activated K^{+} currents). These would no longer be activated when the current of interest is blocked. Finally, one should keep in mind that when two random variables are subtracted from each other, their mean values subtract but their variances add. Thus, subtracting two sets of currents will result in a signal with greater noise, which would make it difficult to apply fitting methods that rely on the properties of current fluctuations.

4 Fitting the Data

The objective is to find the kinetic mechanism that best explains the experimental data, but also agrees with prior knowledge. As discussed, a kinetic mechanism is fully defined by its topology, given by the number of states and the connectivity matrix of possible transitions, and by the parameters quantifying rate constants and voltage dependence. Finding the topology and finding the parameters are generally approached as separate problems, and our focus here is on estimating parameters for a given topology. A computational procedure for finding the best parameters combines a parameter optimization engine (the optimizer) and an algorithm that calculates how well the model explains the data (the cost function, or the goodness of fit). The optimizer starts with a set of initial values and iteratively explores the parameter space, according to a defined search strategy, until it finds a set of parameters that maximizes the goodness of fit. For each point sampled in the parameter space, the optimizer calls the estimation algorithm to evaluate the goodness of fit. Typically, the optimizer is data- and model-blind, although it can be tweaked for a particular problem. A variety of general optimization algorithms that have been described in the literature [60] and are available in numerical libraries can be applied to ion channels. In contrast, the function that calculates the goodness of fit is very specialized and can be quite complicated.

4.1 Goodness of Fit

In general, how well the model explains the data can be defined in different ways, depending on the data and the model. For a deterministic time series contaminated by measurement noise, the goodness of fit is typically given by the sum of squared errors, S, between the experimental data points and the fit curve:

$$S = \sum_t (y_t - f_t(\mathbf{M}, \mathbf{K}))^2, \tag{21}$$

where y_t is the data point measured at time t and f_t $(\mathbf{M}, \mathbf{K})$ is the calculated value at time t, given a structural model $\mathbf{M}$ and a set of parameters $\mathbf{K}$. The *best fit* parameters are those that correspond to the lowest S value that could be reached.

Curve fitting is not the ideal method for data generated by ion channels or other stochastic processes. These data are not defined by a deterministic function that can be calculated for every time point. Instead, they are a stochastic sequence of channel states, contaminated with random measurement noise. Nevertheless, this stochastic sequence is generated by the ion channel according to a probability distribution, which is determined by the kinetic mechanism [32]. For single-channel data, this probability distribution can be used to calculate the likelihood of the data, L [36]:

$$L = p(\mathbf{Y}|\mathbf{M}, \mathbf{K}), \tag{22}$$

where p is the conditional probability of the data sequence $\mathbf{Y}$, given a model topology $\mathbf{M}$ and a set of parameters $\mathbf{K}$. The best fit parameters correspond to the highest L value that could be reached. In practice, the logarithm of the likelihood function is used instead of the likelihood itself, because L may reach intractably small or large values.

Ideally, macroscopic currents should also be approached as a stochastic process, using a likelihood-based goodness of fit. A variety of mathematical and computational algorithms have been designed to calculate the likelihood of a macroscopic current [44, 46, 47, 50], all making various approximations to speed up the computation. Ultimately, the fastest but theoretically the least accurate approximation that can be made is to completely ignore the stochastic nature of the macroscopic current. Essentially, the goodness of fit in this case is calculated as the sum of squared errors between the experimental data and the calculated macroscopic current, I_t:

$$S = \sum_t (y_t - I_t)^2. \quad (23)$$

This approximation is most suitable when the analyzed current is generated by many channels, when stochastic fluctuations are small relative to the mean value and comparable to the measurement noise. All other methods that make more accurate assumptions exploit in some way the fluctuations of the current, and theoretically should produce more accurate or more precise parameter estimates. However, as discussed above, many experimental data are not clean enough for noise analysis and the mean of the current may be the only reliable source of information. This condition describes well the macroscopic currents generated by voltage-gated ion channels in whole-cell patch-clamp experiments. In the following section, we assume that the goodness of fit is calculated as the sum of squared errors, S.

4.2 Computing the Cost Function

A variety of voltage-clamp protocols can be applied to determine the kinetic mechanism, as illustrated in Fig. 2. For each data set that is included in the fitting procedure, the estimation algorithm must calculate the goodness of fit. When the cost function is the sum of squared errors, S, then the mean current I_t must be calculated for every point in the data. Essentially, the algorithm must simulate a macroscopic current in response to the same voltage-clamp protocol as was used to record the data, given the set of parameters proposed by the optimizer in that iteration. For two-pulse protocols, such as those shown in Fig. 2, the simulated current sequence must also be processed in the same way as the experimental data. For example, the experimental recovery from inactivation (Fig. 2d) is calculated, as a function of time and recovery potential, as the ratio between the peak current obtained with the test pulse and the peak current obtained with the conditioning pulse. Although it might be tempting, it is a bad idea to calculate the theoretical recovery from inactivation using the sum of non-inactivated state occupancies. Instead, it should be calculated as for the experimental curve: first, simulate the response of the model to the two-pulse protocol, then, from this simulation, calculate the ratio of the two peaks.

I_t is most efficiently computed recursively, using Eq. 8 to calculate $\mathbf{P}_{t+1}$ from $\mathbf{P}_t$, where t and $t + 1$ refer to consecutive samples. The computation is initialized with $\mathbf{P}_0$, which is calculated as the equilibrium state occupancies that correspond to the holding voltage. The entire sequence of operations can be summarized as follows:

$$\begin{aligned}
\mathbf{P}_{\text{eq}} &= \mathbf{1} \cdot \mathbf{B}_{0, V_{\text{H}}}, \\
\mathbf{P}_0 &= \mathbf{P}_{\text{eq}}, \\
\mathbf{P}_1 &= \mathbf{P}_0 \times \mathbf{A}_{\delta t, V_1}, \\
&\cdots \\
\mathbf{P}_t &= \mathbf{P}_{t\text{-}1} \times \mathbf{A}_{\delta t, V_t}, \\
I_t &= \mathbf{P}_t \times \mathbf{I}_{V_t}, \\
S_t &= \left(y_t - I_t\right)^2,
\end{aligned} \tag{24}$$

where $\mathbf{1}$ is the normalized unity vector, with each element equal to $1/N_{\text{S}}$; $\mathbf{B}_{0, V_{\text{H}}}$ is the spectral matrix corresponding to λ_0 and calculated for a voltage equal to the holding potential, V_{H}; $\mathbf{A}_{\delta t, V_t}$ is the transition probability matrix calculated for δt and a voltage equal to the command potential at time t, V_t; S_t is the squared error at time t. Finally, $\mathbf{I}_{V_t}$ is a vector with dimension of current, with each element equal to the maximum current that would be generated if all the channels resided in that state:

$$\mathbf{I}_{V_t} = \mathbf{g} \times (V_t - V_{\text{R}}) \times N_{\text{C}}. \tag{25}$$

When the command voltage changes during a protocol sequence, the spectral matrix $\mathbf{B}_0$ and the transition probability matrix $\mathbf{A}_{\delta t}$ are replaced with the matrices calculated for that voltage value. As discussed, instead of the command voltage, one could use the actual voltage measured at the membrane, when available, or a voltage corrected for errors caused by the uncompensated series resistance.

The total sum of squared errors, S, is the sum of squared errors for all data points used in the analysis. S could be divided in components corresponding to individual data sets, each multiplied by a weighting factor:

$$S = \sum_i w_i \times S_i. \tag{26}$$

These weighting factors can be chosen empirically, to establish the relative contribution of each data component to the cost function.

4.3 Model Parameters

For a given model topology, the unknown parameters to be determined are the rate constant factors k_{ij}^0 and k_{ij}^1. However, the macroscopic current depends on additional quantities: the unitary conductance, $\mathbf{g}$, and the total number of channels, N_{C}. Calculating

I_t in the cost function requires these quantities. Normally, for a given ion channel type, the unitary conductance has the same value in every recording and can be estimated directly from single-channel data, or via noise analysis from macroscopic currents [53, 61]. Although N_C can also be determined through noise analysis, it takes a different value in each experimental preparation and it cannot always be known. One possibility is to make N_C a parameter to be estimated [44]. If the data used in the fit were obtained from multiple experiments, then there will be multiple N_C parameters, one for each preparation. The downside with this approach is a potentially large increase in the dimensionality of the parameter space, which would slow down the optimizer. Another possibility is to normalize the current in each data set to the local maximum value. The disadvantage in this case is a greater ambiguity in the estimated kinetic parameters. Furthermore, it can be problematic to analyze fluctuations. With some models, distinct combinations of rate constants and channel count values can generate the same macroscopic current in response to a voltage-clamp protocol [44]. However, in principle this ambiguity could be resolved by adding more protocols to the fit.

4.4 Prior Knowledge

Including prior knowledge in the model is necessary: although it restricts the freedom of the optimizer to search for parameters, it ensures that the parameters that best explain the new data are also in agreement with previous experiments and theory. Prior knowledge can be encoded in the topology of the model, but also in the kinetic parameter values. For example, the number and sequence of distinct conformational states that can be assumed by voltage sensors is defined by topology, while their degree of cooperativity is defined by rate constants. One could also encode in the model a hypothesis about the kinetic mechanism, and test it by fitting the data with this model. If the optimizer can find parameters that explain the data well, then the hypothesis is *potentially* correct, and vice versa.

A good example of including prior knowledge in the model is the study of inactivation in Nav channels by Kuo and Bean [62]. Phenomenologically, the steady-state and transient inactivation properties of Nav currents appear to be strongly voltage sensitive, as initially determined and modeled by Hodgkin and Huxley [21]. Yet, in more recent studies, very little electrical charge was detected to move during inactivation [26, 59, 63–66]. While Hodgkin and Huxley postulated that activation and inactivation are independent processes in Nav currents, Kuo and Bean proposed instead an allosteric coupling of inactivation to activation, which makes inactivation *apparently* but not *intrinsically* voltage dependent. Their model is shown in Fig. 3.

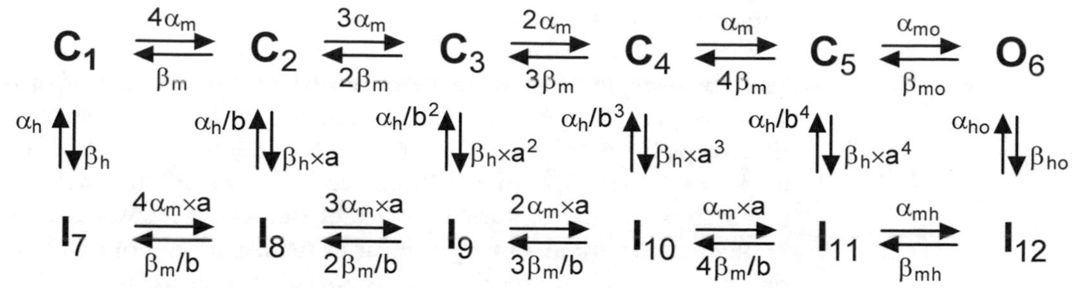

$$
\begin{array}{ccccccccccc}
C_1 & \underset{\beta_m}{\overset{4\alpha_m}{\rightleftharpoons}} & C_2 & \underset{2\beta_m}{\overset{3\alpha_m}{\rightleftharpoons}} & C_3 & \underset{3\beta_m}{\overset{2\alpha_m}{\rightleftharpoons}} & C_4 & \underset{4\beta_m}{\overset{\alpha_m}{\rightleftharpoons}} & C_5 & \underset{\beta_{mo}}{\overset{\alpha_{mo}}{\rightleftharpoons}} & O_6 \\
\alpha_h \uparrow\downarrow \beta_h & & \alpha_h/b \uparrow\downarrow \beta_h \times a & & \alpha_h/b^2 \uparrow\downarrow \beta_h \times a^2 & & \alpha_h/b^3 \uparrow\downarrow \beta_h \times a^3 & & \alpha_h/b^4 \uparrow\downarrow \beta_h \times a^4 & & \alpha_{ho} \uparrow\downarrow \beta_{ho} \\
I_7 & \underset{\beta_m/b}{\overset{4\alpha_m \times a}{\rightleftharpoons}} & I_8 & \underset{2\beta_m/b}{\overset{3\alpha_m \times a}{\rightleftharpoons}} & I_9 & \underset{3\beta_m/b}{\overset{2\alpha_m \times a}{\rightleftharpoons}} & I_{10} & \underset{4\beta_m/b}{\overset{\alpha_m \times a}{\rightleftharpoons}} & I_{11} & \underset{\beta_{mh}}{\overset{\alpha_{mh}}{\rightleftharpoons}} & I_{12}
\end{array}
$$

Fig. 3 Representing ion channel kinetic mechanisms with state models. This model has been formulated for Nav channels [62]. States $C_1 \ldots C_5$ and O_6 represent the non-inactivated channel, whereas $I_7 \ldots I_{12}$ are inactivated states. O_6 is the only conducting (open) state. The pathway either from C_1 to C_5 or from I_7 to I_{12} corresponds to the activation of the four voltage sensors, assumed to be equal and independent. This assumption is denoted by the 4:3:2:1 or 1:2:3:4 ratios in the factors multiplying the α_m or β_m rates. The C_5 to O_6 transition corresponds to the opening of the channel. The model allows the channel to inactivate without opening, from any of the closed states $C_1 \ldots C_5$. However, the channel is most likely to inactivate from the open state O_6 or when more voltage sensors are activated (e.g., from C_5). This property is implemented via the allosteric factors *a* and *b*, which control the equilibrium and transition frequency between closed and inactivated states. The $I_7 \ldots I_{11}$ transition rates also include the allosteric factors *a* and *b*, to satisfy microscopic reversibility. When the channel reaches the open state O_6 upon membrane depolarization, it is quickly and completely absorbed into the inactivated states I_{12} and I_{11}. Inactivation from closed states happens more slowly, which gives the channel a chance to open before it inactivates during an action potential [54]. The only rates with significant voltage sensitivity are α_m and β_m. The allosteric coupling between activation and inactivation can explain the apparent voltage sensitivity of inactivation but also the minimal electrical charge detected to move within the channel during inactivation

4.5 Model Constraints and Free Parameters

Knowledge—or hypotheses—about rate constants can be implemented either as a set of mathematical constraints or as a penalty term added to the cost function. Generally, constraints are defined as an invertible transformation, *c*, between a set of *model* parameters **K** and a set of *free* parameters **F**:

$$\mathbf{F} = c(\mathbf{K}), \tag{27}$$

$$\mathbf{K} = c^{-1}(\mathbf{F}). \tag{28}$$

The model is defined by the **K** parameters, which are subject to the constraints implemented by *c*. In contrast, the optimizer is model-blind and operates with the **F** parameters, which are “free” to take any value in the $(-\infty, +\infty)$ range. The optimizer searches in the free parameter space and finds a set $\mathbf{F}^*$ that is converted, via the c^{-1} transformation, into a set $\mathbf{K}^*$ that best explains the data. The search is initialized with a set of free parameters $\mathbf{F}^0$, obtained via the *c* transformation from an initial set of model parameters $\mathbf{K}^0$.

If no constraints are formulated, *c* is the identity transformation. In this case, the number of free parameters is equal to the number of model parameters. However, at least one type of constraint has to be applied, which is to keep the rate constant factors k_{ij}^0 greater than zero. This constraint can be implemented easily by using ln k_{ij}^0 as the free variable, which can take any value in the $(-\infty, +\infty)$ range, but restricts k_{ij}^0 to $(0, +\infty)$. This transformation keeps the numbers of free and model parameters equal. However, any other type of constraint reduces the number of free parameters by one. A variety of useful model constraints can be implemented with c as a simple linear transformation. For example, one can constrain a rate to have a constant value, or two rates to have a constant product, or one can enforce microscopic reversibility. Detailed explanations of how to implement linear constraints are available in the literature [39, 44, 67].

Sometimes, a model parameter is allowed to take any value in a defined range, as implemented via the c transformation, but some of these values may be considered "better" than others. For example, a rate constant estimated at 100,000 s^{-1} by the (model-blind) optimizer, although physically acceptable, could be considered quite unlikely when obtained from data sampled at 1 kHz. In this case, a penalty term can be factored into the cost function S^*:

$$S^* = S \times p(\mathbf{K}), \tag{29}$$

where the penalty $p(\mathbf{K})$ is a function of model parameters $\mathbf{K}$. In principle, constraints implemented via the c transformation can also be formulated as a penalty, by making $p(\mathbf{K})$ equal to one when $\mathbf{K}$ satisfies the constraints, or equal to a very large number when not. However, the advantage of having a reduced number of free parameters would be lost, and defining $p(\mathbf{K})$ is not trivial.

Fundamentally, any assumption about the kinetic mechanism results in adding one or more computational constraints to the model, and thus reduces the number of free parameters. While having fewer parameters makes the fitting easier, the model will also have less flexibility in explaining the data. For example, the model proposed by Kuo and Bean [62], although making a radical departure from the classical dogma, makes the simplifying assumption that the channel has identical and independent voltage sensors, which dramatically limits the number of free parameters. This assumption has later been invalidated by clever biophysical experiments [68]. Another study relaxed the constraint that the inactivation rates in the Kuo and Bean model are voltage independent and found that a small but finite charge can explain their data better [54]. Ideally, one should always start with a well-constrained model that has relatively few free parameters and gradually remove the constraints until the fit no longer improves.

4.6 Searching for Best Parameters

Let's take an intuitive look at how the optimizer searches for parameters. If we had a model that has only two parameters, we could imagine a 2D surface that represents on the z axis the error of the model relative to the data, as a function of the two parameter values represented on the x and y axes. The greater the error, the greater the z value. Hopefully, somewhere on this surface there is a single point where the error is the lowest. That is the solution that the optimizer has to find, corresponding to the set of best parameters. If we could apply a grid over the parameter space and calculate the error at every node, we could simply identify the parameter values where the error is lowest. As a further refinement, we could apply a smaller and finer grid on that point and improve the estimate.

Unfortunately, a grid search is prohibitive, because the number of free parameters can be large and the cost function can be computationally expensive. For example, for only two parameters, a 100×100 search grid requires 10,000 evaluations of the cost function. For some of these parameter values, the cost function may even be impossible to calculate, due to numerical instability, particularly for large k_{ij}^{1} values. For ten free parameters, which is not an unusually large number, we would need 100^{10} evaluations. Assuming an optimistic 1 ms computation time per cost function evaluation, this search would take a long, long, long time. Essentially, the optimizer is in the situation of a tourist trapped in a multidimensional universe, having to find the best restaurant in the city, in complete darkness and without map or smart phone, just guided by smell.

Clearly, the optimizer must use a clever and more efficient strategy. A solution is to mimic the effects of gravity. If we place a ball on our error surface, the ball will start rolling downhill, eventually settling at the bottom. Like our tourist, the ball is in darkness: it does not see the whole map, but it's simply taken by gravity down the local gradient. An optimization algorithm can use the same search strategy, following the error function down its gradient in the multidimensional parameter space. Because of noise, the error is never zero at the bottom of the surface. Instead, the search is terminated when the gradient is zero for all parameters.

We had good experience with the Davidon-Fletcher-Powell method [69], implemented in code as *dfpmin* [70]. Compared to other search strategies, such as *simplex* [71], *dfpmin* is very quick and efficient. As with any gradient-based method, *dfpmin* requires the gradients of the cost function with respect to each parameter. In some cases, the gradients can be calculated analytically [44]. When not, the gradients could be approximated numerically by evaluating the cost function at two points in the parameter space that are separated by a very small distance, for each parameter. Due to potential numerical errors in the calculation, this distance must be chosen carefully, to be sure that the cost function does actually change over that small distance.

4.7 Interpreting Fitting Results

An example of fitting macroscopic currents with an ion channel kinetic mechanism is shown in Fig. 4. Although the fit is clearly very good, the best parameters found by the optimizer should be taken with a grain of salt. First, it's possible that the parameters are not the very best, but only a local solution. Following the smell of food, our tourist will eventually find a restaurant, yet a much better one may be just around the corner. How does he find the very best place

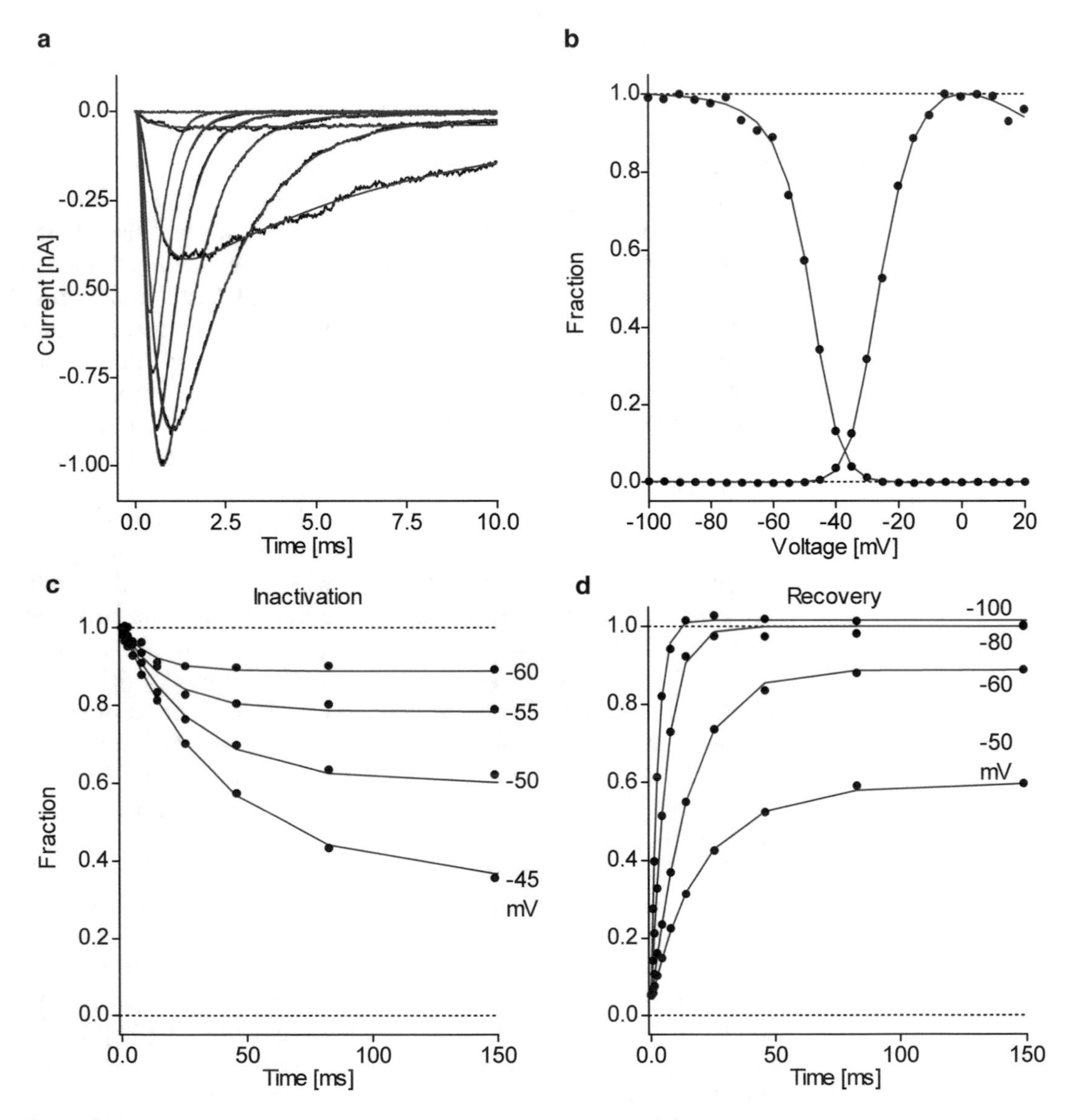

Fig. 4 Fitting the data. This is an example where macroscopic data obtained with several voltage-clamp protocols, as discussed in Fig. 2, were pooled together and fitted with the kinetic mechanism shown in Fig. 3, using a computational algorithm that minimized the sum of squared errors. The response of the model to the same voltage-clamp protocols as used to record the data is represented by the *red* trace, which corresponds to the best parameters found by the optimizer. Adapted from [54]

to eat, without trying them all? Indeed, finding the global minimum is a difficult problem in optimization [72]. A poor man's global search strategy is simply to restart the optimizer at different points in the parameter space and take the overall best. While there is no theoretical guarantee, the more restarting points tried, the more likely it is that the solution is truly global. Alternatively, one could use algorithms specifically designed for global search, which are slower but more exhaustively explore the parameter space and can even search across model topologies [73, 74].

A second issue to consider is the theoretical (a priori) identifiability of the model, which has two related aspects [75]. First, there may be distinct model topologies that explain the data equally well [76, 77]. Second, for a given model structure, there may be multiple parameter sets that are equally good [44], either as a continuum or as discrete points in the parameter space. As a result, even when the optimizer finds the global best, there may be other solutions that are just as good (although the corresponding physical models may not be equally plausible). The theoretical identifiability of a model does not depend on the quality of experimental data. Instead, it depends on the voltage-clamp protocols that were included in the analysis and on the mathematical criteria that were used to calculate the goodness of fit. Essentially, both of these narrow the range of equivalent solutions in the parameter space, which can lead to unique solutions. For example, if we tried to fit a stationary macroscopic current using the sum of squared errors as the goodness of fit, there would be a large continuum of solutions that could all explain equally well the data. However, if we then switched to a maximum likelihood method, the range of equivalent solutions in the parameter space would be narrower, because a solution must explain not only the mean value of the current but also the properties of fluctuations. If we further add a voltage step protocol, the range of equivalent parameters would narrow even more. Finally, including prior knowledge implemented as model constraints or as penalty would further restrict parameter space and increase model identifiability.

How do we know when a parameter solution is unique? A simple empirical technique can be used to determine whether a continuum of equivalent solutions exists. Starting from the solution found by the optimizer, one parameter can be slightly perturbed, and then constrained to a constant value. The resulting parameter set should give a higher error. Then, we restart the optimizer and test whether it is able to reach the same goodness of fit as before, by adjusting the other parameters to compensate for the change in the one that was constrained. If the optimizer returns to the same goodness of fit, though with a different set of parameters, we have proof that a continuum of solutions exists around that point. In this case, one could either add constraints to make the model simpler or add stimulation protocols to make the data more informative.

Another important issue is the practical (*a posteriori*) identifiability of the model, which depends on the properties of the data used in analysis [78]. Each time a new data set is recorded and analyzed, the parameter estimates and the cost function will be different. Because of stochastic fluctuations and experimental noise and artifacts, these quantities are not deterministic, but vary from trial to trial. Moreover, even a good model would never give a perfect fit. The variance of a parameter estimate depends on how much information is contained in the data about that parameter, relative to noise and fluctuation levels. A parameter that has large variance can take a broad range of values without changing the fit significantly, and thus cannot be trusted much. Ultimately, it comes down to how many times the channel visits a state, and how long-lived that state is. For states that are rarely visited or are very long-lived, there will be little information in the data about the transitions connecting that state to others. For example, in the limit case of a single-channel recording, a state simply cannot be identified if it is never visited in a data set. In general, certain features of the model that are theoretically observable in practice may be buried in the noise and hidden or distorted by artifacts. To reduce parameter variance, one can include more data in the analysis, but can also optimize the voltage-clamp protocols. The variance in the cost function is a factor to consider when different models are tested. If two models are comparable, their cost function probability distributions may overlap, which means that, statistically, the wrong model will sometimes give a better fit than the correct model [38, 79].

What if the best fit is still bad? This could be an indication that the topology of the model is wrong or that the optimization starting point was not appropriate and the optimizer has been trapped in a local minimum. As a solution, the optimization can be restarted with different initial values. If this doesn't improve the fit, one could try different model topologies. One should be aware, however, that when the data included in the fit are collected with multiple voltage-clamp protocols, it becomes more difficult for the optimizer to explain all the data collectively. To improve the fit, there are simple changes that can be made to a model, such as inserting a state or connecting or disconnecting two states. Parameter estimates should always be inspected. For example, if the optimizer estimates a rate constant at one million per second, that could indicate something wrong with the model or with the data. Or, if rate constants that lead to an end state are very small while the opposite rates are very large, this could indicate a lack of evidence in the data for that state. Another example is when the rate constants connecting two states have very large values, which could indicate that the two states are in fact just one.

5 Testing Models in Live Neurons

As shown in the previous section, it is possible to find an ion channel model that explains voltage-clamp data well and gives insight into the biophysical mechanism of the channel. However, in an excitable cell, there are many ion channel types that work together to generate specific patterns of firing activity. A cell is a complex system where multiple components interact nonlinearly [80]. In contrast, voltage-clamp experiments isolate the channel of interest from this system and test it with predefined voltage waveforms. It is quite possible that some features of the kinetic mechanism that are critically important to the function of the cell may not be revealed in the voltage-clamp data and may not be captured by the model. Ideally, the model should also be tested functionally, in a cellular context.

A powerful tool for studying the function of voltage-gated ion channels in live neurons is dynamic clamp [81–83]. The principle is to pharmacologically block the channel of interest and then functionally replace it with an injected current, dynamically calculated on the basis of a kinetic model [18]. As a first-order approximation, where we ignore the potential regulatory function of the permeant ions, Ca^{2+} in particular [84], the neuron makes no distinction between the native current and the model-based current, which are not necessarily carried by the same ions. Then, if the model is accurate, the neuron would exhibit the same firing pattern as with the actual channel. The sensitivity of the firing pattern to channel properties and the contribution of that particular current to spiking can be easily studied by varying the properties of the model and manipulating the model-based current in real time. The major advantage of this hybrid experimental-computational approach is that a channel can be investigated within a live cell without any knowledge about other conductances or cell properties.

5.1 Solving the Model in Real Time

Dynamic clamp can be understood within the context of a cellular model. To make it easier to explain the concepts, we make several simplifying assumptions: (1) besides voltage-independent leak channels, the neuron contains only Nav and Kv channels, with kinetic mechanisms described by Markov models; (2) the neuron has a single compartment and the membrane is isopotential; and (3) the model corresponds to an ideal whole-cell recording (zero access resistance, no pipette capacitance, etc.) [17]. The state of the model as a function of time is completely described by three variables: the membrane potential V, and the state occupancies of the Nav and Kv channels, $\mathbf{P}_{\mathrm{Na}}$ and $\mathbf{P}_{\mathrm{K}}$. These state variables evolve in time according to the following ordinary differential equations:

$$C\frac{\mathrm{d}V}{\mathrm{d}t} = -(I_{\mathrm{Na}} + I_{\mathrm{K}} + I_{\mathrm{leak}}) + I_{\mathrm{app}}, \tag{30}$$

$$\frac{d\mathbf{P}_K}{dt} = \mathbf{P}_K \times \mathbf{Q}_K, \quad (31)$$

$$\frac{d\mathbf{P}_{Na}}{dt} = \mathbf{P}_{Na} \times \mathbf{Q}_{Na}, \quad (32)$$

where I_{app} is the current injected into the neuron through the patch-clamp pipette. To run a computer simulation of our model neuron, we would have to integrate these equations with an ODE solver. A real neuron "integrates" a similar set of differential equations, just more complex, to account for multiple cellular compartments, ion channel stochasticity, etc.

In voltage clamp, I_{app} is in principle equal to the sum of all ionic currents, so as to keep V equal to the command voltage and dV/dt equal to a predefined value (e.g., zero for a constant voltage step, or a finite value for a voltage ramp). In this sense, I_{app} becomes a measure of the total ionic currents active in the cell. In current clamp, I_{app} is typically used to test the firing properties of a neuron under a range of conditions. For example, I_{app} can be a constant value to bias the membrane potential or it can be a predefined waveform that mimics excitatory or inhibitory synaptic input. In dynamic clamp, I_{app} is not predefined. Instead, I_{app} is calculated in real time, as a function of the membrane potential V and some other quantities.

How can dynamic clamp be used to test a voltage-gated ion channel model in a live neuron? Let's consider the case of Nav channels. First, we pharmacologically block the channel, with TTX in this case. As the Nav current was eliminated, the equations "integrated" by the cell simplify to just two:

$$C\frac{dV}{dt} = -(I_K + I_{leak}) + I_{app}, \quad (33)$$

$$\frac{d\mathbf{P}_K}{dt} = \mathbf{P}_K \times \mathbf{Q}_K. \quad (34)$$

Next, we replace the blocked current with a current generated by a Nav model that is solved on the computer. Effectively, we now have a hybrid biological-computational model that has the same set of ODEs, but two equations are "integrated" by the cell, and one is integrated on the computer:

$$C\frac{dV}{dt} = -(I_K + I_{leak}) + I_{app}, \ \text{integrated by the neuron} \quad (35)$$

$$\frac{d\mathbf{P}_K}{dt} = \mathbf{P}_K \times \mathbf{Q}_K, \ \text{integrated by the neuron} \quad (36)$$

$$\frac{\mathrm{d}\mathbf{P}_{\mathrm{Na}}}{\mathrm{d}t} = \mathbf{P}_{\mathrm{Na}} \times \mathbf{Q}_{\mathrm{Na}}, \text{ integrated on the computer} \quad (37)$$

where the injected current, I_{app}, is now equal to the negative current generated by the Nav model, $-I_{\mathrm{Na}}^{\mathrm{C}}$, plus a constant component, I_{inj}, that can be used to apply current steps or for other functions:

$$I_{\mathrm{app}} = -I_{\mathrm{Na}}^{\mathrm{C}} + I_{\mathrm{inj}}. \quad (38)$$

The channel model is solved on the computer over discrete time steps, using the recursive method:

$$\mathbf{P}_{\mathrm{Na},\ t+\delta t} = \mathbf{P}_{\mathrm{Na},t} \times \mathbf{A}_{\mathrm{Na,V}}, \quad (39)$$

where $\mathbf{A}_{\mathrm{Na,V}}$ is the transition probability matrix calculated for a given membrane potential V. $\mathbf{A}_{\mathrm{Na,V}}$ can be precalculated over a voltage range (e.g., −100 to +100 mV, every 0.1 mV) and stored in a look-up table. As illustrated in Fig. 5, the model is solved in a real-time computational loop, where every iteration corresponds to one integration step. For each iteration, V is read from the amplifier through the digital acquisition card (DAQ). Depending on V, the appropriate $\mathbf{A}$ matrix is selected from the look-up table and used to update $\mathbf{P}_{\mathrm{Na}}$. Then, $I_{\mathrm{Na}}^{\mathrm{C}}$ is recalculated and injected into the neuron via I_{app}. This loop must execute fast enough so that the voltage at the membrane does not change significantly within one iteration, which would invalidate both the selected $\mathbf{A}_{\mathrm{Na,V}}$ and the injected

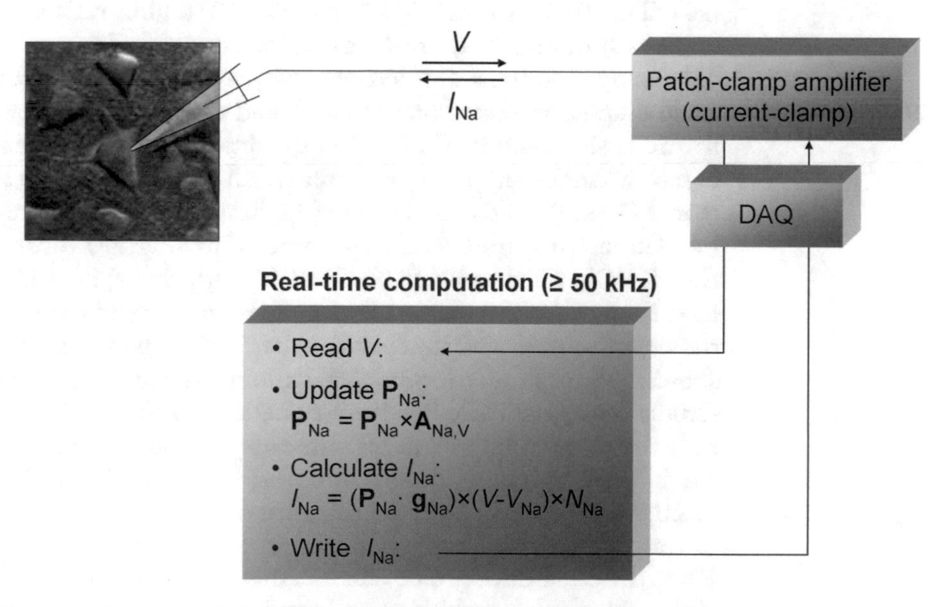

Fig. 5 Testing ion channel models in live neurons with dynamic clamp. As illustrated here, a Nav current is blocked with TTX and replaced with a model-based current, which is calculated in real time on the basis of a model and injected into the neuron via the patch-clamp pipette [18]

current I_{Na}^{C}. The update rate should match the maximum rate of voltage change, which is typically the rising phase of the action potential. An update every 20 μs (50 kHz), or faster, is generally adequate. Once an update interval is chosen, every iteration of the loop should be completed precisely within that time. To ensure predictable time steps with minimum variability, the code should run with real-time priority on the computer.

5.2 Equipment and Software

While dynamic clamp can be performed under a variety of electrophysiology paradigms, we focus here on whole-cell patch-clamp experiments in neurons [17]. Thus, in addition to the equipment and software used for patch clamp, one also needs a digital acquisition card, a dedicated computer, and specialized software for real-time computation. Ideally, the patch-clamp amplifier should feature "true" current clamp. We had good results with HEKA's EPC 9 and 10 amplifiers, as well as with Molecular Devices' Multiclamp 700B. EPC 10 is more convenient, because it allows summation of external and internal current in current-clamp mode. In contrast, the Multiclamp amplifier has only one input connection for applied current, which is normally used by the external DigiData digitizer. A solution in this case is to use an electronic summation circuit or a mechanical switch box. Although we don't have first-hand experience with other instruments, there are several commercially available patch-clamp amplifiers that feature true current clamp, e.g., those made by A-M Systems, NPI, and Warner Instruments.

Although some patch-clamp amplifiers already include an internal (EPC 10) or external (Multiclamp 700B) digitizer, these are not necessarily optimized for real-time feedback acquisition, where, in a very short time (tens of microseconds), a sample is read from the analog input, processed on the CPU, and another sample is written to the analog output. With a few exceptions (e.g., the hardware-based dynamic-clamp device commercialized by Cambridge Electronic Design), all dynamic-clamp applications use digitizers made by National Instruments. At the time of writing, we recommend the NI PCIe-6351 or NI PCIe-6361 (slightly faster) boards, which have been optimized for very low latency. One should be aware that the manufacturer typically specifies the maximum rate for buffered acquisition, not for real-time applications. Transferring one single sample across the PCIe bus has a finite latency that limits dramatically the throughput rate in real-time acquisition. For example, the maximum rate that we could obtain with a NI PCIe-6361 card is ~220,000 acquisition cycles per second, even though the board can acquire two million samples per second in buffered mode. Nevertheless, a throughput rate like this is excellent, being comparable with the bandwidth of the patch-clamp amplifier in current clamp.

Historically, the first dynamic-clamp programs were coded in some flavor of real-time Linux [85–87]. At the time, obtaining acceptable real-time performance under the MS Windows operating system—or any other non-real-time OS—was simply not possible. On such a system, user programs can be interrupted at random by other programs or by the operating system itself. Another limitation at the time was the driver provided by National Instruments for programming their boards, which was incredibly slow for real-time applications (~1,000 cycles per second, according to our tests). However, the situation has completely changed over the last ten years, with the development by National Instruments of new digitizers and optimized drivers, with the advent of multicore processors and an improved PCIe bus, and with the general increase in CPU speeds. Today, dynamic-clamp software can be run in MS Windows with excellent real-time performance, on par with what is achieved under real-time Linux.

Dynamic-clamp programs are available for both real-time Linux [88] and MS Windows [18, 89–91]. For the more biophysically inclined user, we recommend our own implementation of dynamic clamp in the QuB software, which runs under MS Windows (http://milesculabs.biology.missouri.edu/QuB). The major advantages are integration with a variety of ion channel modeling algorithms, a powerful scripting language for customized models and protocols, and sophisticated methods for solving Markov models of ion channels, deterministically or stochastically. The only method that can be used to solve large Markov models accurately is the matrix method described in this chapter, which is available in our software. Once a few quantities are precalculated and stored in look-up tables, very large Markov models can be solved using only vector–matrix multiplications, which can be executed very quickly on modern CPUs or on graphics processors (GPUs). For example, we were able to run models with as many as 26 states at 50 kHz or faster [18]. The matrix method is also very stable and accurate, even with long sampling intervals, which is generally not the case with methods that rely exclusively on ODE solvers to advance the state probabilities. In particular, integration with the Euler method, which is practically the only one that is fast enough for real-time applications, is bound to fail with even small Markov models [18].

In principle, any desktop computer can be used for dynamic clamp. However, for high-performance applications (large models and high-throughput rates), we recommend a fast computer that is used exclusively to run the dynamic-clamp engine. We had the best results with multicore Intel Xeon CPUs, installed in dual-processor server-grade systems. For example, the computer that we currently use in the lab has two Intel Xeon E5-2667 v2 8-core processors, clocking at 3.3 GHz, and runs Windows 7 Pro 64-bit. Our system was built by ASL, Inc., but many other computer integrators sell configurable systems. Of all the components, the most critical are the CPU and the motherboard.

5.3 Preparing and Running a Dynamic-Clamp Experiment

Setting up a dynamic-clamp experiment involves a few steps. First, the voltage monitor output of the patch-clamp amplifier should be connected to one of the analog inputs of the National Instruments digitizer, while the external current input of the amplifier should be connected to one of the analog outputs of the digitizer. Next, one needs to zero the offsets and calibrate the scaling factors between the amplifier and digitizer, for both input and output. The calibration procedure depends on the specific dynamic-clamp software but the idea is to make sure that the membrane voltage value read into the dynamic-clamp software is the same as the value reported by the patch-clamp amplifier. Likewise, the external current reported by the amplifier should match the current sent by the dynamic-clamp software. In our experience, there are always slight offsets of a few mV in membrane potential and a few pA in injected current, between the amplifier and the digitizer. These offsets must be compensated for in the software. In particular, one should be careful that the amplifier receives no unwanted external current when I_{app} is equal to zero, as even a small current of a few pA can alter the firing pattern of a neuron. Most amplifiers have adjustable gain in current clamp (e.g., 1 pA/mV). The smallest gain should be selected that still allows the injection of the largest current that might be predicted by the model. For example, a model-based Nav current ranges from a few pA in the interspike interval, small but sufficient to influence neuronal firing properties, to several nA during an action potential.

The pipette capacitance should be reduced as much as possible by coating with Sylgard or other agents. In our experience, the residual capacitance should be no more than 5–6 pF, otherwise ringing may occur in dynamic clamp when large currents are injected, particularly with Nav currents during action potentials. Once a patch is obtained, the typical artifact estimation and compensation procedures should be applied for series resistance and pipette capacitance, as well as for membrane capacitance. Then, upon switching to current clamp, the pipette capacitance compensation should be slightly reduced (10–20 %) to avoid ringing, while series resistance should be compensated 100 %.

An example of a dynamic-clamp experiment is shown in Fig. 6, adapted from a previous study on serotonergic Raphé neurons in the brainstem [54]. In that study, the voltage-gated sodium current was investigated with voltage clamp to determine its kinetic mechanism and tested with dynamic clamp to determine its contribution to the neuronal firing. These neurons exhibit a pattern of spontaneous spiking with low frequency (1–5 Hz) and broad action potentials (4–5 ms), as shown in Fig. 6a, top traces. Other properties that are visible during a current step injection are a slow reduction in spiking frequency, a broadening of the action potential shape, and a reduction in spike height. A slow kinetic process was also observed in the voltage-clamp data [54, 92–94], which has

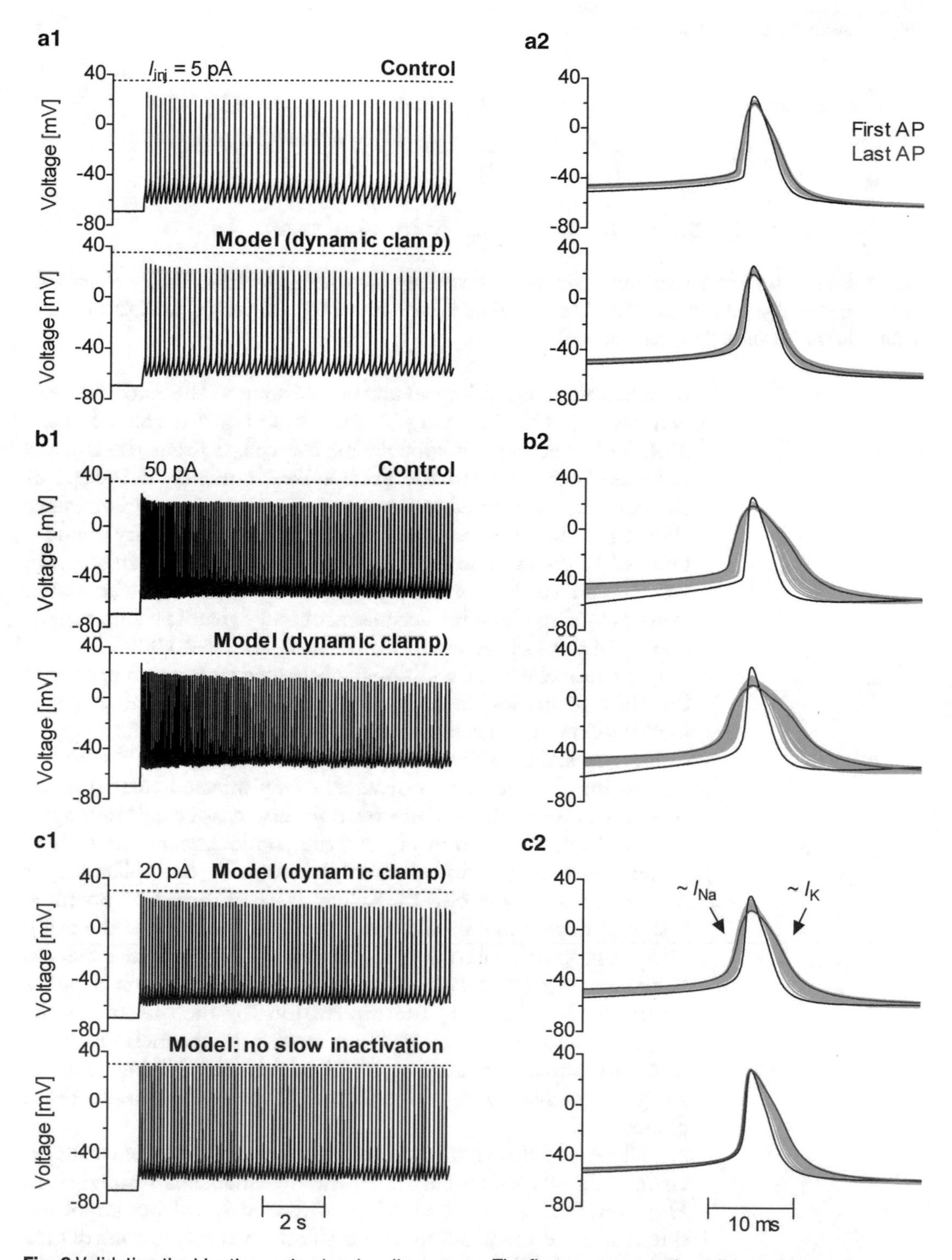

Fig. 6 Validating the kinetic mechanism in a live neuron. The figure compares the spiking patterns of a neuron under control conditions and with the Nav current blocked with TTX and replaced with a model-based current injected with dynamic clamp. A Nav model based on the kinetic mechanism shown in Fig. 7 was able to explain well the firing properties, including the slow adaptation in action potential shape and frequency (**a** and **b**, *lower traces*, and **c**, *upper traces*). In contrast, the Nav model shown in Fig. 3 could not reproduce the slow adaptation (**c**, *lower traces*). Adapted from [54]

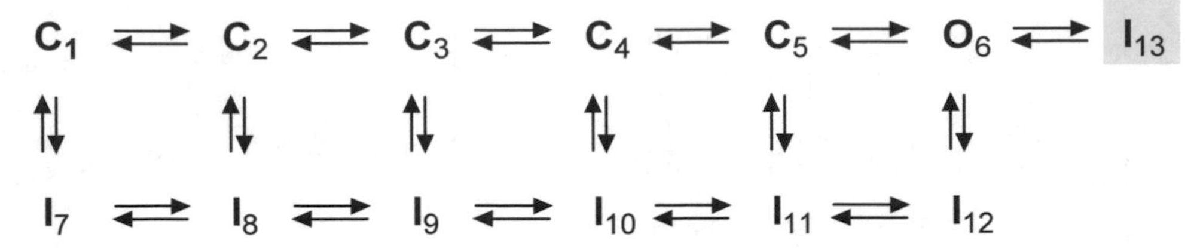

Fig. 7 A Nav channel kinetic mechanism obtained from voltage-clamp data and validated in live neurons. An inactivated state (I_{13}) with special properties was added to the Kuo and Bean model [62] to explain voltage-clamp data and neuronal firing behavior [54]

been explained by adding an inactivated state to the Kuo and Bean Nav model, as shown in Fig. 7. After blocking Nav channels with TTX and injecting this model into the cell, a firing pattern was obtained that exhibited the same spike frequency and shape as the control, maintained under a range of injected current values (Fig. 6a, b, lower traces). For this experiment, a unitary conductance of 10 pS was used and the total number of Nav channels, N_C, was chosen so as to match the maximum slope of the action potential rising phase between control and dynamic-clamp experiments. In most of the examined cells, N_C was $\approx$ 20,000.

The model with an additional inactivated state also reproduced the slow adaptation in frequency and action potential shape. A logical follow-up question is whether the adaptation in firing properties is caused by the slow inactivation detected within the kinetic mechanism. If true, then a dynamic-clamp-injected current based on a model that lacks slow inactivation should not result in adaptation. Indeed, as shown in Fig. 6c, this model generates a spiking pattern that clearly differs from the control: the spike height remains constant and only the falling phase of the action potential is slowed down. One should remember that the slope of the rising phase is proportional to the total number of Nav channels that are available to generate current, whereas the falling phase depends mostly on Kv channels. The explanation for the observed spike shape adaptation is that more and more Nav channels enter the slow inactivated state after each action potential, leaving progressively fewer channels available to contribute current to the rising phase.

Clearly, in this case the kinetic model obtained from voltage-clamp data was well validated by the dynamic-clamp experiment. However, the keen observer will have noticed a small but important difference: the action potential starts more abruptly in control than with the model (Fig. 6a2, b2). In fact, this is not a shortcoming of the kinetic model but a technical limitation of the brain slice preparation, where neurons maintain their processes intact. Although Nav channels are distributed throughout the cell in the soma, dendrites, and axon [95, 96], the axonal initial segment [97] has special properties that make it the site of action potential initiation

[98–100]. From the axonal initial segment, the action potential backpropagates to the soma, causing the abrupt onset [101–104]. In contrast, the model-based current is injected strictly in the soma, which becomes the site of action potential initiation. This configuration resembles the case of dissociated neurons that have lost their axon and exhibit similarly smoothly rising action potentials. Nevertheless, this abrupt action potential onset can be reproduced in a dynamic-clamp experiment by adding a virtual axonal compartment that generates its own spike and thus contributes additional current to the rising phase of the somatic action potential [18]. Overall, it is quite remarkable that basic spiking properties are reproduced so well by a soma-injected model-based current.

References

1. Hille B (2001) Ion channels of excitable membranes. Sinauer Sunderland, MA
2. Frank HY, Yarov-Yarovoy V, Gutman GA, Catterall WA (2005) Overview of molecular relationships in the voltage-gated ion channel superfamily. Pharmacol Rev 57:387–395
3. Bezanilla F (2000) The voltage sensor in voltage-dependent ion channels. Physiol Rev 80:555–592
4. Sigworth FJ (1994) Voltage gating of ion channels. Q Rev Biophys 27:1–40
5. Yellen G (1998) The moving parts of voltage-gated ion channels. Q Rev Biophys 31:239–295
6. Armstrong CM, Hille B (1998) Voltage-gated ion channels and electrical excitability. Neuron 20:371–380
7. Hodgkin AL, Huxley A, Katz B (1952) Measurement of current–voltage relations in the membrane of the giant axon of Loligo. J Physiol 116:424–448
8. Bean BP (2007) The action potential in mammalian central neurons. Nat Rev Neurosci 8:451–465
9. Richie J (1973) Energetic aspects of nerve conduction: the relationships between heat production, electrical activity and metabolism. Prog Biophys Mol Biol 26:147–187
10. Connors BW, Gutnick MJ (1990) Intrinsic firing patterns of diverse neocortical neurons. Trends Neurosci 13:99–104
11. Marder E, Goaillard J-M (2006) Variability, compensation and homeostasis in neuron and network function. Nat Rev Neurosci 7:563–574
12. Prinz AA, Bucher D, Marder E (2004) Similar network activity from disparate circuit parameters. Nat Neurosci 7:1345–1352
13. Harris-Warrick RM, Marder E (1991) Modulation of neural networks for behavior. Annu Rev Neurosci 14:39–57
14. Getting PA (1989) Emerging principles governing the operation of neural networks. Annu Rev Neurosci 12:185–204
15. Grillner S (2003) The motor infrastructure: from ion channels to neuronal networks. Nat Rev Neurosci 4:573–586
16. Harris-Warrick RM (2002) Voltage-sensitive ion channels in rhythmic motor systems. Curr Opin Neurobiol 12:646–651
17. Marty A, Neher E (1995) Tight-seal whole-cell recording. In: Sakmann B, Neher E (eds) Single-channel recording, 2nd edn. Springer, New York, pp 31–52
18. Milescu LS, Yamanishi T, Ptak K, Mogri MZ, Smith JC (2008) Real-time kinetic modeling of voltage-gated ion channels using dynamic clamp. Biophys J 95:66–87
19. Izhikevich EM, Edelman GM (2008) Large-scale model of mammalian thalamocortical systems. Proc Natl Acad Sci U S A 105:3593–3598
20. Armstrong CM (2006) Na channel inactivation from open and closed states. Proc Natl Acad Sci U S A 103:17991–17996
21. Hodgkin AL, Huxley AF (1952) A quantitative description of membrane current and its application to conduction and excitation in nerve. J Physiol 117:500–544
22. Catterall WA, Raman IM, Robinson HP, Sejnowski TJ, Paulsen O (2012) The Hodgkin-Huxley heritage: from channels to circuits. J Neurosci 32:14064–14073
23. Hoshi T, Zagotta WN, Aldrich RW (1994) Shaker potassium channel gating. I: transitions near the open state. J Gen Physiol 103:249–278

24. Zagotta WN, Hoshi T, Aldrich RW (1994) Shaker potassium channel gating. III: evaluation of kinetic models for activation. J Gen Physiol 103:321–362
25. Zagotta WN, Hoshi T, Dittman J, Aldrich RW (1994) Shaker potassium channel gating. II: transitions in the activation pathway. J Gen Physiol 103:279–319
26. Vandenberg CA, Bezanilla F (1991) A sodium channel gating model based on single channel, macroscopic ionic, and gating currents in the squid giant axon. Biophys J 60:1511–1533
27. Schoppa NE, Sigworth FJ (1998) Activation of shaker potassium channels. III. an activation gating model for wild-type and V2 mutant channels. J Gen Physiol 111:313–342
28. Schoppa NE, Sigworth FJ (1998) Activation of Shaker potassium channels. II. kinetics of the V2 mutant channel. J Gen Physiol 111:295–311
29. Schoppa NE, Sigworth FJ (1998) Activation of shaker potassium channels. I Characterization of voltage-dependent transitions. J Gen Physiol 111:271–294
30. Rothberg BS, Magleby KL (2000) Voltage and Ca^{2+} activation of single large-conductance Ca^{2+} –activated K^{+} channels described by a two-tiered allosteric gating mechanism. J Gen Physiol 116:75–99
31. Box GE, Draper NR (1987) Empirical model-building and response surfaces. John Wiley & Sons, New York
32. Colquhoun D, Hawkes AG (1995) The principles of the stochastic interpretation of ion-channel mechanisms. In: Sakmann B, Neher E (eds) Single-channel recording, 2nd edn. Springer, New York, pp 397–482
33. Eyring H (1935) The activated complex in chemical reactions. J Chem Phys 3:107–115
34. Sigg D (2014) Modeling ion channels: past, present, and future. J Gen Physiol 144:7–26
35. Colquhoun D, Hawkes AG (1995) A Q-matrix cookbook. In: Sakmann B, Neher E (eds) Single-channel recording, 2nd edn. Springer, New York, pp 589–633
36. Hawkes A, Jalali A, Colquhoun D (1990) The distributions of the apparent open times and shut times in a single channel record when brief events cannot be detected. Philos Trans R Soc Lond B Biol Sci 332:511–538
37. Ball FG, Sansom MS (1989) Ion-channel gating mechanisms: model identification and parameter estimation from single channel recordings. Proc R Soc Lond B Biol Sci 236:385–416
38. Csanady L (2006) Statistical evaluation of ion-channel gating models based on distributions of log-likelihood ratios. Biophys J 90:3523–3545
39. Qin F, Auerbach A, Sachs F (1996) Estimating single-channel kinetic parameters from idealized patch-clamp data containing missed events. Biophys J 70:264–280
40. Qin F, Auerbach A, Sachs F (2000) A direct optimization approach to hidden Markov modeling for single channel kinetics. Biophys J 79:1915–1927
41. Qin F, Auerbach A, Sachs F (2000) Hidden Markov modeling for single channel kinetics with filtering and correlated noise. Biophys J 79:1928–1944
42. Venkataramanan L, Sigworth F (2002) Applying hidden Markov models to the analysis of single ion channel activity. Biophys J 82:1930–1942
43. Lampert A, Korngreen A (2014) Markov modeling of ion channels: implications for understanding disease. Prog Mol Biol Transl Sci 123:1–21
44. Milescu LS, Akk G, Sachs F (2005) Maximum likelihood estimation of ion channel kinetics from macroscopic currents. Biophys J 88:2494–2515
45. Irvine LA, Jafri MS, Winslow RL (1999) Cardiac sodium channel Markov model with temperature dependence and recovery from inactivation. Biophys J 76:1868–1885
46. Celentano JJ, Hawkes AG (2004) Use of the covariance matrix in directly fitting kinetic parameters: application to GABAA receptors. Biophys J 87:276–294
47. Moffatt L (2007) Estimation of ion channel kinetics from fluctuations of macroscopic currents. Biophys J 93:74–91
48. Stepanyuk AR, Borisyuk AL, Belan PV (2011) Efficient maximum likelihood estimation of kinetic rate constants from macroscopic currents. PLoS One 6:e29731
49. Sigworth FJ (1985) Open channel noise. I. noise in acetylcholine receptor currents suggests conformational fluctuations. Biophys J 47:709–720
50. Stepanyuk A, Borisyuk A, Belan P (2014) Maximum likelihood estimation of biophysical parameters of synaptic receptors from macroscopic currents. Front Cell Neurosci 8:303–317
51. Qin F, Li L (2004) Model-based fitting of single-channel dwell-time distributions. Biophys J 87:1657–1671
52. Anderson C, Stevens C (1973) Voltage clamp analysis of acetylcholine produced end-plate current fluctuations at frog neuromuscular junction. J Physiol 235:655–691

53. Sigworth F (1980) The variance of sodium current fluctuations at the node of Ranvier. J Physiol 307:97–129
54. Milescu LS, Yamanishi T, Ptak K, Smith JC (2010) Kinetic properties and functional dynamics of sodium channels during repetitive spiking in a slow pacemaker neuron. J Neurosci 30:12113–12127
55. Bar-Yehuda D, Korngreen A (2008) Space-clamp problems when voltage clamping neurons expressing voltage-gated conductances. J Neurophysiol 99:1127–1136
56. Stuart G, Spruston N, Sakmann B, Häusser M (1997) Action potential initiation and backpropagation in neurons of the mammalian CNS. Trends Neurosci 20:125–131
57. Milescu LS, Bean BP, Smith JC (2010) Isolation of somatic Na^+ currents by selective inactivation of axonal channels with a voltage prepulse. J Neurosci 30:7740–7748
58. Heinemann SH (1995) Guide to data acquisition and analysis. In: Sakmann B, Neher E (eds) Single-channel recording, 2nd edn. Springer, New York, pp 53–91
59. Armstrong CM, Bezanilla F (1974) Charge movement associated with the opening and closing of the activation gates of the Na channels. J Gen Physiol 63:533–552
60. Fletcher R (1987) Practical methods of optimization. John Wiley & Sons, Chichester
61. Heinemann SH, Conti F (1992) Nonstationary noise analysis and application to patch clamp recordings. Methods Enzymol 207:131–148
62. Kuo CC, Bean BP (1994) Na^+ channels must deactivate to recover from inactivation. Neuron 12:819–829
63. Bezanilla F, Armstrong CM (1977) Inactivation of the sodium channel. I sodium current experiments. J Gen Physiol 70:549–566
64. Sheets MF, Hanck DA (1995) Voltage-dependent open-state inactivation of cardiac sodium channels: gating current studies with Anthopleurin-A toxin. J Gen Physiol 106:617–640
65. Sheets MF, Hanck DA (2005) Charge immobilization of the voltage sensor in domain IV is independent of sodium current inactivation. J Physiol 563:83–93
66. Armstrong CM, Bezanilla F (1977) Inactivation of the sodium channel. II. gating current experiments. J Gen Physiol 70:567–590
67. Colquhoun D, Dowsland KA, Beato M, Plested AJ (2004) How to impose microscopic reversibility in complex reaction mechanisms. Biophys J 86:3510–3518
68. Cha A, Ruben PC, George AL Jr, Fujimoto E, Bezanilla F (1999) Voltage sensors in domains III and IV, but not I and II, are immobilized by Na^+ channel fast inactivation. Neuron 22:73–87
69. Fletcher R, Powell MJ (1963) A rapidly convergent descent method for minimization. Comput J 6:163–168
70. Press WH, Teukolsky S, Vetterling W, Flannery B (1992) Numerical recipes in C: the art of scientific computing. Cambridge University, Cambridge, UK
71. Nelder JA, Mead R (1965) A simplex method for function minimization. Comput J 7:308–313
72. Horst R, Pardalos PM (1995) Handbook of global optimization. Kluwer Academic Publishers, London
73. Menon V, Spruston N, Kath WL (2009) A state-mutating genetic algorithm to design ion-channel models. Proc Natl Acad Sci U S A 106:16829–16834
74. Gurkiewicz M, Korngreen A (2007) A numerical approach to ion channel modelling using whole-cell voltage-clamp recordings and a genetic algorithm. PLoS Comput Biol 3:e169
75. Cobelli C, DiStefano JJ 3rd (1980) Parameter and structural identifiability concepts and ambiguities: a critical review and analysis. Am J Physiol 239:R7–R24
76. Kienker P (1989) Equivalence of aggregated Markov models of ion-channel gating. Proc R Soc Lond B Biol Sci 236:269–309
77. Bruno WJ, Yang J, Pearson JE (2005) Using independent open-to-closed transitions to simplify aggregated Markov models of ion channel gating kinetics. Proc Natl Acad Sci U S A 102:6326–6331
78. Horn R (1987) Statistical methods for model discrimination. applications to gating kinetics and permeation of the acetylcholine receptor channel. Biophys J 51:255–263
79. Colquhoun D, Hatton CJ, Hawkes AG (2003) The quality of maximum likelihood estimates of ion channel rate constants. J Physiol 547:699–728
80. Koch C, Segev I (2000) The role of single neurons in information processing. Nat Neurosci 3:1171–1177
81. Sharp AA, O'Neil MB, Abbott LF, Marder E (1993) Dynamic clamp: computer-generated conductances in real neurons. J Neurophysiol 69:992–995
82. Tan RC, Joyner RW (1990) Electrotonic influences on action potentials from isolated ventricular cells. Circ Res 67:1071–1081

83. Robinson HP, Kawai N (1993) Injection of digitally synthesized synaptic conductance transients to measure the integrative properties of neurons. J Neurosci Methods 49:157–165
84. Kaczmarek LK (2006) Non-conducting functions of voltage-gated ion channels. Nat Rev Neurosci 7:761–771
85. Butera RJ Jr, Wilson CG, DelNegro C, Smith JC (2001) A methodology for achieving high-speed rates for artificial conductance injection in electrically excitable biological cells. IEEE Trans Biomed Eng 48:1460–1470
86. Christini DJ, Stein KM, Markowitz SM, Lerman BB (1999) Practical real-time computing system for biomedical experiment interface. Ann Biomed Eng 27:180–186
87. Dorval AD, Christini DJ, White JA (2001) Real-time linux dynamic clamp: a fast and flexible way to construct virtual ion channels in living cells. Ann Biomed Eng 29:897–907
88. Lin RJ, Bettencourt J, White J, Christini DJ, Butera RJ (2010) Real-time experiment interface for biological control applications. Conf Proc IEEE Eng Med Biol Soc 2010:4160–4163
89. Kullmann PH, Wheeler DW, Beacom J, Horn JP (2004) Implementation of a fast 16-bit dynamic clamp using LabVIEW-RT. J Neurophysiol 91:542–554
90. Yang Y, Adowski T, Ramamurthy B, Neef A, Xu-Friedman MA (2015) High-speed dynamic-clamp interface. J Neurophysiol 113:2713–2720
91. Nowotny T, Szűcs A, Pinto RD, Selverston AI (2006) StdpC: a modern dynamic clamp. J Neurosci Methods 158:287–299
92. Mickus T, Hy J, Spruston N (1999) Properties of slow, cumulative sodium channel inactivation in rat hippocampal CA1 pyramidal neurons. Biophys J 76:846–860
93. Fleidervish IA, Friedman A, Gutnick M (1996) Slow inactivation of Na^+ current and slow cumulative spike adaptation in mouse and guinea-pig neocortical neurones in slices. J Physiol 493:83–97
94. Goldfarb M, Schoorlemmer J, Williams A, Diwakar S, Wang Q, Huang X, Giza J, Tchetchik D, Kelley K, Vega A (2007) Fibroblast growth factor homologous factors control neuronal excitability through modulation of voltage-gated sodium channels. Neuron 55:449–463
95. Vacher H, Mohapatra DP, Trimmer JS (2008) Localization and targeting of voltage-dependent ion channels in mammalian central neurons. Physiol Rev 88:1407–1447
96. Lai HC, Jan LY (2006) The distribution and targeting of neuronal voltage-gated ion channels. Nat Rev Neurosci 7:548–562
97. Palay SL, Sotelo C, Peters A, Orkand PM (1968) The axon hillock and the initial segment. J Cell Biol 38:193–201
98. Baranauskas G, David Y, Fleidervish IA (2013) Spatial mismatch between the Na^+ flux and spike initiation in axon initial segment. Proc Natl Acad Sci U S A 110:4051–4056
99. Kole MH, Ilschner SU, Kampa BM, Williams SR, Ruben PC, Stuart GJ (2008) Action potential generation requires a high sodium channel density in the axon initial segment. Nat Neurosci 11:178–186
100. Colbert CM, Johnston D (1996) Axonal action-potential initiation and Na^+ channel densities in the soma and axon initial segment of subicular pyramidal neurons. J Neurosci 16:6676–6686
101. Palmer LM, Stuart GJ (2006) Site of action potential initiation in layer 5 pyramidal neurons. J Neurosci 26:1854–1863
102. Shu Y, Duque A, Yu Y, Haider B, McCormick DA (2007) Properties of action-potential initiation in neocortical pyramidal cells: evidence from whole cell axon recordings. J Neurophysiol 97:746–760
103. Stuart G, Schiller J, Sakmann B (1997) Action potential initiation and propagation in rat neocortical pyramidal neurons. J Physiol 505:617–632
104. Yu Y, Shu Y, McCormick DA (2008) Cortical action potential backpropagation explains spike threshold variability and rapid-onset kinetics. J Neurosci 28:7260–7272

Chapter 14

Recording and Hodgkin-Huxley Kinetic Analysis of Voltage-Gated Ion Channels in Nucleated Patches

Mara Almog and Alon Korngreen

Abstract

The patch-clamp technique is widely used to measure and characterize ion currents flowing through the cell membrane. Various configurations can be used according to the research hypothesis. A relatively new configuration of the patch-clamp technique is the outside-out nucleated patch, which allows measuring somatic currents in voltage-clamp recordings due to a reduction in technical issues. Characterization of ionic currents is mostly based on the Hodgkin-Huxley model, a detailed mathematical and biophysical model of membrane excitability. This chapter describes voltage-clamp experiments step by step, from preparation and performing outside-out nucleated patch experiments through to the Hodgkin-Huxley analysis of the recorded data.

Key words Nucleated patch, Outside-out patch, Voltage-gated channels, Patch clamp, Kinetic analysis, Hodgkin-Huxley model

1 Introduction

The invention of the patch-clamp technique [1] made it possible to record ionic currents from single voltage-gated ion channels and, using the whole-cell variant of the technique, from ensembles of channels. A quick search through PubMed reveals that since 1995 the number of publications containing "patch clamp" in their title or abstract has been large and roughly constant. About one third of these publications are related, according to PubMed, to voltage-gated ion channels. In the 1990s the acute brain slice preparation emerged as the major in vitro preparation for investigating neurons in the central nervous system (CNS) [2]. This preparation, in combination with differential interference contrast optics and digital imaging [3], allowed repeated and accurate measurements selectively from visually identified neurons in many regions of the CNS. It is, therefore, not surprising that the number of papers describing properties of ion channels in the CNS soared in the mid-1990s.

Alon Korngreen (ed.), *Advanced Patch-Clamp Analysis for Neuroscientists*, Neuromethods, vol. 113,
DOI 10.1007/978-1-4939-3411-9_14,

This chapter focuses on some technical aspects of this field of research, dealing with using the patch-clamp technique to record from voltage-gated channels. This technique controls the membrane voltage and measures the membrane current [4–6]. During an experiment the experimenter uses a patch-clamp amplifier to set the voltage for holding the membrane potential (i.e., "holding voltage" or "holding potential"). The voltage-clamp electrode connected to the amplifier measures the membrane voltage and feeds an input to the amplifier. The amplifier calculates the "error signal" between the voltage command (V_c) and the membrane potential (V_m) (subtracts the membrane potential from the voltage command ($V_c - V_m$)), amplifies it, and sends the output to the electrode. This transient negative feedback current to the cell reduces the "error signal" to zero and maintains the desired membrane voltage.

Currents recorded from voltage-gated channels using the voltage-clamp technique are widely analyzed using the modern application of the Hodgkin-Huxley analysis [7] developed by Cole and Marmont and improved by Hodgkin, Huxley, and Katz in the 1940s [8]. The measured membrane current (I_m) is the sum of the ionic current (I_i) and the capacity current (C):

$$I_{\mathrm{m}} = I_{\mathrm{i}} + C\frac{\mathrm{d}V}{\mathrm{d}t}. \tag{1}$$

During an experiment, while the membrane voltage is held constant, the $\mathrm{d}V/\mathrm{d}t$ is zero, leading to a zero capacitive current. An exception occurs when the voltage is changed rapidly from one steady state to another. After completing the voltage change the membrane current (I_m) equals the ionic current (I_i).

In the voltage-clamp technique the current injected into the cell spreads radially and decreases with a space constant (λ). In nonspherical cells the injected current decays with distance from the injection point, leading, in the best case, to a smaller change in membrane voltage and, in the worst case, to no change at all. To overcome the space-clamp error and measure membrane current (I_m) the cell should be iso-potential, so that all the channels in the membrane experience the same voltage.

Using "patch-clamp" microelectrodes to inject current at a single point controls the voltage of spherical cells, especially small cells, creating a good space clamp. If the cell body has extensions (e.g., axons and dendrites), voltage clamping the membrane leads to measuring a distorted current due to the inability to control the membrane voltage [9–11]. The outside-out nucleated configuration overcomes space-clamp errors, since this configuration can be considered equivalent to a small spherical cell in whole-cell configuration.

The configuration of the patch-clamp technique chosen for an experiment must be appropriate for answering the research hypothesis and minimize technical issues (e.g., space clamp, signal-to-noise ratio, duration of the recording, stability). Outside-out nucleated patches are a simple and straightforward approach that easily overcomes some of the technical issues arising during recordings of voltage-gated and ligand-gated ion currents. For this reason nucleated patches have been used to study macroscopic ligand-gated ion channels [12–14], voltage-gated ion currents [15–19], proton-gated ion channels [20], membrane capacitance charging current [21], and passive membrane properties [22].

2 Extracting a Nucleated Patch

The pipette (with a relative low resistance of 4–6 MΩ), filled with the intracellular solution and under slightly positive pressure, is lowered into the bath solution, positioned close to the cell body, and slowly moved towards it to contact the cell membrane (Fig. 1a). When the pipette touches the cell membrane, the positive pressure in the pipette pushes away the cell membrane, creating a concave vesicle (Fig. 1a). Suction is now required to seal the cell membrane to the pipette (membrane-to-glass seal), and a negative holding potential improves the seal. Achieving a gigaohm seal leads to the cell-attached configuration.

Now applying additional suction to the membrane inside the pipette ruptures it, leading to the whole-cell configuration. Further negative pressure is applied to the pipette (100–200 mBar above ambient pressure, depending on the diameter of the patch pipette and neuron size). This draws the nucleus towards the pipette. A slow and smooth retraction of the pipette then brings with it a large portion of the somatic membrane covering the nucleus (Fig. 1b–d). Once the nucleus is free from the cell, the plasma membrane closes behind it (Fig. 1e, f) forming an almost spherical and very large outside-out patch (diameter about 5–8 μm, Fig. 1e, f). At this point it is useful to reduce the pressure applied to the pipette to a few mBar above ambient pressure.

The nucleated configuration has several advantages over the conventional outside-out patch. First, the current measured is larger due to the relatively larger surface (usually 100–1200 pA but values may vary widely beyond this). Currents of this size allow pharmacological manipulations and the approach has been successfully used to investigate the properties of various voltage- and ligand-gated channels. Second, the structural support given by the nucleus makes nucleated patches more stable than regular outside-out patches and therefore they are more durable (from minutes to hours) [14]. Third, the outside-out nucleated patch can be considered a small spherical cell in the whole-cell

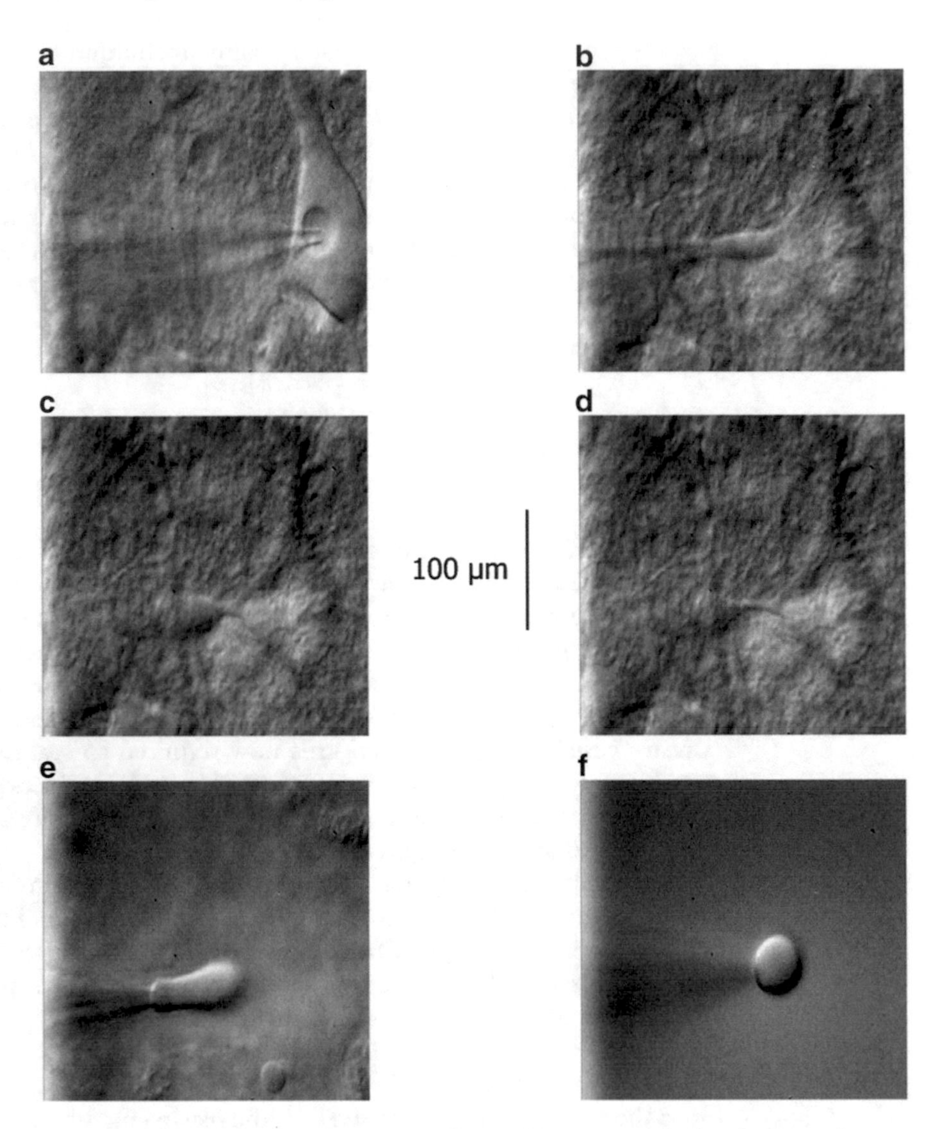

Fig. 1 Forming an outside-out nucleated patch. (**a**) The positive pressure in the pipette pushes away the cell membrane, creating a concave vesicle. (**b–d**) After reaching a gigaohm seal and rupturing the membrane inside the pipette the pipette is carefully withdrawn from the cell body. (**e**, **f**). The cell membrane extracted together with the nucleus forms a small sphere. Modified with permission from [35]

configuration (Fig. 1f). This avoids space-clamp errors and there is little filtration due to the membrane capacitance, allowing very rapid recording of voltage-gated ion currents. Nonetheless, one should keep in mind that the nucleated patch is an outside-out patch. Care should, therefore, be taken to determine the extent of channel run-down (Sect. 3.8).

3 Data Acquisition

3.1 Filter and Sampling Rate

Biological events are continuous or analog signals, which can be measured as discrete signals or digitized using an analog-digital (A/D) converter. The filter and sampling rate of the A/D converter must be selected to accurately fit the dynamics of the desired signal and its noise characteristics. Also to be considered are the RC components of the cell membrane and the electrode series resistance (R_s) that function together as an RC filter. Selecting the highest sampling rate (100–300 kHz) produces an enormous amount of data that does not necessarily contain relevant information. The minimum sampling rate that accurately represents an analog waveform is twice the bandwidth. Derived from the Nyquist-Shannon theorem, this is called the Nyquist frequency. Selecting a sampling rate lower than the Nyquist frequency distorts high frequency signals (aliasing). Aliasing can be prevented by filtering high frequency signals using a low-pass filter (frequency not higher than half the sampling frequency). Voltage-clamp recordings, including from nucleated patches, require a high frequency filter of at least 20 kHz to measure macroscopic ion currents, leading to a minimal sample frequency of 50 kHz. (For further clarification see Chap. 3 by Meents and Lampert.)

3.2 Capacitance Compensation

Two sources of capacitance appear during voltage-clamp recordings: (1) the pipette and holder capacitance ("fast" capacitance, due to the fast time constant of the current); (2) the membrane capacitance ("slow" capacitance). The fast capacitance is compensated by adjusting its amplitude and the time constant (some amplifiers offer an automatic compensation feature). This step is essential for estimating the series resistance (R_s). Compensation of the slow capacitance is essential in whole-cell and outside-out patches (including nucleated patches). This capacitance is compensated by applying an exponentially shaped voltage through a small capacitor in the amplifier. The sum of two exponentials compensates both the fast and the slow capacitance. The voltage-clamp recording must be amplified by at least ×5–10 for best resolution of the currents, the amplification requiring compensation of the fast and the slow capacitances to prevent saturating the amplifier.

3.2.1 Pipette Coating and Fire Polishing

Patch-clamp amplifiers can typically compensate pipette capacitance up to 10 pF. However, the pipette capacitance of pipettes made from glass with a high dialectic constant can exceed the compensation range of the amplifier, preventing good compensation. This problem can be solved by coating about 1 mm from the shank toward the end of the tip with an elastomer, such as Sylgard No. 184 resin, which reduces the fast capacitance and pipette noise. After coating, the pipette tip must be fire-polished. This step is essential to enable gigaohm seal formation and to prevent the tip from penetrating the cell membrane. Fire polishing is not necessary with some cell types.

3.3 Series Resistance Compensation

The membrane potential in nucleated patches during voltage-clamp recordings is controlled by the potential fed through the electrode. Ideally, this allows complete control of the membrane potential after compensating the access resistance between the pipette and the interior of the cell. This resistance is called the series resistance (R_s) because the access resistance and the pipette are in series with the current recording. The R_s stems partly from the pipette resistance but mostly from the broken membrane particles created by forming the nucleated configuration. R_s size determines the compensating voltage command.

R_s slows the charging of the membrane capacitance (C_m), leading to a slow charging time constant (hundreds of microseconds). This is too long to examine rapid voltage-gated ion currents, such as Na^+ currents. The combination of R_s and C_m also functions as a bandwidth current filter. The R_s reduces the voltage imposed on the cell membrane, distorting the currents recorded. Consequently, to maintain the voltage error below 1 %, the R_m must be two orders of magnitude larger than the R_s. As R_s compensation is very sensitive to changes in the pipette capacitance during the recordings, it is very important to first compensate the fast capacitance (pipette capacitance) (Sect. 3.2).

3.4 Leak Subtraction

Any ion current measured under voltage clamp includes leakage currents and membrane capacitive currents. The membrane capacitive current carries important information (the membrane area of the cell can be derived from it), but in most experiments this information is irrelevant. As the capacitive and leakage currents are linear and not voltage dependent, they can be subtracted, either online or off-line, from the ion current (leak subtraction). Another method of subtracting the leakage and capacitive currents and avoiding noise enhancement is adding specific ion channel blockers that completely eliminate the ion current being measured. The signal recorded after blocker application can then be subtracted from the ion current recorded without blocker.

3.5 Liquid Junction Potential

A liquid junction potential develops at the tip of the pipette during electrophysiological experiments when the pipette solution differs from the bath solution, as in outside-out nucleated patches. This leads to different mobilities of the ions. Not correcting the liquid junction potential leads to a significant error in membrane potential, ranging from 1 to 2 mV for pipette solutions containing KCl to more than 10 mV for pipette solutions containing K-gluconate or K-methylsulfate. Unfortunately, in most studies not only is the liquid junction potential not compensated but its value is rarely mentioned.

3.5.1 Measuring of Junction Liquid Potential

Measuring the liquid junction potential requires a simple series of steps [23]: (1) the recording bath is filled with the pipette solution to be used during the experiment; (2) the pipette, containing the same pipette solution, is placed in the bath; (3) the potential is measured in the current-clamp condition (it should be close to zero) and is set to zero; (4) the solution in the bath is replaced by the extracellular bath solution (e.g., artificial cerebrospinal fluid) without moving the pipette; (5) the potential is measured again and now gives the liquid junction potential; (6) to test for reversibility the bath solution can be replaced with the pipette solution. As long the bath solution does not change, especially its chloride ion concentration, the liquid junction potential will remain steady during the experiment. However, in some cases it is necessary to change the bath solution during the experiment, leading to changes in the liquid junction potential. These cases require a salt bridge. As Barry and Diamond [24] pointed out, it is be simpler to locally change the solution using a puffer pipette thus preventing potential changes at the reference electrode.

In outside-out nucleated and whole-cell patches the liquid junction potential is corrected according to the following equation

$$V_{\mathrm{m}} = V_{\mathrm{p}} - V_{\mathrm{LJP}}, \qquad (2)$$

where V_{m} is the corrected membrane potential, V_{p} is the pipette voltage, and V_{LJP} is the liquid junction potential [23]. The liquid junction potential can be calculated theoretically using the Henderson formula for liquid junction potential [25]. Although the calculated liquid junction potential is extremely accurate, it should be verified experimentally, especially if the solution contains high concentrations of divalent ions [25].

3.6 Kinetic Protocols

Macroscopic currents are the sum of many microscopic ion current events (single ion channel events). The kinetics parameters of the macroscopic currents (e.g., time constants, conductance, and voltage dependence) can be measured using pulse protocols. These must be designed such that the measured kinetics parameters reflect the kinetic parameters of microscopic ion currents. A pulse protocol usually includes steps of specified voltage and duration. Voltage ramps or sine waves, etc., may be of advantage for phase-sensitive measurements.

To investigate the kinetics of a voltage-gated ion channel, one can alternate each segment of the voltage steps or the duration to create a series of pulse protocols. To ensure the channel is in its resting state before application of a voltage step, each step must be preceded and followed by a strong hyperpolarization for some tens of milliseconds.

3.7 Data Averaging

To increase the signal-to-noise ratio in voltage-clamp experiments, one repeatedly records the current yielded by the same pulse protocol (the experimental conditions must remain stable throughout the recording). This is particularly important when recording small ion currents (about a few tens to hundreds of picoampere current). The averaged data is smoother because the noise is reduced by a factor of $\sqrt{n}$, where n is the number of runs of the pulse protocol. Some of the recording software currently available averages the data online and stores either only the averaged data or the raw and the averaged data, whereas other software stores only the raw data, leaving data averaging for off-line analysis software.

3.8 Run-Up and Run-Down

The activity of many types of ionic channels declines with time in both outside-out (including nucleated patches) and inside-out patches. This phenomenon, called "run-down," is also observed in whole-cell patches, possibly due to washout and dephosphorylation of endogenous cytoplasmic factors. Run-down is time dependent, often fast, but is not voltage dependent. For example, Ca^{2+} voltage-gated channels show complete shunting of the channel within a few minutes (Fig. 2a), without changes in the gate current [15, 26, 27] (Fig. 2b). Adding ATP and the protease inhibitor calpastatin to the intracellular solution can completely prevent run-down or, at least, prolong the process [26, 28, 29].

Several types of ion channels show enhancement of the ion channel activity or "run-up" during the first minutes of measurement (Fig. 2a, b; [15, 30, 31]). This phenomenon can be confused with facilitation (fast or slow), especially in Ca^{2+} voltage-gated channels, which display both processes [30]. Run-up and run-down are also observed in Na^{+} voltage-gated channels recorded in excised patches (Fig. 2c), but this appears sensitive to the age of the animal. Run-up is observed in relatively young animals (p14) (smooth line, Fig. 2c), whereas older animals (p38) show run-down over time (dashed line, Fig. 2c). Therefore, before measuring ion currents, it is necessary to examine whether run-up or/and run-down occur and to optimally design the experiment, for example, compromising on the age of the animal (if Na^{+} channels are being recorded) or on the recording duration (number of protocols to run, as with Ca^{2+} channels).

4 Hodgkin-Huxley Analysis

According to the Hodgkin-Huxley model [7], each ion channel contains physical "gates" that modulate the ion flow through the channel. The gates may be either open (conductive) or closed ($C \rightleftarrows O$). The probability (p) of a gate being open is voltage dependent and ranges from 0 to 1. The fraction of a channel population in its conductive state can be represented as:

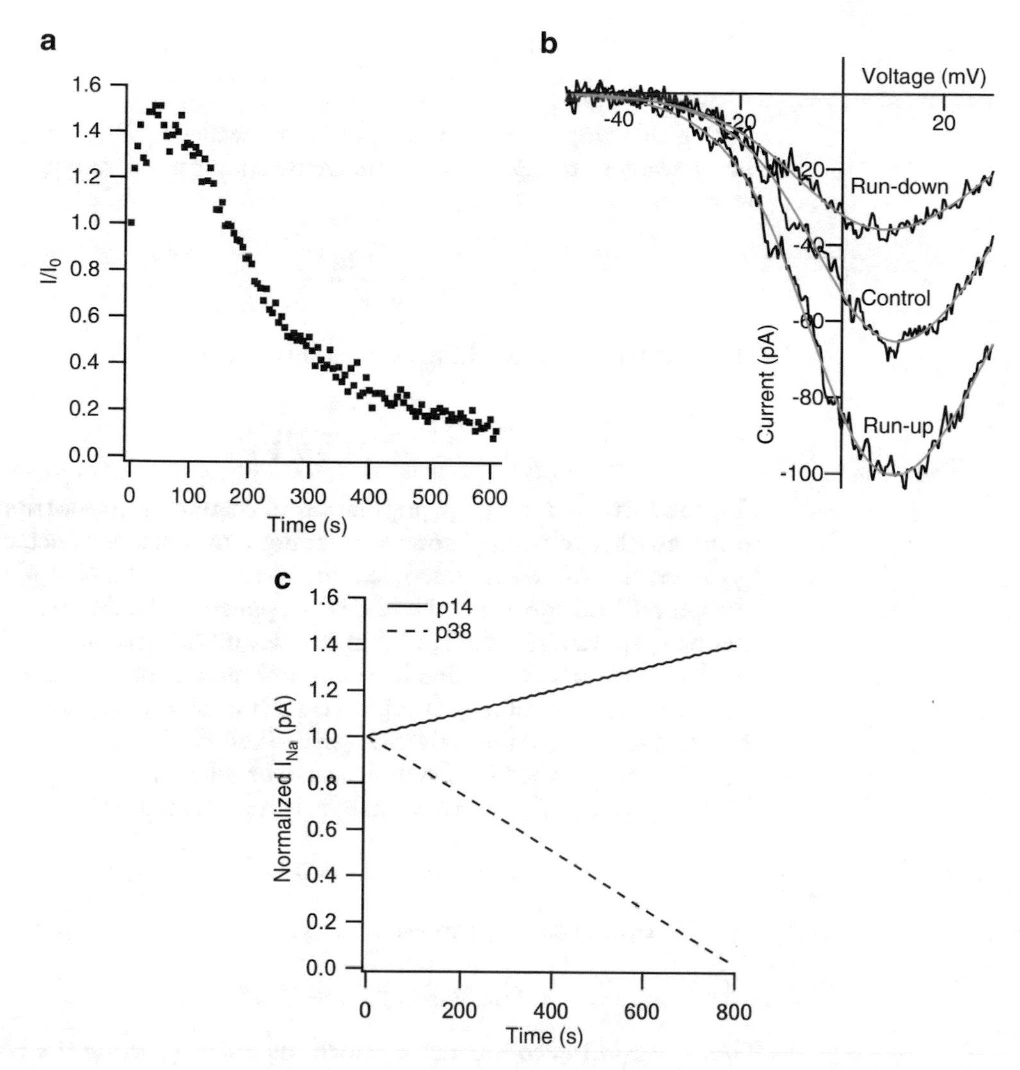

Fig. 2 Run-up and run-down of ion channels in nucleated patches. (**a**) Normalized Ca^{2+} current in a nucleated patch from the first trace recorded ($t = 0$) plotted as function of time (from [15, 30, 31]). (**b**) Activation curves of Ca^{2+} currents obtained at $t = 0$ (control), $t = 47$ s (run-up), and $t = 273$ s (run-down) in the experiment shown in (**a**). The *smooth lines* are the fit of a Boltzmann function (from [15, 30, 31]). (**c**) A linear regression fitted to an averaged Na^+ current (not shown) normalized to the first trace recorded ($t = 0$) as function of time. The *smooth line* shows a run-up for young animals (p14) and the *dashed line* shows an extreme run-down for old animals (p38) with current vanishing after 800 s

$$1 - p \underset{\beta(V)}{\overset{\alpha(V)}{\rightleftharpoons}} p \quad (3)$$

where α is the transition rate between the nonconductive state (C) and the conductive state (O) and β is the transition rate between the conductive state (O) and the nonconductive state (C). The transition rates are voltage dependent and have units of s^{-1}. In the Hodgkin-Huxley model the gate states obey first-order kinetics:

$$\frac{dp}{dt} = \alpha(V)(1 - p) - \beta(V)\,p. \tag{4}$$

During clamping the membrane potential reaches equilibrium over time (steady state, $dp/dt = 0$). At steady state the above equation becomes:

$$p_{t\to\infty} = \frac{\alpha(V)}{\alpha(V) + \beta(V)}. \tag{5}$$

The time course for reaching this equilibrium is:

$$\tau(V) = \frac{1}{\alpha(V) + \beta(V)}. \tag{6}$$

The conductance for a large population of channels is proportional to the number of open channels, reaching a maximal conductance ($\overline{g}$) when all channels are open, leading to the equation: $G = \overline{g}\prod p$. The generalized equations above can be applied to the different ion channels, for example, the Hodgkin-Huxley model describes the K^+ and Na^+ channels embedded in the membrane of the squid giant axon. In their equations p is replaced with a variable representing the gate type—m for the activation gate of the K^+ channel, n for the activation gate of the Na^+ channel, and h for the inactivation gate of the Na^+ channel. The K^+ conductance is represented as:

$$G_K = \overline{g}_K\, p_m^4 = \overline{g}_K\, m^4 \tag{7}$$

and Na^+ conductance is represented by:

$$G_{Na} = \overline{g}_{Na}\, p_n^3\, p_h = \overline{g}_{Na} n^3 h \tag{8}$$

The power in the conductance equations is set by fitting the conductance curve of each ion channel, using equations with different power values. This procedure yielded a power of 3 for the n gates (for Na^+ channel) and power of 4 for the m gates (for K^+ channels).

4.1 Ion Current Analysis

The ion current recordings can be analyzed using analysis and graphic software like Origin, Matlab, or IGOR. Using averaging increases the signal-to-noise ratio making for a smoother signal. For further analysis the current must be normalized to a standard reference response, such as the maximal current recorded or the maximal conductance calculated. This allows comparison among different patches and eliminates the effects of run-up and/or rundown of the ion current during the recording.

4.1.1 I–V Curves

Functional parameters cannot be extracted from the raw data. A common method for examining the voltage behavior of voltage-gated ion channels at steady state is plotting current–voltage curves. This gives maximal conductance, slope, ion reversal potential, voltage dependence, threshold activation, and ion selectivity.

Activation *I–V* Curve

The current–voltage curve of an ion current recorded using an activation pulse protocol (Fig. 3a) can be calculated by measuring the peak current at each voltage step and plotting it against the increasing voltage (Fig. 3b), after correcting for the liquid junction potential (see Sect. 3.5). Traditionally, the next step is converting the *I–V* curve to a *G–V* curve (Fig. 3c). The ion conductance is a function of the number of open channels and is calculated according to Ohm's law ($I = G(V_m - V_{rev})$). The ion conductance at each voltage step is derived by dividing the recorded ion current by the driving force, defined as the difference between the membrane voltage (V_m) and the ion reversal potential (V_{rev}). The ion reversal potential can be extracted from the *I–V* curve, being the voltage at which the ion current changes direction (where the curve crosses the *x*-axis). If the ion current does not cross the *x*-axis, the reversal potential can be extrapolated from the curve. Next, a sigmoidal or Boltzmann function is fitted to the *G–V* curve, describing the voltage-dependent activation of an ion channel:

$$G = G_{min} + \frac{G_{max} - G_{min}}{1 + e^{\left(\frac{V_m - V_{1/2}}{k}\right)}} \quad (9)$$

where G is the conductance, G_{min} and G_{max} are the minimum and maximal conductances, respectively, V_m is the membrane potential, $V_{1/2}$ is the voltage at which the conductance is half-maximal, and k is the slope of the curve, reflecting the charge movement during the gating transition (Fig. 3c, smooth line).

This approach may introduce large errors in the estimated value of the conductance as the voltage approaches the positive ion reversal potential (e.g., Na^+, Ca^{2+}). Thus, another straightforward method is to combine the two equations (Ohm's law and the Boltzmann equations) into one [10] and fit it directly to the *I–V* curve:

$$\frac{I}{I_{max}} = G_{max}\left(\frac{1}{1 + e^{-\left(\frac{V_m - V_{1/2}}{k}\right)}}\right)^a (V_m - V_{rev}) \quad (10)$$

where I/I_{max} is the current normalized to its maximal value, V_m is the membrane potential, $V_{1/2}$ is the voltage at which the conductance is half-maximal, k is the slope of the curve, and a is the number of activation gates (Fig. 3d, smooth line). This equation has produced better results than the traditional analysis protocol, avoiding errors in estimating the conductance value.

The Hodgkin-Huxley model assumes that activation ion conductance is proportional to the *n*th power activation gate which obeys a first-order Boltzmann equation [7]. "*a*" represents the number of activation gates in each ion channel and differs in value among channels. To fit the rising phase of the K^+ conductance from

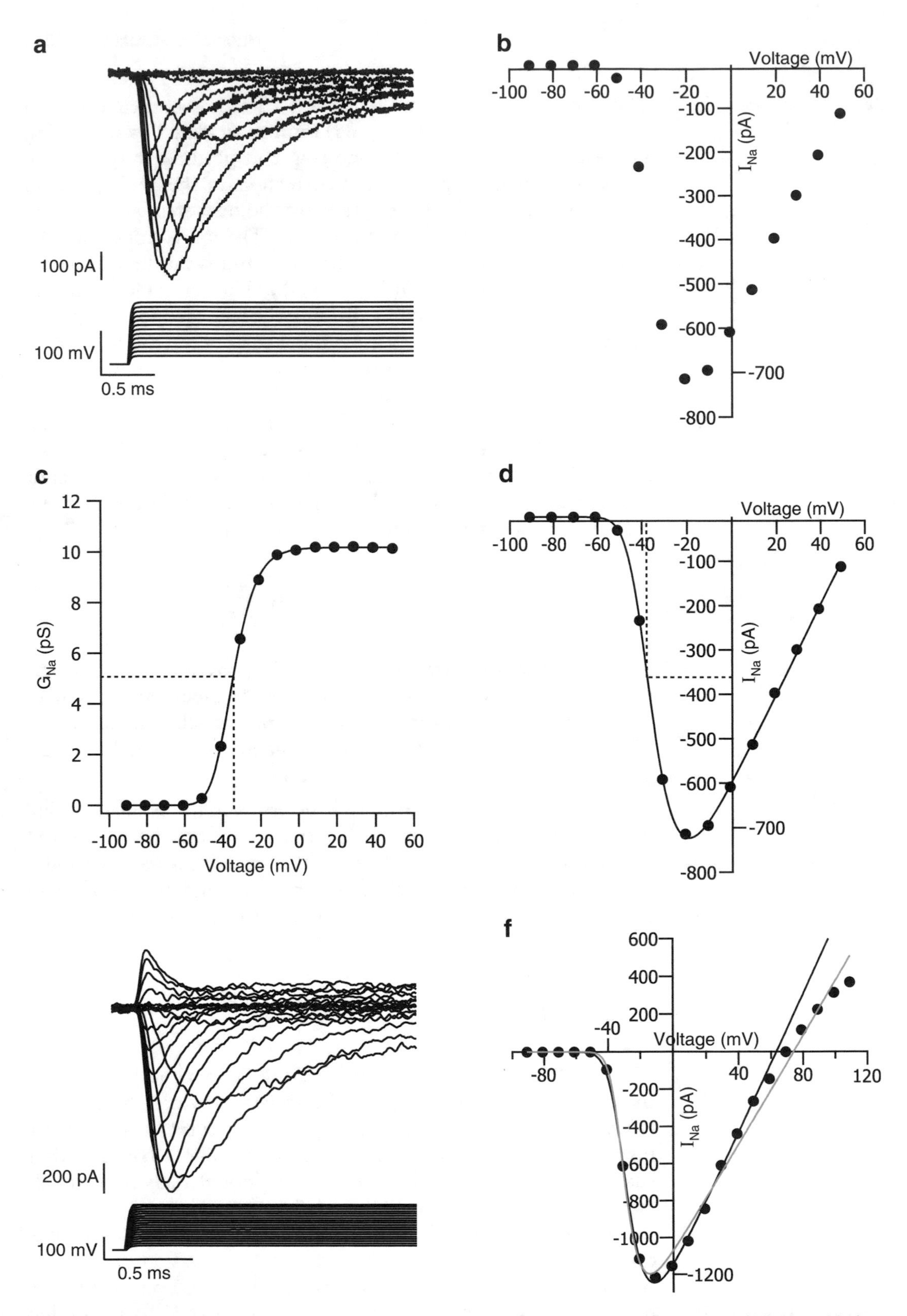

Fig. 3 Activation analysis of ion currents. (**a**) Na^+ currents elicited by an activation protocol displayed below. (**b**) Peak current plotted versus voltage, yielding an *I–V* activation curve. The Na^+ reversal potential can be extrapolated from the curve. (**c**) The Na^+ conductance curve was calculated by dividing the *I–V* curve by the driving force. A third-order Boltzmann function was fitted to the conductance curve (*smooth line*). The *dashed*

zero to a finite value, the model assumes $a = 4$. The rising phase of the Na^+ conductance can be described by $a = 3$ [7]. Traditionally, $a = 2$ is used for Ca^{2+} conductance.

I–V curves can also reflect further behaviors, such as current rectification, i.e., an ion current passes more easily in one direction through the membrane than in the other. This property is shown by a subtype of K^+ channels, the inward-rectifier K^+ channel, through which current passes inward more easily than outward and vice versa for the outward-rectifier K^+ channels. Na^+ channels show the opposite rectification behavior (Fig. 3e, f). Fitting the entire *I–V* curve may result in a distorted fit due to rectification of the current (grey line, Fig. 3f). Thus, for an accurate fit the Boltzmann equation should be fitted to the part of the curve before the rectification begins (black line, Fig. 3f).

Instantaneous *I–V* Curve

The tail current peak measured using a deactivation pulse protocol can be plotted as a function of the voltage changes after correcting the liquid junction potential (dashed line, Fig. 4a). This yields an instantaneous *I–V* curve (Fig. 4b). Typically, the curve is linear for all channels, indicating a voltage dependence of the channel gate rather than voltage dependence of the channel conductance [32].

Inactivation *I–V* Curve

Most of the voltage-gated ion channels (Na^+, Ca^{2+}, K^+) close during a voltage step (inactivation) [33]. The steady-state inactivation peak current can be measured using a steady-state inactivation pulse protocol and plotted versus the increasing voltage after correcting for the liquid junction potential (Fig. 5a). The current at the different voltages can be normalized to the maximal current. The curve can be fitted with a first-order Boltzmann equation according to the Hodgkin-Huxley model [7] (Fig. 5b, smooth line):

$$I/I_{\max} = \frac{1}{1 + e^{\left(\frac{V_m - V_{1/2}}{k}\right)}} \tag{11}$$

where $I/I_{\max}$ is the current normalized to its maximal value, V_m is the membrane potential, $V_{1/2}$ is the voltage at which the conductance is half-maximal, and k is the slope of the curve.

Fig. 3 (continued) lines indicate the voltage required to activate half the channel population ($V_{1/2}$). (**d**) The activation *I–V* curve in (**a**) was fitted with a combined Ohm's law and a third-order Boltzmann equation. The *dashed lines* indicate $V_{1/2}$. (**e**) Inward and outward Na^+ currents elicited using an activation protocol displayed below. (**f**) The *I–V* curve shows rectification of the outward Na^+ current. The curve was fitted partially (without the rectification) to obtain a best fit (*smooth black line*). The *smooth grey line* is the fit of the entire curve

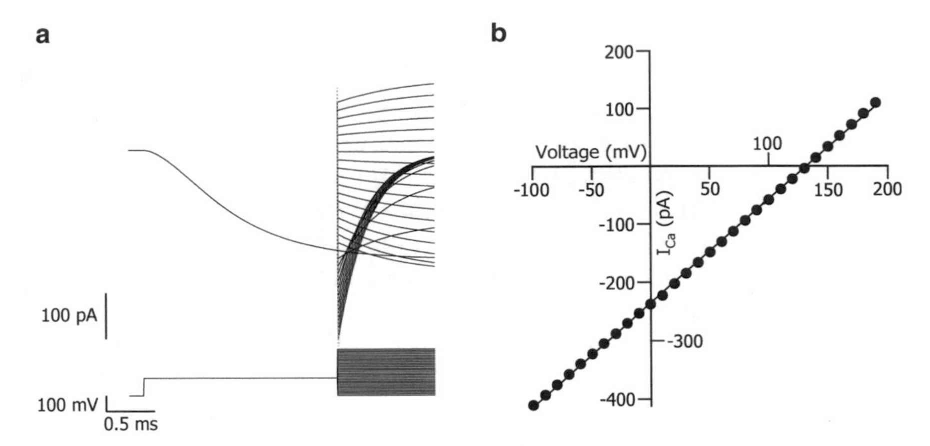

Fig. 4 Instantaneous *I–V* curve. (**a**) The kinetics of Ca^{2+} currents in nucleated patches was inserted into a Ca^{2+} channel model. The Ca^{2+} tail currents were simulated by a deactivation pulse protocol displayed below. (**b**) The peak tail current 10 μs from the briefly depolarizing step (*dashed grey line* in **a**) was plotted against increasing voltage, yielding a linear relationship

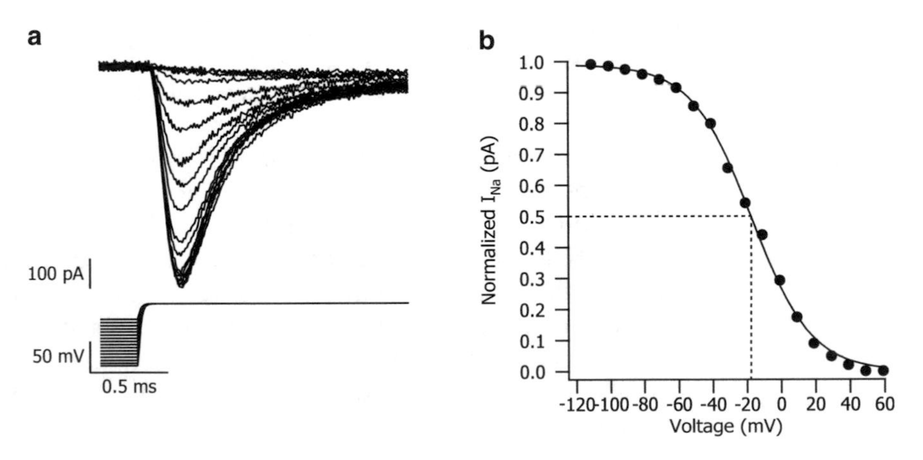

Fig. 5 Steady-state inactivation analysis. (**a**) Na^+ steady-state current inactivation elicited using a steady-state inactivation pulse protocol displayed below. (**b**) The peak current normalized to its maximal amplitude plotted against the pre-pulse voltage, yielding an *I–V* inactivation curve. A first-order Boltzmann equation was fitted (*smooth line*). The *dashed lines* indicate $V_{1/2}$

4.1.2 Time Constants (τ)

Ion currents can be described in terms of kinetic parameters reflecting the time dependence of the conductance, i.e., the time constant (τ). Each ion channel shows a time constant for each kinetics state (activation, deactivation, inactivation) that differs in its duration. If time constants differ, either between channels (e.g., K^+ and Na^+ channels) or between channels subtypes (e.g., T- and L-type Ca^{2+} channels), a specific ion current can be isolated kinetically.

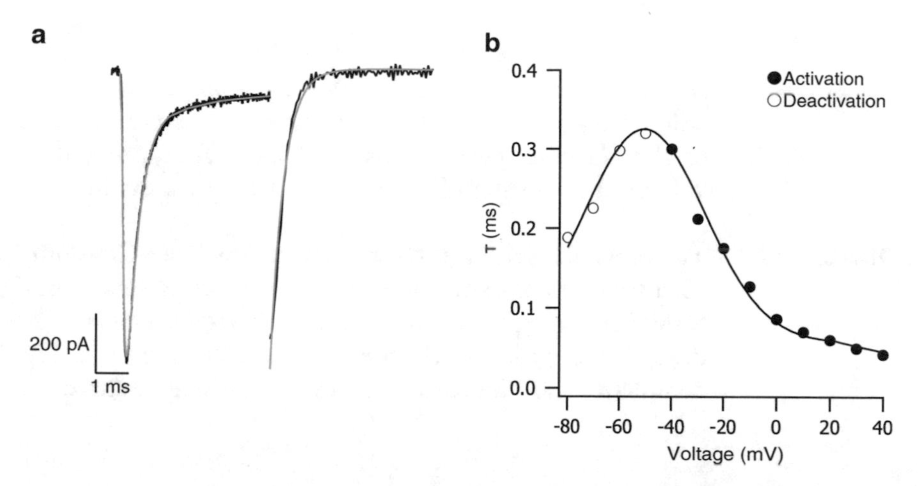

Fig. 6 Analysis of activation kinetics. (**a**) Activation (*left*) and deactivation (*right*) fitting of a third-order Hodgkin-Huxley model (*grey line*). (**b**) Activation (*filled circle*) and deactivation (*open circle*) time constants determined from traces like those in (**a**). The *smooth line* is the Gaussian function fitted to the curve

Activation Time Constant

The activation ion currents can be fitted using the following equation [34]:

$$I_0 + I_\infty \cdot (1 - \exp(-(t - t_0)/\tau_\mathrm{m}))^a \exp \cdot (-(t - t_0)/\tau_\mathrm{h}) \quad (12)$$

where t is time, I_∞ is the steady-state current, I_0 is the current at $t = 0$, τ_m and τ_h are the activation and inactivation time constants, respectively, of the exponential relaxation, and a is the number of activation gates in the model (Fig. 6a, right). The activation (τ_m) and inactivation (τ_h) time constants can be extracted from the declining and the rising phases of the ion current at the different voltages.

The activation time constant (τ_m) can be plotted against the increasing voltages to obtain the ion channel kinetics (●, Fig. 6b). The curves obtained from the activation ion currents are restricted to a narrow voltage range. Completing the activation and inactivation curves requires further analysis.

To complete the activation time constant curve the time constant must be extracted at more hyperpolarized voltages. This can be done by fitting Eq. (13) to the rising phase of the tail currents recorded using the deactivation pulse protocol (Fig. 6a, left) and

$$I_0 + I_\infty \cdot \exp\left(-\frac{at}{\tau}\right) \quad (13)$$

where t is time, I_∞ is the steady-state current, I_0 is the current at $t = 0$, τ is the deactivation time constant, and a is the number of activation gates in the model. The deactivation time constant can be added to the activation time constant curve (○, Fig. 6b). The time constant curve displays a bell-shaped voltage dependence, thus it can be fitted with a Gaussian function:

$$C_1 + C_2 \cdot \exp\left(-\left(\frac{V_{\mathrm{m}} - C_3}{C_4}\right)^n\right) \quad (14)$$

where C_1 is the voltage-independent time constant, C_2 is the height of the Gaussian peak, V_{m} is the membrane potential, C_3 is the voltage at the center of the peak, and C_4 is the standard deviation.

Inactivation Time Constant

The previous section (section "Activation Time-Constant") briefly discussed how to extract inactivation time constants from the rising activation currents at relatively depolarized voltages. This procedure gives a partial inactivation time constant curve (●, Fig. 7c). To complete the inactivation time constant curve at more

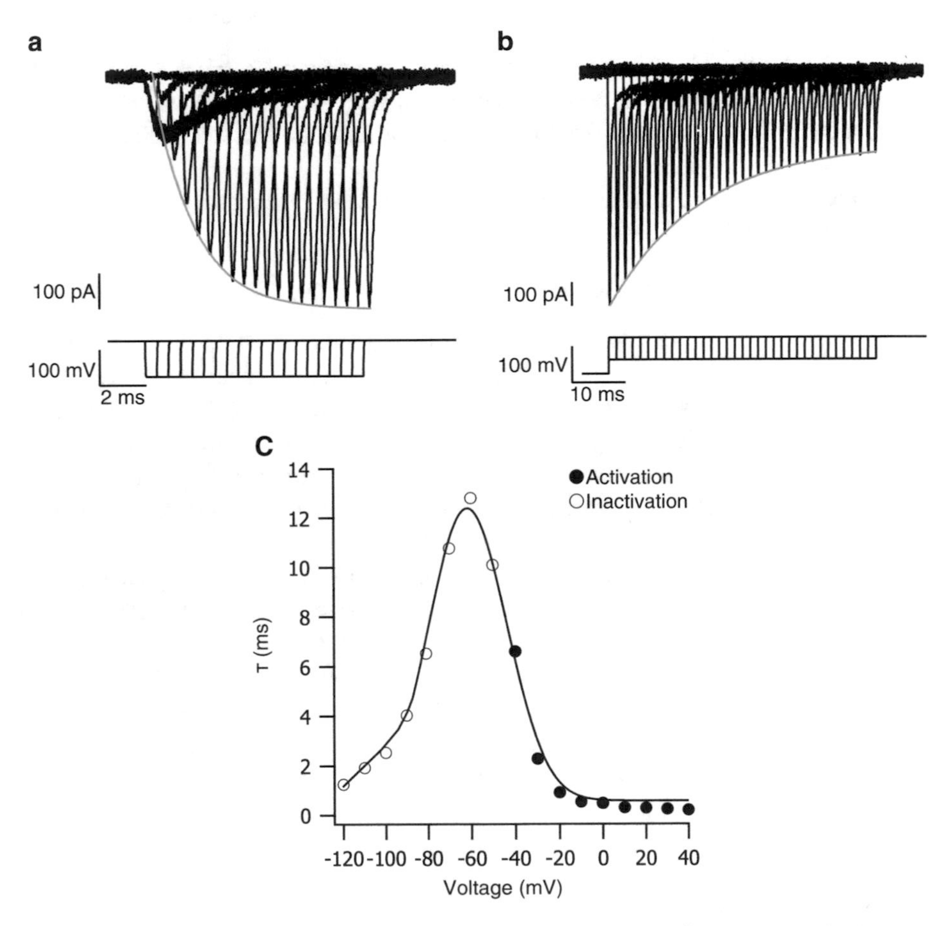

Fig. 7 Analysis of inactivation kinetics. (**a**) Recovery from inactivation Na^+ currents elicited using recovery from an inactivation protocol at −100 mV displayed *below*. An exponential equation was fitted to the declining phase of the peak current that recovered from inactivation (*smooth grey line*). (**b**) Inactivation onset of Na^+ currents elicited using an onset of inactivation protocol at −60 mV displayed at the *bottom*. An exponential equation was fitted to the rising phase of the peak current (*smooth grey line*). (**c**) The Na^+ inactivation time constant calculated from the rising phase of the activation currents (*filled circle*) and from the recovery inactivation and the onset inactivation currents (*open circle*) plotted as a function of voltage. The *smooth line* is the Gaussian fit

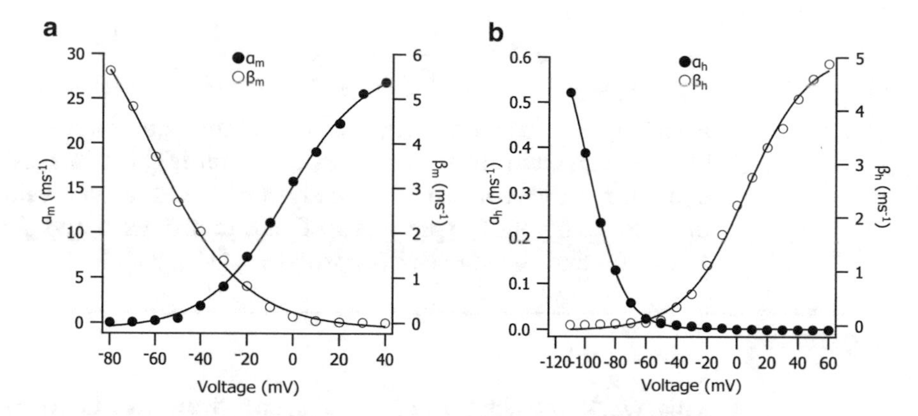

Fig. 8 Activation and inactivation rate constants. The forward (α) and the backward (β) rate constants for the Na^+ channel activation (**a**) and inactivation (**b**) are plotted versus voltage

hyperpolarized voltages, one must analyze the currents yielded by the recovery from inactivation and onset of inactivation pulse protocols. An exponential equation is fitted either to the declining phase of the current that recovered from inactivation (Fig. 7a) or to the rising phase of the inactivation current at different voltages (Fig. 7b). The fitting gives an inactivation time constant that can be added to the curve as a function of the voltage (○, Fig. 7c). The time constant curve displays a bell-shaped voltage dependence; thus it can be fitted with a Gaussian function as in the activation time constant curve (section "Activation Time-Constant").

The Ca^{2+} channel is exceptional in having two types of inactivation, Ca^{2+}- and voltage-dependent inactivation. The inactivation time constant measured from the rising phase of the Ca^{2+} current obtained by an activation pulse protocol contains both inactivation currents. For this reason, an extracellular solution containing Ba^{2+} ions rather than Ca^{2+} ions is preferred when examining Ca^{2+} channels. The inactivation shape with Ca^{2+} solution differs from that obtained with Ba^{2+} solution, with the inactivation phase of the Ca^{2+} current steeper than the activation of the Ba^{2+} current. This difference can arise from the absence of Ca^{2+}-dependent inactivation in the Ba^{2+} solution, the remaining inactivation being only voltage dependent. Therefore, to understand the inactivation kinetics of Ca^{2+} channels, if possible, it seems reasonable to calculate two inactivation time constant curves, one with the Ca^{2+} solution and a second with the Ba^{2+} solution.

4.1.3 Calculating the Transition Rates α and β

The analyzed data can be presented as α and β curves. According to the equations of p_∞ and τ (Eqs. (3)–(6)), the rate constant can be displayed by:

$$\alpha(V) = \frac{p_\infty(V)}{\tau(V)}, \quad \beta(V) = \frac{1 - p_\infty(V)}{\tau(V)} \quad (15)$$

where p_∞ is calculated using the Boltzmann equation (Sect. 4.1.1) and τ is extracted from the recorded currents (Sect. 4.1.2). These equations easily convert the steady-state conductance values and time constants into expressions of rate constants α and β for the activation and the inactivation processes (Fig. 8).

Acknowledgments

This work was supported by a grant from the German-Israeli Foundation to AK (#1091-27.1/2010).

References

1. Neher E, Sakmann B, Steinbach JH (1978) The extracellular patch clamp: a method for resolving currents through individual open channels in biological membranes. Pflugers Arch 375(2):219–228
2. Edwards FA, Konnerth A, Sakmann B, Takahashi T (1989) A thin slice preparation for patch clamp recordings from neurones of the mammalian central nervous system. Pflugers Arch 414(5):600–612
3. Stuart GJ, Dodt HU, Sakmann B (1993) Patch-clamp recordings from the soma and dendrites of neurons in brain slices using infrared video microscopy. Pflugers Arch 423(5-6):511–518
4. Sigworth F (1995) Design of the EPC-9, a computer-controlled patch-clamp amplifier. 1. Hardware. J Neurosci Methods 56 (2):195–202
5. Johnston D, Wu SM-s (1995) Foundations of cellular neurophysiology. MIT Press, Cambridge, MA
6. Sigworth FJ, Affolter H, Neher E (1995) Design of the EPC-9, a computer-controlled patch-clamp amplifier. 2. Software. J Neurosci Methods 56(2):203–215
7. Hodgkin AL, Huxley AF (1952) A quantitative description of membrane current and its application to conduction and excitation in nerve. J Physiol 117(4):500–544
8. Hodgkin AL, Huxley AF, Katz B (1952) Measurement of current-voltage relations in the membrane of the giant axon of Loligo. J Physiol 116(4):424–448
9. Schaefer AT, Helmstaedter M, Schmitt AC, Bar-Yehuda D, Almog M, Ben-Porat H, Sakmann B, Korngreen A (2007) Dendritic voltage-gated K^+ conductance gradient in pyramidal neurones of neocortical layer 5B from rats. J Physiol 579(Pt 3):737–752
10. Schaefer AT, Helmstaedter M, Sakmann B, Korngreen A (2003) Correction of conductance measurements in non-space-clamped structures: 1. Voltage-gated K^+ channels. Biophys J 84(6):3508–3528
11. Bar-Yehuda D, Korngreen A (2008) Space-clamp problems when voltage clamping neurons expressing voltage-gated conductances. J Neurophysiol 99(3):1127–1136
12. Chen X, Whissell P, Orser BA, MacDonald JF (2011) Functional modifications of acid-sensing ion channels by ligand-gated chloride channels. PLoS One 6(7):e21970
13. Kim NK, Robinson HP (2011) Effects of divalent cations on slow unblock of native NMDA receptors in mouse neocortical pyramidal neurons. Eur J Neurosci 34(2):199–212
14. Sather W, Dieudonne S, MacDonald JF, Ascher P (1992) Activation and desensitization of N-methyl-D-aspartate receptors in nucleated outside-out patches from mouse neurones. J Physiol 450:643–672
15. Almog M, Korngreen A (2009) Characterization of voltage-gated Ca^{2+} conductances in layer 5 neocortical pyramidal neurons from rats. PLoS One 4(4), e4841
16. Bekkers JM (2000) Properties of voltage-gated potassium currents in nucleated patches from large layer 5 cortical pyramidal neurons of the rat. J Physiol 525(Pt 3):593–609
17. Jung SC, Eun SY (2012) Sustained K^+ outward currents are sensitive to intracellular heteropodatoxin2 in CA1 neurons of organotypic

cultured hippocampi of rats. Kr J Physiol Pharmacol 16(5):343–348

18. Khurana S, Liu Z, Lewis AS, Rosa K, Chetkovich D, Golding NL (2012) An essential role for modulation of hyperpolarization-activated current in the development of binaural temporal precision. J Neurosci 32(8):2814–2823
19. Korngreen A, Sakmann B (2000) Voltage-gated K^+ channels in layer 5 neocortical pyramidal neurones from young rats: subtypes and gradients. J Physiol 525(Pt 3):621–639
20. Lin YC, Liu YC, Huang YY, Lien CC (2010) High-density expression of Ca^{2+}-permeable ASIC1a channels in NG2 glia of rat hippocampus. PLoS One 5(9)
21. Gentet LJ, Stuart GJ, Clements JD (2000) Direct measurement of specific membrane capacitance in neurons. Biophys J 79 (1):314–320
22. Veruki ML, Oltedal L, Hartveit E (2010) Electrical coupling and passive membrane properties of AII amacrine cells. J Neurophysiol 103 (3):1456–1466
23. Neher E (1992) Correction for liquid junction potentials in patch clamp experiments. Methods Enzymol 207:123–131
24. Barry PH, Diamond JM (1970) Junction potentials, electrode standard potentials, and other problems in interpreting electrical properties of membranes. J Membr Biol 3 (1):93–122
25. Barry PH, Lynch JW (1991) Liquid junction potentials and small cell effects in patch-clamp analysis. J Membr Biol 121(2):101–117
26. Costantin JL, Qin N, Waxham MN, Birnbaumer L, Stefani E (1999) Complete reversal of run-down in rabbit cardiac Ca^{2+} channels by patch-cramming in Xenopus oocytes; partial reversal by protein kinase A. Pflugers Arch 437(6):888–894
27. Josephson IR, Varadi G (1996) The beta subunit increases Ca^{2+} currents and gating charge movements of human cardiac L-type Ca^{2+} channels. Biophys J 70(3):1285–1293
28. Belles B, Hescheler J, Trautwein W, Blomgren K, Karlsson JO (1988) A possible physiological role of the Ca^{2+}-dependent protease calpain and its inhibitor calpastatin on the Ca^{2+} current in guinea pig myocytes. Pflugers Arch 412 (5):554–556
29. Romanin C, Grosswagen P, Schindler H (1991) Calpastatin and nucleotides stabilize cardiac calcium channel activity in excised patches. Pflugers Arch 418(1-2):86–92
30. Ikeda SR (1991) Double-pulse calcium channel current facilitation in adult rat sympathetic neurones. J Physiol 439:181–214
31. Scamps F, Valentin S, Dayanithi G, Valmier J (1998) Calcium channel subtypes responsible for voltage-gated intracellular calcium elevations in embryonic rat motoneurons. Neuroscience 87(3):719–730
32. Hodgkin AL, Huxley AF (1952) The components of membrane conductance in the giant axon of Loligo. J Physiol 116(4):473–496
33. Hodgkin AL, Huxley AF (1952) The dual effect of membrane potential on sodium conductance in the giant axon of Loligo. J Physiol 116(4):497–506
34. Connor JA, Stevens CF (1971) Voltage clamp studies of a transient outward membrane current in gastropod neural somata. J Physiol 213 (1):21–30
35. Gurkiewicz M, Korngreen A (2006) Recording, analysis, and function of dendritic voltage-gated channels. Pflugers Arch 453(3):283–292

Chapter 15

Creating and Constraining Compartmental Models of Neurons Using Experimental Data

Stefanos S. Stefanou, George Kastellakis, and Panayiota Poirazi

Abstract

In order to understand the information processing at the level of individual nerve cells, detailed information is required about the complex interactions between the anatomical structure of the neurons, their physiological properties, and synaptic input. Embodying such information in a formal theoretical model is useful as a predictor as it can simulate the electrical behavior of neurons in ways or conditions that may not be possible experimentally. In order for a model neuron to perform as closely as a real neuron, it needs to be constrained against all the available experimental data. The calibration of the parameters in each compartment requires a careful examination of the relevant literature, experimental evidence, as well as the intuition of the experimenter. However the hand tuning of model parameters can be extremely tricky due to the increased complexity of today's compartmental neuron models. Consequently the choices of structural model and the level of detail at which these models are constructed are critical considerations and because of our limited knowledge of the system the predictive value of each model is also limited. Therefore, a neuronal model is as good predictor as it was created to be. In this chapter we will address how a realistic conductance-based compartmental model is constructed and what are the common restrictions that prevent a model from reproducing the desired biological dynamics.

Key words Cable theory, Compartmental model, Morphology, Ion channels, Software, Active properties, Passive properties, Synapse, Optimisation, Neural network

1 Introduction

1.1 Equivalent Circuits

Compartmental models are conductance-based models of neurons composed of multiple spatial compartments, which can reproduce the electrical behavior of the cell to a high level of accuracy. In order to mimic the signal propagation properties of a real cell, the morphology of a neuron is broken down to a set of cylinders, which are connected to each other through axial resistances in a tree-like structure as shown in Fig. 1. Each of these cylinders can have different lengths and diameters, and different, but uniform, membrane properties. The main assumption in the compartmental approach is that every cylinder can be treated as an isopotential element; the essentially continuous structure of a neuron can thus

Alon Korngreen (ed.), *Advanced Patch-Clamp Analysis for Neuroscientists*, Neuromethods, vol. 113,
DOI 10.1007/978-1-4939-3411-9_15, © Springer Science+Business Media New York 2016

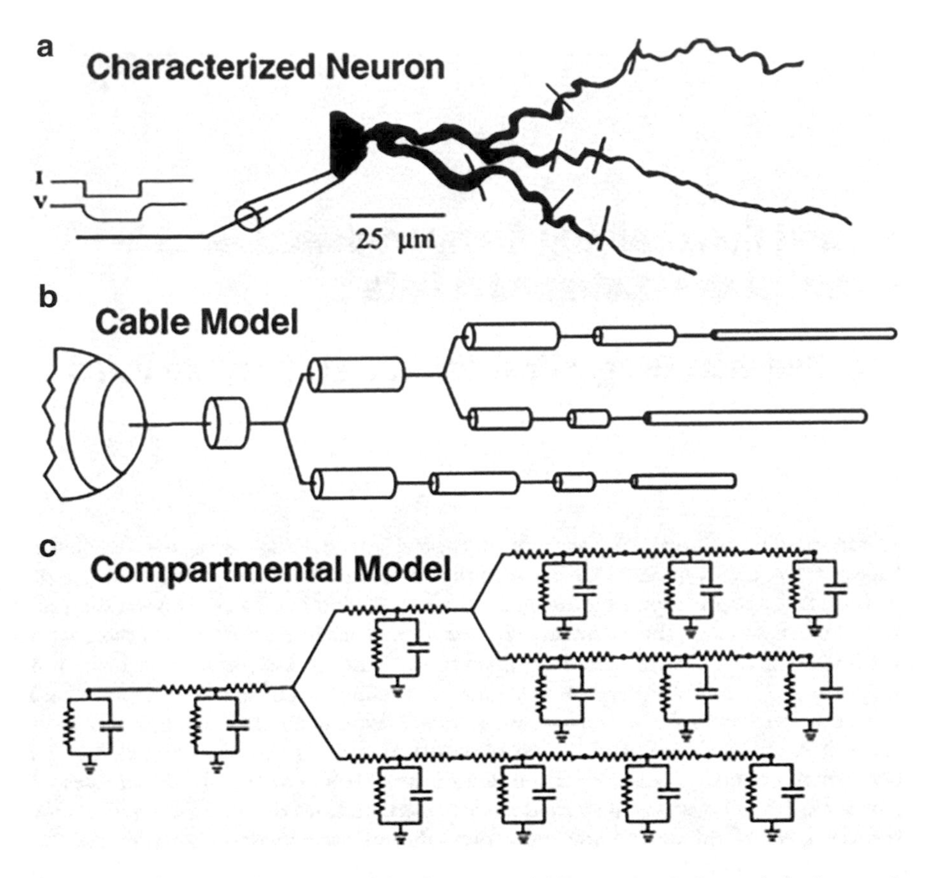

Fig. 1 Creating morphologically detailed models of neurons. (**a**) Reconstructed morphologies can be acquired from stained neurons. Creation of a model neuron starts with compartmentalization of the characterized morphology to isopotential cylinders. (**b**) The cable model specifies the necessary cylindrical isopotential compartments that will be implemented in the model. (**c**) Each cylindrical compartment is modeled by its equivalent electrical circuit, which has passive properties. Figure adapted from [60]

be approximated by a linked assemblage of such isopotential elements. This approach simplifies the mathematics while posing no restriction on the membrane properties of each compartment or the geometry that the model neuron can simulate. This seminal equivalent cylinder model was introduced by Wilfrid Rall [1] and was based on observations that Golgi-stained motor neurons in the cat appeared to be consonant with the constraints required to collapse the complex branched dendrites of a neuron into a single equivalent cylinder. Such a model made it possible to develop analytical expressions derived from cable theory and proved to be sufficient to study the spread of signals in lengthy neurites.

In each neuronal compartment, the voltage across the cell membrane V_m (Fig. 2) depends on both the passive properties of the membrane (membrane capacitance C_m and input resistance R_m), as well as on voltage-dependent currents (e.g., G_k in Fig. 2) and external current input. Compartments are connected with each

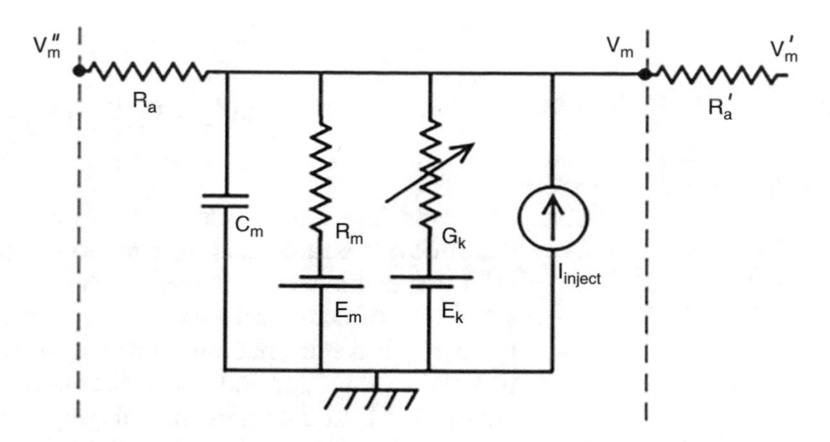

Fig. 2 The equivalent circuit of a compartment that is part of a neuronal cable. The voltage across the membrane (V_m) depends on the combination of passive membrane properties as well as on the mixture of ionic mechanisms that populate the membrane surface. Synaptic input is modeled via an additional synaptic current (I_{inject}). Figure adapted from [60]

other in series through the axial resistances (R_a, R_a' in Fig. 2). The axial resistance is used to model the movement of electrical charge when there is a voltage differential across successive compartments. The circuit is simple, consisting of a capacitor that simulates the non-conducting bilipid cell membrane and a resistor as an analog of the conducting ionic channels on the membrane. This simplification of a passive membrane is adequate to reproduce the electrical responses tested on real neurons. Despite some initial errors and misinterpretations, models of passive neuronal cell membranes were able to make predictions about the propagation of electric currents in the neuron around the late 1950s [1–3].

The theoretical foundation of compartmental modeling is the Cable Theory, which was originally developed to study the propagation of signals in telegraphic cables [4]. The solution to the problem of the electric current flow in and out of a cylindrical core conductor was initially applied to study the electrical conductive properties of neuronal cells [5]. The formulation of the cable theory enabled the creation of compartmental models of neurons. Its basis lays on the creation of an electrical circuit that performs equivalently as a cylindrical part of neuronal cell membrane. According to cable theory, the spatial spread of electrical signals over time is modeled with the following nonlinear differential equation:

$$\frac{l^2 \partial^2 V_j}{R_a \partial x^2} = C_m \frac{\partial V_j}{\partial t} + \frac{V_j}{R_m} + I_j. \tag{1}$$

In practice, neuronal simulators make use of a Taylor-series approximation of this equation, which yields, for the case of the circuit shown in Fig. 2:

$$C_{\mathrm{m}}\frac{\mathrm{d}V_{\mathrm{m}}}{\mathrm{d}t}=\left[\frac{E_{\mathrm{m}}-V_{\mathrm{m}}}{R_{\mathrm{m}}}\right]+\sum_{\mathrm{k}}[(E_{\mathrm{k}}-V_{\mathrm{m}})G_{\mathrm{k}}]+\left[\frac{V'_{\mathrm{m}}-V_{\mathrm{m}}}{R'_{\mathrm{a}}}\right]+\left[\frac{V''_{\mathrm{m}}-V_{\mathrm{m}}}{R_{\mathrm{a}}}\right]+I_{\mathrm{inject}}. \quad (2)$$

This modeling framework can be extended to include branched dendritic tree structures. Cable theory, in general, provides a powerful framework which allows for the accurate phenomenological simulation of neural cells with very complex dendritic and axonal processes. It is not uncommon to use thousands of compartments to simulate extended and complex dendritic trees such as the Purkinje cell [6]. In the following paragraphs, we discuss the practical aspects of creating accurate neuronal models of neurons and using them to simulate their electrophysiological behavior as explained above.

2 Modeling Dendritic Morphology

Neurons vary widely in their morphological characteristics from region to region, and this variability is believed to affect the processing of incoming synaptic signals [7]. Consequently, in order to create neuron models that closely resemble a real neuron one must assume or use an actual neuronal morphology. The use of anatomically correct morphological reconstructions of the desired cell type is significant for the later steps of constraining the passive properties of the model. Errors in reconstructions, e.g., in diameter, can propagate to deviations in estimated cell capacitance, passive membrane resistance and passive axial resistance compared to the true parameter values of the real neuron.

Seminal modeling work by Rall [1] was based in part on observations of Golgi-stained neurons using light microscopy. The rapid Golgi technique allows the full staining of individual neurons in fixed tissue. However, it is not recommended for neural modeling because of the unavailability of physiological data from the stained neurons. Radioactive or fluorescent tracers have also been used [8, 9], but were limited in the sense that they revealed only parts of the dendritic tree that was needed, e.g., for dendritic localization during recordings. The use of light microscopy is limited by the wavelength of the light. In typical conditions a thin line ~0.1 μm will be imaged as a 0.3 μm (limit), leading to errors in the subsequent modeling. Additionally, profiles that expand vertically (towards the imaging axis) can be missed altogether. Advanced optical techniques like laser scanning microscopy [10] or confocal microscopy [11] permit reliable reconstruction of neuronal structures. Although the resolution of confocal microscopy is about 1.4 times higher than a light microscope [12], tracing the whole neuron at high resolution (e.g., for fine dendrites) requires to extend over hundreds or thousands of confocal stacks. Today the standard for creating 3-dimensional reconstructions of real cells is biocytin labeling by diffusion via the whole-cell pipette. This technique allows for simultaneous recording and provides a higher contrast than most available dyes and an overall durable staining [13]. The reconstruction for light or confocal microscopy is usually performed automatically by specialized software such as Neurolucida, which creates a 3d morphology representation in an automated way. To convert Neurolucida files to morphology file formats used by compartmental simulators the CVapp is recommended, available from the www.neuromorpho.org database of neuronal morphologies.

As mentioned before, when creating a realistic compartmental model one needs to create a structure of connected cylindrical compartments that morphologically matches the real cell. However, in practice, the construction of compartmental models requires simplifications of the original morphology which are employed in order to make simulations faster by reducing the number of equivalent cylinders that need to be simulated. Because each compartment is regarded, by definition, as an isopotential element, its dimensions must be chosen to minimize the computational error introduced by this assumption. Consequently the length of each compartment must be chosen to be a fraction of the characteristic length (λ), thus ensuring that the signal decay within a compartment can be negligible. Failure to implementing a detailed morphology correctly can seriously limit the number of phenomena that the model can reproduce, or lead to wrong conclusions. Dendritic spike initiation or dendritic synaptic integration, for example, cannot be simulated in very simple morphological models.

On the other hand, modeling the complete morphology of the neuron is computationally intensive, as the simulation time increases with the number of compartments. It is thus sometimes useful to find the simplest possible form of a model that performs well phenomenologically. A reduced morphology, retaining the characteristics needed for the model to reproduce the experimental data, reduces the number of model parameters that need to be defined and validated while also increasing computation speed. Reducing a number of passive dendritic trees to an equivalent passive cable is straightforward [14, 15], but not always possible for active dendrites. This is due to the different local impedance of the reduced dendrite. Other simplifications can be made in active models, e.g., collapsing the detailed morphology of spines on the compartment they are attached to. A morphologically reduced model can be used just to reproduce the complex phenotype of a real neuron [16] rather than recreating a detailed model neuron in silico where there is the need for all the variables to be directly related to experimentally measurable quantities.

3 Modeling Passive Cable Properties

Modeling a membrane patch that contains only passive ionic conductance is a simplistic, but fundamental way of understanding active membranes. Passive channels do not change their permeability to ions in response to changes in the membrane potential, thus resulting in a linear or sublinear response of the membrane to incoming synaptic stimuli. Passive neuron models can be used as an electrical skeleton for the active conductances to be added later. Additionally, they can be used to reconstruct synaptic currents or conductances that had been filtered along the neuronal tree, because neuronal morphology acts as a low-pass filter. Finally, a passive neuronal model can be used as a simple model lacking the complicated construction and validation of the active ones. The modeling of passive channels can also be used to simulate experimental findings as in [17]. Each channel has selective permeability to some ions. The total current density flowing in a neurite through a channel is the sum of the contributions from the individual ions.

To construct a model of a neuron, it is necessary first to characterize its passive electrical properties. This may not always be possible, due to lack of experimental data or due to, e.g., accumulation of variability or errors. Additionally, it is not easy to isolate the effect of passive conductances, because under normal conditions the active conductances dominate over the passive ones. Due to these limitations, there is a margin for error in estimating the model's passive constants.

Passive membrane capacitance per unit area (C_m), passive membrane resistance per unit area (R_m), and passive membrane cytoplasmic resistivity (R_i) are frequently referred to as the specific passive parameters. Passive membrane capacitance per unit area is unlikely to vary much along the neuron. However passive membrane resistance and its reciprocal R_i are more likely to change along the length of a neuron. Estimating the specific passive parameter constants (R_m, R_i, C_m) is considered as the "inverse estimation problem" [18] and, along with the morphology, determines the electrotonic properties of the neuron. Generally, the assumption that these constants are uniform along the compartments can be made, reducing the values that need to be fitted to just three. To obtain the correct values of passive parameters one needs to initially infer the above constants and then optimize them to fit electrophysiological recordings. In order to exclude the effect of active, voltage-gated conductances, it is typical to block them using pharmacological agents. Ideally one would optimize recordings to get a linear (passive) response [19]. It is considered best practice to reduce the complexity of the morphology to obtain good estimates of the unknowns that can be refined later [18].

A safe method for constraining a passive model is to first acquire electrophysiological recordings from a single neuron (preferably the same neuron whose morphology is used). Initially one can measure neuron responses to long pulses of current injected at the soma. Pulses of both polarities and in different holding voltages around the resting membrane potential are preferred [20, 21]. After checks for nonlinear responses like rectifying currents, sag (membrane undershoot), and creep (membrane overshoot), following with pharmacological blocking if observed, one can record again from different points of the neuron and repeat the checks for linear scale as before. The fitting of the model to the above recordings can be achieved either manually tweaking the specific passive parameter constants or utilizing an automated method in conjunction with a direct comparison error function like least-squares fitting. (The different error functions and optimizing methods are discussed later in the text.) Other passive properties of the cell, such as its input resistance, should also be reproduced by the model neuron. This would reflect the direct relation of the model passive constants with the real neuron, as a result of the careful calibration. It is possible to use direct voltage recordings, as opposed to voltage clamp to estimate the passive parameters of the membrane. This approach has the advantage that it eliminates any additional noise introduced by the voltage-clamp amplifier and the need to compensate for the series resistance. Tight-seal whole-cell recordings using patch-clamp pipettes can be used to assess the membrane time constant directly from a passive response [19].

4 Modeling Active Membrane Mechanisms

Even if basic assumptions can be extracted from a well-constrained, morphologically detailed passive model, the electrophysiological behavior of real neurons is dominated by active conductances. The active channels which are found in different concentration gradients across the surface of neuronal cells and their dendritic arbors underlie all sorts of nonlinear phenomena which render neurons powerful computational devices. In the equivalent cylinder of Fig. 2, active channels are modeled by a battery (reversal potential E_k) in series with a variable conductance (maximal conductance g_k). These channels can open and close in response to membrane potential changes or in response to specific chemical neurotransmitters.

Completing a passive model by adding active mechanisms gives the modeler great flexibility. The resulting detailed model after validation constitutes the respective in silico model of a real neuron. This can be used along with real neurons to confirm or reject conceptual hypotheses about the complex interplay of membrane conductances, synaptic mechanisms, and morphology.

More importantly, the model can also be used to perform novel experiments entirely in silico, due to the advantage that all variables are visible and can be controlled directly.

Realistic neuronal models used to study the integration of synaptic input ought to incorporate a complete dendritic morphology and all voltage- and ligand-gated channels. For example, [22] simulated a Prefrontal Cortex cell model which contains 13 different distributed ionic mechanisms representing different types of Na^+ and K^+ channels, voltage-gated Ca^{2+} channels, hyperpolarizing Ca^{2+}-dependent K^+ channels, etc. These conductances are not uniformly present in all compartments of a neuron. It is important to carefully calibrate the values of these channels within physiological limits and to ascertain their distribution based on detailed studies. This can be a daunting task for the modeler, as a lot of this information is either missing or not available for all different neuronal types.

If the model contains channels that are controlled by specific ions, the simulation of ionic concentrations inside the cell is also crucial. For example the intracellular calcium concentration controls the activation of certain Ca^{2+}-activated K^+ channels and is crucial for the initiation of phenomena believed to underlie synaptic plasticity. Usually four processes that affect intracellular calcium concentration are modeled: (a) the entry of Ca^{2+} via various ionic channels, I_{ca}, (b) the diffusion of calcium throughout the cell, (c) the action of intracellular calcium-binding proteins (buffers), and (d) the efflux of Ca^{2+} via membrane-bound pumps. Intracellular calcium concentration can be modeled in a phenomenological way [23]. Thus the model will contain fewer parameters than a biophysical model and be much easier to constrain. If such a model is not enough to model the dynamics of the system, an ad hoc model can be used [24].

The spatial distribution of conductances is of great importance for accurate model generation. For example, the distribution of the A-type, D-type, and M-type potassium channels can change as a function of the distance from the soma [25]. However obtaining the conductance gradients for all the membrane mechanisms is not feasible so the modeler is assuming uniform distribution or can experiment with different gradients and how they fit to experimental data.

The first step in constructing an active model using a strongly calibrated passive model is the review of the relevant literature for reported values of each current in order to establish upper and lower bounds for the maximum conductance of each active membrane mechanism of the specific neuron type. Scouting the experimental literature for biophysical property values can often be an arduous task. Physiological parameters show great variability between experiments, different animals, and different brain regions and depend on experimental manipulations. Reviews of experimental literature regarding specific mechanisms are often invaluable and sometimes a direct experience with the neuron can be vital for the modeler. An ideal model should integrate data from the same neuron type, region, animal, and age. However, it is proven unfeasible to obtain all the data in a single recording, or wait for data to become available. In such a case one needs to improvise integrating data from the most closely related animal species and/or neuron types. In the opposite case, where a multitude of data is available, experimentation is advised too. For example channel current kinetics fitted over a single experiment instead over the mean of recordings might give better results due to the voltage-dependent activation and inactivation that is presumably lost in averaged responses [6]. It is only very recently that attempts have been made to systematically catalogue the biophysical properties *en masse*. The online database NeuroElectro (neuroelectro.org) attempts to extract electrophysiological property values from the literature for all the neuronal types and catalogue them in a centralized database. This provides an invaluable tool that allows the visualization of such properties and the comparison between different types of cells.

Often, the complexity resulting from a morphologically and biophysically detailed model can be relaxed at some points in order for the simulation to be computationally efficient. Synaptic contacts that have the same isopotential postsynaptic target and the same presynaptic model neuron can be combined to a single synapse with greater weight for efficiency. It is also considered best practice to include the axon hillock in the model, as the action potentials are initiated there, especially if the modeler wants to reproduce back-propagating action potentials [26]. Although a detailed approach would require to explicitly model the spines that cover the majority of dendrites, one can add two extra compartments (to simulate the spine and its neck) at every synapse, if just the Ca^{2+} concentration is of interest. The rest of the spines can be collapsed to the compartment they are attached to for computational efficiency.

Parameterizing (e.g., standardizing channel kinetics by scaling due to different experimental temperatures) by validating against experimental data and subsequently testing the model is an essential but not always simple task. Accumulated electrophysiological data can be divided into two datasets, one used to constrain the model and one used to test it. The behaviors of the model that need to be reproduced must be defined and quantified. If an automatic parameter search method is to be used for the validation process, one should formalize the desirable behavior by an error function. Generally, the error function must be estimated after the tuning. In such a way the validation will not change, but useful insights about the importance of different parameters and model robustness can be obtained.

5 Synaptic Mechanisms

Simulating chemical synaptic input is an essential component of biophysically realistic modeling. A great advantage of compartmental modeling is that it allows the uneven distribution of different synaptic mechanisms in the dendritic tree. The spatiotemporal arrangement of synaptic inputs can lead to different modes of synaptic integration, which greatly affect the output of the cell [27–29]. Thus, it is important for the synaptic stimulation protocols used in compartmental simulations to be realistic, and to correspond to patterns of activity that are possible to exist in the brain, both for excitatory and inhibitory synapses.

Most excitatory connections in the cortex terminate on small protrusions, called spines, which serve to compartmentalize chemical synapses, and can also affect the local voltage response of the dendrites. While most compartmental models do not model the spine structure for reasons of computational efficiency, some models have explicitly studied the excitability properties of dendritic spines, simulating spines with a series of compartments [30, 31]. Both AMPAα-Amino-3-hydroxy-5-methyl-4-isoxazolepropionic acid (AMPA) (α-Amino-3-hydroxy-5-methyl-4-isoxazolepropionic acid) and NMDA (*N*-methyl-D-aspartate) excitatory receptor dynamics are usually modeled, as well as GABA (Gamma-Aminobutyric acid) receptor dynamics. In order to model realistic synaptic responses, voltage-clamp recordings (EPSCs and IPSCs) are required. Of particular importance are the recorded miniature EPSCs and IPSCs (mEPSCs and mIPSCs respectively), which provide the minimal, quantal current generated, presumably by the release of a single vesicle from the presynaptic terminal.

Multiple approaches can be taken to model the postsynaptic response at chemical synapses. It is possible to create detailed kinetic models of receptors, which account for all possible states of the receptor using probabilistic Markov models of ion channels. These models can be used to study the synaptic response in greater detail. For example, a kinetic model of the NMDA receptor has been used in [32] to study the effects of Mg^{2+} blockade kinetics in synaptic plasticity [33]. In practice however, this method is slow and is often replaced by simpler and computationally more efficient models. The temporal profile of synapse activation can be modeled by simple alpha functions or double exponentials. A more realistic method for modeling synaptic transmission uses analytical solutions to derive a model that accounts for summation and saturating responses and is also computationally efficient [32]. By tuning the parameters of this model, it is possible to simulate the action of both excitatory and inhibitory receptors.

Although synaptic excitation is modeled with some degree of detail, axonal synaptic transmission is usually not explicitly modeled. Instead, axons are considered as simple delay lines, and action potentials are modeled as unitary events. Thus, presynaptic processes are usually not modeled explicitly, and a common assumption is that the activation of each synapse saturates the postsynaptic receptor density.

6 Model Calibration/Optimization

In order for a model neuron to perform as closely as possible to a real neuron, it needs to be constrained against all the available experimental data for the specific neuron type. The calibration of all parameters in each compartment requires a careful examination of the relevant literature, experimental evidence, as well as the intuition of the experimenter. The concentrations of ionic mechanisms across the neuronal membrane as well as their maximum conductance values need to be calibrated in order to reproduce electrophysiological behavior that is similar to the one obtained from recordings, including local nonlinear phenomena, such as dendritic spikes. Due to the lack of data about many types of neurons, the density of ionic currents is often left as a free parameter in the model.

In order to constrain compartmental models, usually the voltage responses of the cell to in vitro current stimulation with varying simple currents are used. This is often a manual and iterative process which depends on the intuition and experience of the modeler and can be time-consuming. Kinetic models of active conductances are usually modeled using the Hodgkin and Huxley formulation. In order to constrain these models, whole-cell recordings are often used, attempting to match the model response to the recorded waveforms. An example of model fitting to experimental waveforms is given in Destexhe et al. for the thalamic relay neuron [21].

On the other hand, it is possible to constrain compartmental models using extracellularly recorded action potentials. Gold et al. [34] used extracellular recordings from a CA1 cell in order to constrain the parameters of a CA1 model. In order to do this, the authors calculated the extracellular potential based on the ionic currents on the neuronal membrane using the Line Source Approximation algorithm [35]. The authors found that extracellular potentials may be better suited to constrain compartmental models that include nonuniform active conductance densities [34], because the extracellular action potential waveform is more sensitive to the composition of active ionic conductances.

6.1 Automated Model Optimization

As the application of new experimental techniques is generating a continuously growing ensemble of experimentally measured properties for multiple neuronal types, and computational models are becoming increasingly detailed, there is growing demand to create computational methods that automatically constrain the large number of free model parameters by using the abundant computational power that is available nowadays. This is an exciting research area in computational neuroscience, which aims to lead to the automated neuronal model construction. Such Optimization Algorithms provide a way of exploring in an efficient way the vast parameter space of the compartmental model, in order to discover the set of parameters that create a response of the model that matches the experimentally measured response. Typically, the construction of these tuning algorithms can be subdivided into two independent stages: identification of the error function and selection of the optimization algorithm to be used. The error function (or fitness/cost function) quantifies the comparison of the model output with the experimental data. To this end the error function must be chosen to (a) reflect the properties of the data that the model has to reproduce, (b) be relatively fast to calculate, and (c) have a solution space as smooth as possible. Such functions can be feature based [36] (Fig. 3) or direct comparisons of electrophysiological traces, each with advantages and drawbacks [37]. Although the error function heavily influences the speed of convergence and parameter set (solution) chosen, the choice of optimization algorithm will determine the balance between the exploration and the exploitation of the parameter space in a finite time. A brute force method maximizes the exploration, a gradient descend method exploits maximally the local gradient, whereas stochastic methods, such as simulated annealing, and evolutionary algorithms combine the two for a more balanced parameter search in a given time. Generally, any error function can be used with any optimization algorithm. Exceptions are the multi-objective problems [36] and linear regression methods [38].

Brute force methods have been used to infer the unknown spatial distributions of ionic channels and to search for regions of the parameter space which are robust to disturbances [39]. Most model optimization studies use some form of gradient descent to identify the optimal parameters that optimize the performance of the compartmental model with respect to electrophysiological recordings from real neurons. In general, these methods attempt to minimize the error between the simulation response and the response of a real cell to the same stimulation protocol. The protocol usually utilized is the response of the neuron to short current injections under specific conditions. However, the derivation of an optimization method is not a trivial issue. First, due to intrinsic noise, experimental data are characterized by large variability; therefore, selecting a single voltage trace to use as the target for optimization is not easy. Second, there is no commonly accepted method for measuring the difference between simulated and

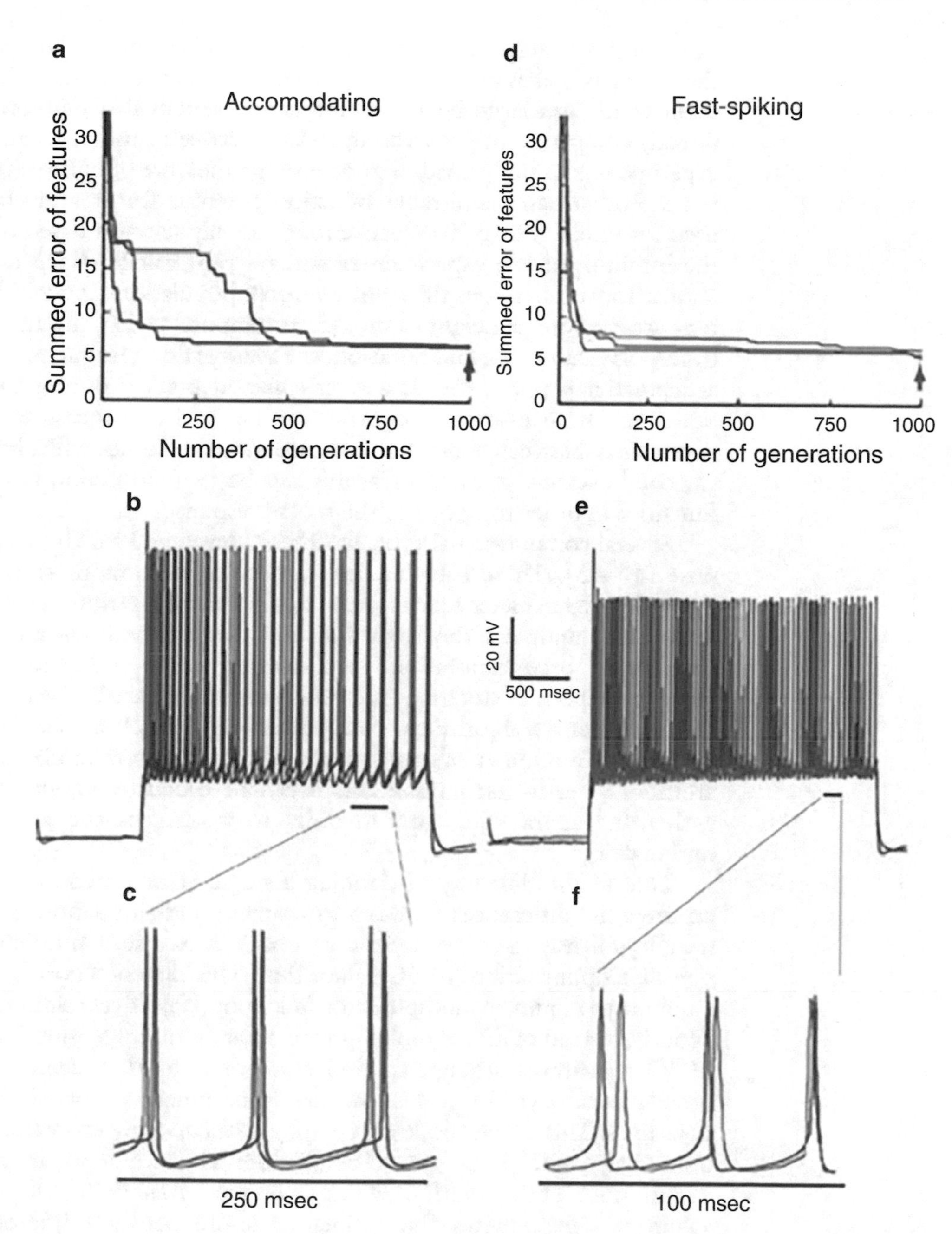

Fig. 3 Fitting of model to experimental traces using a multi-objective optimization procedure. (**a**, **d**) Convergence of summed errors for the accommodating behavior with each generation. In each generation, 300 sets of parameter values (values for maximal ion conductance) are evaluated simultaneously. The error value for the set of parameters that yield the minimal error is plotted. The three lines show the same fitting procedure repeated using different random initial conditions. Optimization was performed for two distinct neuron types: an accommodating and a fast spiking. (**b**, **e**) Comparison between one of the experimental responses (*red/blue*) and the model response (*green* trace) to the same input, using the best set of ion channel conductances obtained at the 1000th generation (point denoted by *red arrow* in **a**). (**c**, **f**) Zoom into the region marked by a *black line* in (**b**) and (**e**) respectively. Adapted from [36]

experimental voltage traces. For example, small temporal shifts in the timing of spikes would lead to high error values, because of the small width and large height of the action potential. Therefore, directly comparing the raw voltage traces between simulation and experiment does not provide a good enough measure of optimality. On the other hand, a number of action potential features can be used for model fitting. These criteria are usually selected based on the intuition of the experimenter and refer to values which are known to vary between different neuronal populations. Such criteria are the voltage height of the AP, its temporal width, the after-hyperpolarization size and duration, the average firing frequency to a depolarizing pulse, the degree of spike adaptation during the pulse, etc. Each of these properties can be used as a measure of discrepancy between simulation and experiment. Additionally, linear combinations of these measures can be used to define error functions in order to optimize the model response.

Several computational tools have been developed for this purpose [40–42]. These tools employ various algorithms to search through the parameter space for the optimal configuration of variables that minimizes the error function. Search methods range from brute force exploration, local random search and gradient descent methods to stochastic methods such as simulated annealing and evolutionary algorithms. Neurofitter [43] is such a tool that implements a number of search methods in order to minimize an arbitrary objective error function. It can be used in conjunction with many neural simulators in order to implement the model optimization.

Due to the difficulty of defining a single error function that captures the differences between experiment and simulation in a meaningful way, a more holistic approach is to use a multiple-objective optimization (MOO) algorithm. This class of algorithms attempts to minimize multiple error functions (objectives) simultaneously, instead of just optimizing with regard to a single function. MOO algorithms attempt to find configurations that *dominate* others, that is, yield equal or smaller error functions for all the objectives. This allows modelers to utilize concurrently error functions that would otherwise be conflicting. The end result is the identification of the solutions which lie on the *Pareto Front*, a set of solutions which identify the optimal trade-offs between different objectives [36]. However, it is imperative to note that utilizing MOO is not an automatic model creation panacea. Usually the variability of the resulting parameters across similar models is large [36, 44], whereas in models where the parameters were constrained in smaller batches (parameter peeling) the variability is much smaller [45, 46].

7 Modeling Networks of Neurons

In addition to building good models of individual neurons, linking molecular and neuronal phenomena to the functional output of brain circuits and to behavior requires detailed knowledge of the dynamics of neuronal networks. Due to experimental constraints, data necessary to create and parameterize these networks have yet to achieve the same level of detail as single neuron models.

Small networks of neurons, particularly those observed in preparations or in vivo, are attractive objects for modeling because all the necessary experimental data of the component neurons that form the network can be gathered easily and then described mathematically. In the case of missing data the modeler can adapt accurate models from similar cells to match the observed responses. As always, the compromises to detail must not affect the network function under study. In some cases lack of important data like conductance gradients of voltage-gated channels might force oversimplifications on the morphology detail of the component neurons also saving an unnecessary strain to the computational resources and focusing the attention on circuit interactions. Since the connectivity of the component neurons define the output of the network, synaptic interactions must be modeled as previously mentioned (see above) and connectivity measurements must be incorporated. Creating and calibrating the network consist only half of the modeling work needed. Equally important is adequate model testing against experimental data (preferably data not used to parameterize the model network). This might not always be a one-time process and the modeler might revert to parameterization for further tuning, leading to a close interplay of modeling and experiment. In addition, if a network is modeled in a more abstract way to be used for insight of its role and the role of similar circuits, one must test the simplifications and how they affect the model performance and interpretation of experimental data in each case.

Following the above a model of a hippocampal network was constructed to study the emerging gamma oscillations [47], a model of an entire cortical column and its interaction with the thalamus [48], microcircuits and larger networks of prefrontal neurons to study the contribution of NMDAR and inhibitory synaptic currents in the emergence of persistent firing observed during working memory [22, 49, 50], as well as the network size requirements (Fig. 4) to support persistent firing [51].

Advances in electrophysiological recordings and imaging are expected to aid the development of more accurate network models in the very near future. With the introduction of the Blue Brain Project [52], the BRAIN Initiative [53], and the Human Brain Project [54] creation of large-scale networks of compartmental models is becoming feasible, for example reconstructing a large

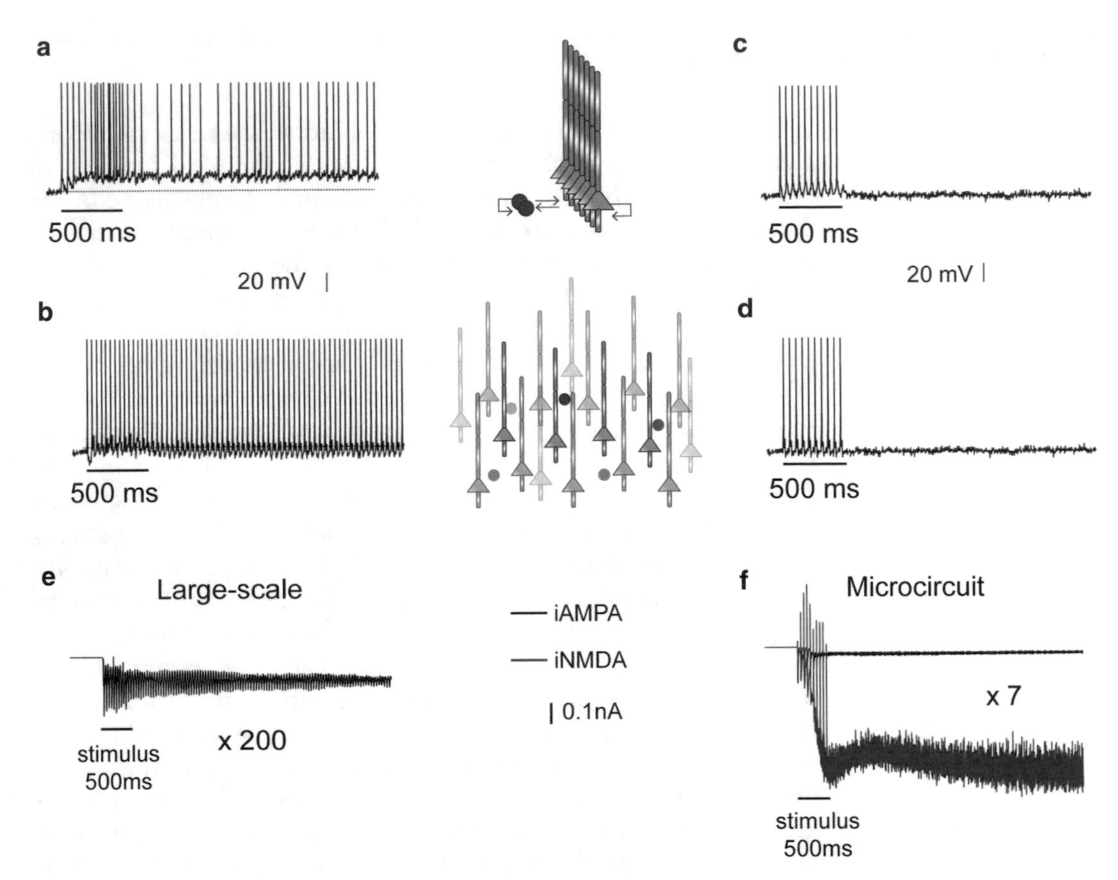

Fig. 4 Depolarizing plateau underlies persistent activity only in the presence of NMDA receptors. Distinct differences in persistent firing of a microcircuit and a much larger network in response to a short stimulus. (**a**) Persistent activity after stimulation at the proximal dendrites of the microcircuit. Note the generation of a depolarizing somatic plateau potential. (**b**) Persistent activity after stimulation at the proximal dendrites of a large-scale network while blocking NMDA receptors and increasing the latency of excitatory synaptic transmission to 40 $\pm$ 10 ms. Note the absence of the depolarizing plateau potential. (**c**) Persistent activity fails to emerge in the microcircuit when NMDA receptors are blocked and the latency of excitatory synaptic transmission is increased to 40 $\pm$ 10 ms, as in (**b**). (**d**) Persistent activity fails to emerge in the large-scale network when NMDA receptors are blocked and the latency of excitatory synaptic transmission is reduced to 1.7 $\pm$ 0.9 ms. (**e**) Average net AMPA current a pyramidal neuron receives from all other pyramidal neurons in the large-scale network in the trials that led to persistent activity. (**f**) Average net AMPA current (*black* trace) and the net NMDA current (*red* trace) a pyramidal neuron receives from all other pyramidal neurons in the microcircuit in the trials that led to persistent activity. Bar indicates stimulus presentation. Adapted from [51]

network of an entire cortical barrel column to unveil structure--function relationships by correlating dendritic morphology and thalamocortical connectivity [55, 56].

8 Bringing It All Together

Computational neuroscience is an established field which has provided numerous insights into the function of neurons and their compartments, and in many cases has preceded and instructed experimental research [57–59]. An accurate compartmental model for a neuron can be an invaluable tool for studying the behavior of a particular cell type. The advantage of compartmental simulations is that they give to the experimenter the ability to manipulate all the properties of the cell. This allows the exploration of neuronal behavior in cohorts that are not yet accessible to experimental investigations. Hopefully, the increase in computational resources available to the modeler will enable detailed models of single neurons and neuronal networks to evolve further, even correspond to larger brain areas, rendering possible behavioral hypotheses testing, prediction of brain disorders, and development of appropriate treatments.

References

1. Rall W (1959) Branching dendritic trees and motoneuron membrane resistivity. Exp Neurol 1:491–527
2. Rall W (1957) Membrane time constant of motoneurons. Science 126:454
3. Rall W (1960) Membrane potential transients and membrane time constant of motoneurons. Exp Neurol 2:503–532
4. Thomson W (1854) On the Theory of the Electric Telegraph. Proceedings of the Royal Society of London (1854–1905) 7:382–399
5. Coombs JS, Eccles JC, Fatt P (1955) The specific ionic conductances and the ionic movements across the motoneuronal membrane that produce the inhibitory post-synaptic potential. J Physiol 130:326–374
6. De Schutter E, Bower J (1994) An active membrane model of the cerebellar Purkinje cell. I. Simulation of current clamps in slice. J Neurophysiol 71:375–400
7. Mainen ZF, Sejnowski TJ (1996) Influence of dendritic structure on firing pattern in model neocortical neurons. Nature 382:363–366
8. Lux H, Schubert P, Kreutzberg G (1970) Direct matching of morphological and electrophysiological data in cat spinal motoneurons. In: Andersen P, Jansen JKS (eds) Excitatory synaptic mechanisms. Universitetsforlaget, Norway, pp 189–198
9. Barrett JN, Crill WE (1974) Specific membrane properties of cat motoneurones. J Physiol 239:301–324
10. Rodriguez A, Ehlenberger D, Kelliher K et al (2003) Automated reconstruction of three-dimensional neuronal morphology from laser scanning microscopy images. Methods 30:94–105
11. Schmitt S, Evers JF, Duch C et al (2004) New methods for the computer-assisted 3-D reconstruction of neurons from confocal image stacks. Neuroimage 23:1283–1298
12. Santer RM (1989) Correlative microscopy in biology. Instrumentation and methods. J Anat 162:279, http://www.ncbi.nlm.nih.gov/pmc/articles/PMC1256462/
13. Marx M, Günter RH, Hucko W et al (2012) Improved biocytin labeling and neuronal 3D reconstruction. Nat Protoc 7:394–407
14. Segev I, Rinzel J, Shepherd GM et al (1995) The theoretical foundation of dendritic function. Trends Neurosci 18:512
15. Rall W (1989) Cable theory for dendritic neurons. MIT Press, Cambridge, MA, pp 9–92
16. Pinsky PF, Rinzel J (1994) Intrinsic and network rhythmogenesis in a reduced traub model for CA3 neurons. J Comput Neurosci 1:39–60
17. Abrahamsson T, Cathala L, Matsui K et al (2012) Thin dendrites of cerebellar interneurons confer sublinear synaptic integration and a gradient of short-term plasticity. Neuron 73:1159–1172
18. Rall W, Burke R, Holmes W (1992) Matching dendritic neuron models to experimental data. Physiol Rev 72:S159–S186
19. Major G, Evans JD (1994) Solutions for transients in arbitrarily branching cables: IV. Nonuniform electrical parameters. Biophys J 66:615–633
20. Major G, Evans JD, Jack JJ (1993) Solutions for transients in arbitrarily branching cables: I. Voltage recording with a somatic shunt. Biophys J 65:423–449
21. Destexhe A, Neubig M, Ulrich D et al (1998) Dendritic low-threshold calcium currents in thalamic relay cells. J Neurosci 18:3574–3588
22. Papoutsi A, Sidiropoulou K, Cutsuridis V et al (2013) Induction and modulation of persistent

activity in a layer V PFC microcircuit model. Front Neural Circuits 7:161
23. Traub RD, Llinás R (1977) The spatial distribution of ionic conductances in normal and axotomized motorneurons. Neuroscience 2:829–849
24. De Schutter E, Angstadt JD, Calabrese RL (1993) A model of graded synaptic transmission for use in dynamic network simulations. J Neurophysiol 69:1225–1235
25. Korngreen A, Sakmann B (2000) Voltage-gated K+ channels in layer 5 neocortical pyramidal neurones from young rats: subtypes and gradients. J Physiol 525:621–639
26. Mainen ZF, Joerges J, Huguenard JR et al (1995) A model of spike initiation in neocortical pyramidal neurons. Neuron 15:1427–1439
27. Poirazi P, Brannon T, Mel BW (2003) Arithmetic of subthreshold synaptic summation in a model CA1 pyramidal cell. Neuron 37:977–987
28. Losonczy A, Magee JC (2006) Integrative properties of radial oblique dendrites in hippocampal CA1 pyramidal neurons. Neuron 50:291–307
29. Polsky A, Mel BW, Schiller J (2004) Computational subunits in thin dendrites of pyramidal cells. Nat Neurosci 7:621–627
30. Shepherd GM, Brayton RK, Miller JP et al (1985) Signal enhancement in distal cortical dendrites by means of interactions between active dendritic spines. Proc Natl Acad Sci U S A 82:2192–2195
31. Gold JI, Bear MF (1994) A model of dendritic spine Ca2+ concentration exploring possible bases for a sliding synaptic modification threshold. Proc Natl Acad Sci 91:3941–3945
32. Destexhe A, Mainen ZF, Sejnowski TJ (1994) An efficient method for computing synaptic conductances based on a kinetic model of receptor binding. Neural Comput 6:14
33. Kampa BM, Clements J, Jonas P et al (2004) Kinetics of Mg2+ unblock of NMDA receptors: implications for spike-timing dependent synaptic plasticity. J Physiol 556:337–345
34. Gold C, Henze DA, Koch C (2007) Using extracellular action potential recordings to constrain compartmental models. J Comput Neurosci 23:39–58
35. Holt GR, Koch C (1999) Electrical interactions via the extracellular potential near cell bodies. J Comput Neurosci 6:169–184
36. Druckmann S, Banitt Y, Gidon A et al (2007) A novel multiple objective optimization framework for constraining conductance-based neuron models by experimental data. Front Neurosci 1:7–18
37. Keren N, Peled N, Korngreen A (2005) Constraining compartmental models using multiple voltage recordings and genetic algorithms. J Neurophysiol 94:3730–3742
38. Huys QJM, Ahrens MB, Paninski L (2006) Efficient estimation of detailed single-neuron models. J Neurophysiol 96:872–890
39. Bhalla U, Bower J (1993) Exploring parameter space in detailed single neuron models: simulations of the mitral and granule cells of the olfactory bulb. J Neurophysiol 69:1948–1965
40. Hines ML, Carnevale NT (2001) NEURON: a tool for neuroscientists. Neuroscientist 7:123–135
41. Rossant C, Goodman DFM, Fontaine B et al (2011) Fitting neuron models to spike trains. Front Neurosci 5:9
42. Friedrich P, Vella M, Gulyás AI et al (2014) A flexible, interactive software tool for fitting the parameters of neuronal models. Front Neuroinform 8:63
43. Van Geit W, Achard P, De Schutter E (2007) Neurofitter: a parameter tuning package for a wide range of electrophysiological neuron models. Front Neuroinform 1:1
44. Hay E, Hill S, Schürmann F et al (2011) Models of neocortical layer 5b pyramidal cells capturing a wide range of dendritic and perisomatic active properties. PLoS Comput Biol 7:e1002107
45. Keren N, Bar-Yehuda D, Korngreen A (2009) Experimentally guided modelling of dendritic excitability in rat neocortical pyramidal neurones. J Physiol 587:1413–1437
46. Almog M, Korngreen A (2014) A quantitative description of dendritic conductances and its application to dendritic excitation in layer 5 pyramidal neurons. J Neurosci 34:182–196
47. Traub RD, Whittington MA, Colling SB et al (1996) Analysis of gamma rhythms in the rat hippocampus in vitro and in vivo. J Physiol 493 (Pt 2):471–484
48. Traub RD, Contreras D, Cunningham MO et al (2005) Single-column thalamocortical network model exhibiting gamma oscillations, sleep spindles, and epileptogenic bursts. J Neurophysiol 93:2194–2232
49. Konstantoudaki X (2014) Modulatory effects of inhibition on persistent activity in a cortical microcircuit model. Front Neural Circuits 8:1–15
50. Durstewitz D, Gabriel T (2007) Dynamical basis of irregular spiking in NMDA-driven prefrontal cortex neurons. Cereb Cortex 17:894–908
51. Papoutsi A, Sidiropoulou K, Poirazi P (2014) Dendritic nonlinearities reduce network size

requirements and mediate on and off states of persistent activity in a PFC microcircuit model. PLoS Comput Biol 10:e1003764

52. Markram H (2006) The blue brain project. Nat Rev Neurosci 7:153–160
53. Alivisatos A, Chun M, Church G (2012) The brain activity map project and the challenge of functional connectomics. Neuron 74:970–974
54. Kandel ER, Markram H, Matthews PM et al (2013) Neuroscience thinks big (and collaboratively). Nat Rev Neurosci 14:659–664
55. Meyer HS, Egger R, Guest JM et al (2013) Cellular organization of cortical barrel columns is whisker-specific. Proc Natl Acad Sci U S A 110:19113–19118
56. Oberlaender M, de Kock CPJ, Bruno RM et al (2012) Cell type-specific three-dimensional structure of thalamocortical circuits in a column of rat vibrissal cortex. Cereb Cortex 22:2375–2391
57. Mel B (1993) Synaptic integration in an excitable dendritic tree. J Neurophysiol 70:1086–1101
58. Poirazi P, Mel B (2001) Impact of active dendrites and structural plasticity on the memory capacity of neural tissue. Neuron 29:779–796
59. Segev I, Rall W (1988) Computational study of an excitable dendritic spine. J Neurophysiol 60:499–523
60. Bower JM, Beeman D (2003) The book of genesis - exploring realistic neural models with the GEneral NEural SImulation System. Genesis. Springer, New York, NY

Index

Alon Korngreen (ed.), *Advanced Patch-Clamp Analysis for Neuroscientists*, Neuromethods, vol. 113,
DOI 10.1007/978-1-4939-3411-9, © Springer Science+Business Media New York 2016

H

I

J

K

L

M

N

O

P

Q

R

S

T

U

V

Printed in the United States
By Bookmasters